Understanding the rheology of concrete

Related titles:

Polymer modified bitumen
(ISBN 978-0-85709-048-5)
This book, the first of its type, provides a comprehensive and in-depth coverage of the science and technology of polymer modified bitumen. After an initial introduction to bitumen and polymer modified bitumen, the book is divided into two parts. Part I focuses on the preparation and properties of a range of polymer modified bitumen, including polymer bitumen emulsions, modification of bitumen with poly(urethanes), waste rubber and plastic and polypropylene fibres. Part II addresses the characterization and properties of polymer modified bitumen. Topics covered include rheology, simulated and actual long term ageing studies, the solubility of bituminous binders in fuels and the use of Fourier transform infrared spectroscopy to study ageing/oxidation of polymer modified bitumen. The book is an essential reference for scientists and engineers from both academia and the civil engineering and transport industries interested in the properties and characterisation of polymer modified bitumen.

Fibrous and composite materials for civil engineering
(ISBN 978-1-84569-558-3)
The application of fibrous materials in civil engineering is an important and interesting development. Steel is one of the most popular composite materials used for structural applications, however many types of steel structures are subjected to corrosion damage. The use of fibrous structures including knitted, woven and braided fabrics are becoming recognised as an attractive reinforcement due to their fibre orientation, conformability and potential of producing shaped reinforcements. This book reviews these fibrous materials with regards to their role in structural reinforcement. Key topics include short fibre reinforced concrete, natural fibre reinforced concrete and high performance fibre reinforced cementitious composites.

Service life estimation and extension of civil engineering structures
(ISBN 978-1-84569-398-5)
This book reviews ways of rehabilitating ageing reinforced concrete structures using fibre-reinforced polymer (FRP) composites and how to assess their remaining service life. The book covers topics such as rehabilitation techniques for service life extension, probabilistic methods for service life estimation, rehabilitation and service life estimation of corroded structures and NDE/NDT of rehabilitated structures.

Details of these and other Woodhead Publishing materials books can be obtained by:

- visiting our web site at www.woodheadpublishing.com
- contacting Customer Services (e-mail: sales@woodheadpublishing.com; fax: +44 (0) 1223 832819; tel.: +44 (0) 1223 499140 ext. 130; address: Woodhead Publishing Limited, 80 High Street, Sawston, Cambridge CB22 3HJ, UK)
- contacting our US office (e-mail: usmarketing@woodheadpublishing.com; tel. (215) 928 9112; address: Woodhead Publishing, 1518 Walnut Street, Suite 1100, Philadelphia, PA 19102-3406, USA)

If you would like to receive information on forthcoming titles, please send your address details to: Francis Dodds (address, tel. and fax as above; e-mail: francis.dodds@woodheadpublishing.com). Please confirm which subject areas you are interested in.

Understanding the rheology of concrete

Edited by
Nicolas Roussel

WOODHEAD
PUBLISHING

Oxford Cambridge Philadelphia New Delhi

Published by Woodhead Publishing Limited,
80 High Street, Sawston, Cambridge CB22 3HJ, UK
www.woodheadpublishing.com

Woodhead Publishing, 1518 Walnut Street, Suite 1100, Philadelphia,
PA 19102-3406, USA

Woodhead Publishing India Private Limited, G-2, Vardaan House,
7/28 Ansari Road, Daryaganj, New Delhi – 110002, India
www.woodheadpublishingindia.com

First published 2012, Woodhead Publishing Limited

British Library Cataloguing in Publication Data
A catalogue record for this book is available from the British Library.

Library of Congress Control Number: 2011939656

ISBN 978-0-08-101645-9 (print)
ISBN 978-0-85709-528-2 (online)

The publisher's policy is to use permanent paper from mills that operate a sustainable forestry policy, and which has been manufactured from pulp which is processed using acid-free and elemental chlorine-free practices. Furthermore, the publisher ensures that the text paper and cover board used have met acceptable environmental accreditation standards.

Typeset by RefineCatch Limited, Bungay, Suffolk

Contents

Contributor contact details

(* = main contact)

Editor and chapters 4 and 11

N. Roussel
IFSTTAR
58 Boulevard Lefebvre
75732 Paris CEDEX 15
France
E-mail: nicolas.roussel@ifsttar.fr

Chapter 1

P. Coussot
Université Paris-Est
Laboratoire Navier (UMR CNRS)
2 Allée Kepler
77420 Champs sur Marne
France
E-mail: philippe.coussot@ifsttar.fr

Chapter 2

G. Ovarlez
Université Paris-Est
Laboratoire Navier (UMR CNRS)
2 Allée Kepler
77420 Champs sur Marne
France
E-mail: guillaume.ovarlez@lcpc.fr

Chapter 3

C. F. Ferraris* and N. S. Martys
National Institute of Standards and Technology
100 Bureau Drive
MS 8615
Gaithersburg MD 20899
USA
E-mail: clarissa@nist.gov

Chapter 5

S. Gauffinet-Garrault
Laboratoire Interdisciplinaire Carnot de Bourgogne
Département I. R. M, Université de Bourgogne
CNRS UMR 5209
Avenue Alain Savary – BP 47870
21078 DIJON CEDEX
France
E-mail: sandrine.gauffinet@u-bourgogne.fr

Chapter 6

X. Chateau
Université Paris-Est
Laboratoire Navier (UMR CNRS)
2 Allée Kepler
77420 Champs sur Marne
France
E-mail: xavier.chateau@gmail.com

Chapter 7

R. J. Flatt*
ETH Zürich
Institute für Baustoffe (IfB)
HIF Building, E 11
Schafmattstrasse 6
8093 Zürich
Switzerland
E-mail: flattr@ethz.ch

I. Schober
Sika Technology AG
Corporate Research
Tüffenwies 16
8048 Zürich
Switzerland
E-mail: schober.irene@ch.sika.com

Chapter 8

K. H. Khayat
Université de Sherbrooke
2500 Boulevard de l'Université
Sherbrooke
Québec J1K 2R1
Canada
E-mail: kamal.khayat@usherbrooke.ca

Chapter 9

S. Grünewald
Department of Design and Construction
Delft University of Technology
Stevinweg 1
PO Box 5048
2600 GA
Delft
The Netherlands
E-mail: S.Grunewald@tudelft.nl

Chapter 10

L. Nyholm Thrane
Danish Technological Institute, Concrete
Gregersensvej
2630 Taastrup
Denmark
E-mail: lnth@dti.dk

Chapter 12

P. Billberg
Swedish Cement and Concrete Research Institute
Drottning Kristinas väg 26
100 44 Stockholm
Sweden
E-mail: peter.billberg@cbi.se

Chapter 13

D. Feys
Université de Sherbrooke
2500 Boulevard de l'Université
Sherbrooke
Québec J1K 2R1
Canada
E-mail: Dimitri.Feys@usherbrooke.ca

Introduction

Throughout its history, concrete has often been called 'liquid stone'. Before turning into an artificial stone, concrete is indeed fluid enough, for a short period, to flow and fill a mould. This property has given the construction industry the ability to cast structural elements such as slabs, beams and columns. Whereas stones are dressed, wood is sawn and steel is shaped at high temperatures, only concrete gives architects the freedom to shape buildings at their will at the building site.

'Liquid' is, however, a very simple word to describe the full complexity of the behaviour of concrete in the fresh state. First, unlike simple liquids such as water or oil, concrete is made of elements of many different sizes (from several nanometres to a couple of centimetres) and of various types (organic or mineral) suspended in water. Each of these elements brings its own contribution to the behaviour of concrete in the fresh state. This contribution is strongly affected not only by mix proportions but also by temperature and the casting process itself. The vast family of industrial cementitious materials present such a variety of behaviours that their classification could seem unattainable. This is even more so for the potential to predict, if only qualitatively, their response in practical processing conditions.

These industrial processing conditions are also very complex. They are transient flows with complex boundary conditions. Lubricating layers may appear, for instance, at the surface of the mould. Moreover, formworks often contain a dense network of reinforcing steel bars, the size of which is close to the size of the coarsest particles in the material. Only a successful casting process is able to guarantee the mechanical bond between the reinforcement bars and concrete and, from a more general point of view, the adequate mechanical and durability performance requirements of the concrete structural element.

Moreover, unlike simple (Newtonian) liquids, the surface does not self-level when concrete is poured in a mould. The material is able to support an amount of the stress generated by gravity without flowing similar to that of mayonnaise, paint or tomato ketchup. In the case of firm concretes, this property is so strong that vibration is often applied to the material to ease the casting process and allow for filling of the formwork. Only recently have concretes able to flow under their own weight without vibration while staying homogeneous (called 'self-compacting concretes' or 'self-consolidating concretes') been developed. This property may, however, prove useful in specific cases such as shotcrete where, after spraying, it is clearly desirable that the material stays on the wall in thick layers rather than flows down it.

Time has also a strong effect on the behaviour of concrete in the fresh state. When left at rest, material consistency increases significantly with time. Some of these changes are reversible; their effects are erased by remixing in a truck for instance or by any type of strong shearing. Other changes, in particular the consequences of the hydration phenomenon, are irreversible and thus contribute to the long-term evolution of material properties (towards the solid state). No matter the origin of the evolution, it may strongly affect the way concrete is cast and the quality of the final structural element. For instance, two successively poured layers of concrete may not form a homogeneous material if the resting time between the pouring of the layers is too long.

It is noticeable in the history of concrete technology that the development of processing methods has been faster than the development of the understanding of the material behaviour in the fresh state. Concrete has indeed been used extensively since the beginning of the twentieth century, whereas the modern science of rheology (i.e. the study of the flow of matter) and the developments in technology which have made it possible have only occurred in the last 50 years.

What are the key objectives of research on the rheology of fresh concrete? Depending on scientific background, an academic answer could be 'the ability to correctly measure and quantify the rheological properties of fresh concrete' or 'the understanding of the correlation between components' proportions and rheological properties in the fresh state' whereas a practitioner would probably answer: 'the ability to predict whether or not a given concrete will correctly fill a given formwork'.

The continuing challenge of understanding particle interactions in cementitious materials in the fresh state up to the engineering prediction of the casting processes lies at the interface of three disciplines: physics, chemistry and fluid mechanics. Physics and chemistry allow correlations to be made between mix design and macroscopic fresh properties. Fluid mechanics allows analysis of the tests used for the measurement of these properties and, in its most advanced form, for the prediction of industrial casting processes.

Keeping the above in mind, we have chosen in this book to gather international experts from the three disciplines. Although their approaches are different, they all have in common the dedication of some or most of their research activities to understanding the fresh properties of cementitious materials. We hope that combining these contributions will form a useful basis for the understanding of the rheology of fresh concrete from mix design to casting processes.

This book is divided into three parts. Following an introductive Chapter 1 on the rheology of complex fluids, chapters 2 to 5 focus on the way the rheological behaviour of cementitious materials can be measured. Chapters 6 to 9 then review the main correlations that have been established between mix design and fresh properties. Finally, chapters 10 to 13 show how the knowledge of fresh properties can allow for the prediction of industrial casting processes. Although chapters 10 to 13 focus mainly on concrete industrial processes, chapters 1 to 9 can be useful for anybody interested in any other cementitious materials.

Chapter 1 goes through the main rheological behaviour types. In particular, the most important behaviour types for cementitious materials (yield stress behaviour and thixotropy) are emphasised along with the importance of distinguishing the main interaction types within the material for evaluating the physical origin of its behaviour. Chapter 2 presents the most used experimental methods for the measurement of the rheological properties of complex suspensions such as fresh cementitious materials. It deals successively with the basis of rheometry, the measurements of the properties of simple yield stress fluids, the measurements of the properties of thixotropic yield, stress fluids, shear localisation and material heterogeneities. Chapter 3 introduces current rheometer technology for measuring the properties of fresh concrete or mortar while Chapter 4 presents some correlations between industrial tests such as the slump test or slump flow test and yield stress. Whereas the previous chapters focus on the properties of fresh cementitious materials within a casting process perspective, Chapter 5 shows how rheometry can also be a valuable tool for the non-destructive study of the evolution of the structure of cement pastes from mixing to setting.

Chapter 6 recalls the main results concerning the effects of particles properties on packing characteristics. The dependence of the overall rheological properties of highly concentrated suspensions of polydisperse particles on the particle size distribution and particle blend composition is investigated. Chapter 7 focuses on the effects of deflocculating polymers on the fluidity of fresh cementitious materials. It first describes the chemical and physical characteristics of these molecules before detailing their modes of actions within a cement suspension. Chapter 8 deals with the high molecular weight water-soluble polymers used as viscosity agents in cementitious materials. After describing their chemical nature, it recalls the various ways they are able to improve the stability of cementitious materials. Chapter 9 focuses on how addition of fibres can affect the rheology of cementitious materials. It also shows how flow-induced fibre orientation can affect the mechanical properties of the hardened material. Chapter 10 gives an overview of the use of numerical tools to simulate the flow of fresh concrete. It presents the three main techniques: homogeneous fluid simulations, modelling of discrete particle flow (distinct element method), and numerical techniques to model particles suspended in a fluid along with examples of application and future trends. Chapter 11 focuses on the origin of the thixotropic behaviour of cementitious materials, before presenting models able the consequences of thixotropy on casting processes. In Chapter 12, existing models for the prediction of formwork pressure as a function of fresh properties of concrete are discussed.

The emphasis is on the effect of the structural build-up of fresh concrete and existing measurement techniques for this crucial parameter. Finally, Chapter 13 deals with the pumping of concrete. After a theoretical description of this specific flow, the influence of the nature of the pumped concrete, the material of the pipe, the total applied pressure, the temperature and bends and reducers are discussed.

Nicolas Roussel

Part I
Measuring the rheological behaviour of concrete

1

Introduction to the rheology of complex fluids

P. COUSSOT, Université Paris-Est, France

Abstract: This chapter considers the main types of rheological behaviour. The basic ways of modelling them (essentially in simple shear) and the physical origins of the behaviour of different material types are explained and typical real materials behaving in that way are mentioned. The most important behaviour types (yielding and thixotropy) for cementitious or concrete materials, the importance of flow regime (or 'physical state') with regards to the time scale of observation and the flow rate for predicting the behaviour type, the importance of distinguishing the main interaction types within the material for evaluating the physical origin of its behaviour are emphasised. Solids, simple fluids, suspensions, slightly non-Newtonian materials, yield stress fluids, thixotropy and viscoelasticity are reviewed.

Key words: solid, liquid, suspension, yield stress fluid, thixotropy.

1.1 Solids

1.1.1 Apparent behaviour

We live in a solid environment made of ground, furniture, cars, etc, which fortunately keep their shape when we use them. If however we apply a force, larger than a critical value, to one of these objects, we reach a situation when they start to strongly deform or even break (actually for a force smaller than this critical value they deform slightly). For two different objects made of the same material this critical force is different, which means it is not an appropriate variable for describing the material's behaviour as it is not intrinsic to it. In fact the appropriate variable is the stress, which is the ratio of the force to its surface of application. Finally the mechanical behaviour of solid materials is basically described in terms of the deformation induced by the stress applied and the critical deformation and critical stress at which the material flows or breaks.

To better understand these concepts of stress and deformation it is useful to consider one of the most frequent and simple situations, namely the 'simple shear'. In this case a piece of material is contained between two parallel solid planes which move relative to each other along one of their directions. The resulting deformation corresponds to the relative motion of parallel planes of materials (Fig. 1.1). The shear stress, τ, is the ratio of the force (F) applied to the section of the piece of material (S): $\tau = F/S$. The deformation, ε, is defined as the ratio of the relative displacement of the solid planes Δx to the thickness H of the piece of material: $\varepsilon = \Delta x/H$.

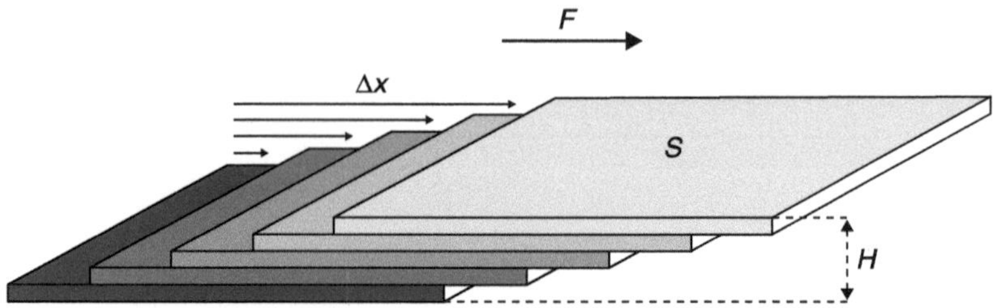

1.1 Principle of a simple shear: relative motion of material planes.

In mechanics we are interested in the constitutive equation of the material, which is a relationship describing the intrinsic behaviour of the material, i.e. independently of the specific conditions to which it is submitted. For a solid undergoing a simple shear, the constitutive equation is mainly described in terms of the relationship between τ and ε, but in some cases the rate of displacement must also be taken into account (see Section 1.7). The simplest solid behaviour is the linear elastic behaviour, in which τ is simply proportional to ε:

$$\tau = G\varepsilon \qquad [1.1]$$

In simple shear the coefficient of proportionality G in this relation is the elastic shear modulus.

For a sufficiently large stress, the material no longer follows this type of behaviour but deforms widely. Here we can distinguish two main types of behaviour: for plastic materials the deformation tends to localise in some specific region within the material; for fragile materials there is a breakage into several parts (Fig. 1.2).

1.1.2 Microstructural origin

Often we do not see what a material is made of and how it is organised at a local scale. Indeed most materials appear smooth or homogeneous at our scale of observation. The elements of materials that are of basic interest for our understanding of their mechanical properties are the largest elements which, by their distribution in space, play a critical role in the behaviour. These elements may be atoms, molecules, polymer chains, cells, clay particles or cement grains, which means that the size of the constitutive elements may range from few nanometres to several centimetres. The basic structure at the origin of a solid's behaviour is the distribution of elements either linked to each other or jammed in a given space. An illustrative example is a vacuum pack of coffee beans: it is apparently solid because the beans are jammed in a limited volume, and its properties depend only on the properties of the beans and their distribution in space.

The simplest material of this type is a solid crystal made of identical atoms or molecules distributed along periodic and symmetrical positions $\mathbf{x}_i$. The specificity

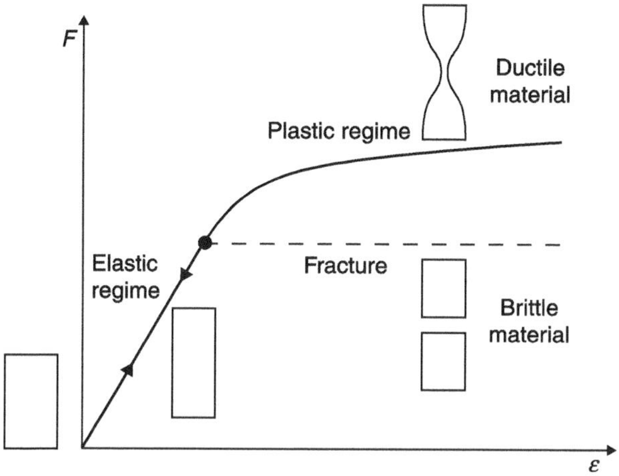

1.2 Two different possible (idealised) behaviour types of solid materials: elastic then fracture or plastic behaviour beyond a critical deformation. The drawings illustrate the different aspects of the material successively in the initial state, for a homogeneous deformation in the elastic regime, then either for a ductile or brittle behaviour.

of such a system is that an atom (situated in $\mathbf{x}$) interacts with its neighbours via strong van der Waals, covalent or ionic forces $\mathbf{F}_i$, and as a function of the distance between the elements $\mathbf{x} - \mathbf{x}_i$, so that this atom has a potential energy (E) defined by $\sum_i \mathbf{F}_i = \nabla E$. In the crystalline state each atom is in a position of force equilibrium (Fig. 1.3), $\Sigma \mathbf{F}_i = 0$, which corresponds to a minimum of the potential energy. As a consequence each atom is in a well of energy from which it can escape only if a sufficiently large force (or a stress) is applied to it. In addition each element is submitted to a thermal agitation which in general induces small amplitude motions of this element around its equilibrium position.

When a low stress is applied, each atom can be only slightly displaced from its initial position, thus reaching a new position of equilibrium: the atoms climb slightly along their potential well. This small displacement of each atom relatively to its neighbour finally induces a limited total displacement of the upper surface relative to the lower surface, and thus a limited deformation of the material. When the stress is released, the atoms fall back to the bottom of their potential well, i.e. they come back to their initial position. This explains the elastic behaviour of the material in the limit of small deformations. For a pure crystal of this type it is possible to relate the basic mechanical properties to the interactions at a local scale between neighbouring atoms. Indeed on one hand the relative displacement of neighbouring atoms induces a sample deformation, on the other hand it requires some energy supply to one atom for it to climb along its potential well. Finally this description explains how the stress increases as a function of the deformation at a local scale, which can be extrapolated at a macroscopic scale.

For a sufficiently large stress some atoms in a crystal can get out of their potential well. This can induce mainly two types of phenomena:

1. A dislocation, i.e. the exit out of their potential well of a limited number of atoms along some planes. In such a case the displacement of this series of atoms is limited. The arrangement is locally broken but finally reformed with other atoms, probably taking advantage of some defects in the initial structure; at a macroscopic scale this corresponds to a plastic behaviour.
2. A breakage, i.e. the exit and separation of all the atoms along some surface through the material, leading to its breakage. After this process the material has in general lost its initial properties because the structure has been irreversibly modified.

Other materials such as glass, some parts of rocks and some polymeric materials are in the glassy state (Fig. 1.3); the atoms or molecules are still densely packed as in the solid state but now the structure is disordered. Generally such materials appear as solids since, like crystals, they only deform slightly when subjected to a stress. However, for such materials, essentially because of the disordered structure, we do not have much information about the relation between the physics at a local scale and their mechanical properties. Indeed, this disorder implies that it is difficult to relate a local displacement to a macroscopic deformation because the

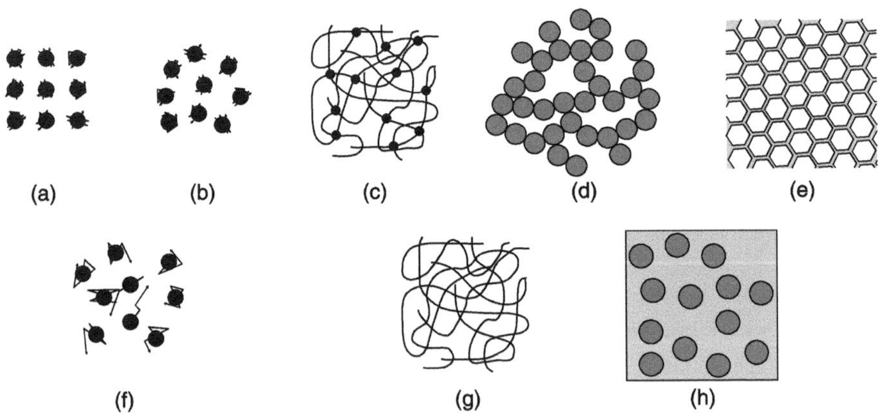

1.3 Different types of material structures associated with solid behaviour: (a) crystalline solid; (b) glass; (c) reticulated polymer; soft solid behaviour; (d) colloidal aggregate; (e) concentrated foam or emulsion. Different types of material structures associated with liquid behaviour: (f) simple molecular liquid; (g) polymer melt or solution; and (h) dilute or semi-dilute suspension in a liquid. For systems (a), (b) and (f) the arrows illustrate the thermal agitation of the elements while the material is macroscopically at rest.

way in which the stress distributes throughout the structure is *a priori* unknown. This is an active field of research.

Due to their formation process most rocks at the surface of the Earth consist of a mixture of small homogeneous crystalline or glassy regions of various types dispersed in a matrix in a glassy state. This complex structure results from the way they have been formed, i.e. via a fast or progressive cooling of magmas containing various species possibly implying fractional crystallisation. There are also many solids, such as soils, woods, etc, with a heterogeneous structure at a local scale. Like most rocks they are made by the aggregation of solid elements. A rheophysical approach to their behaviour requires us to identify the mechanical characteristics of the main components and then to develop a micromechanical approach able to predict the rheological properties of heterogeneous solids.

Finally we have various types of 'soft' solids such as gels and colloids (Fig. 1.3) made of mesoscopic elements dispersed in a liquid. If their concentration is sufficiently high these elements play a crucial role in the main rheological properties of the material; this is proved by the fact that it exhibits properties completely different from those of the interstitial fluid alone. Yet the energy of interaction between the colloidal elements is much lower than the interaction energy between two atoms. For example, while the typical van der Waals interaction energy varies inversely to the sixth power of the distance between atoms (which is very small within a solid), for two colloidal particles it approximately varies inversely to the second power of the distance between particles (which is much larger). This basically explains why the resistance to deformation of a system made of colloidal particles immersed in a simple liquid is much smaller than the resistance of a solid crystal.

In some ideal cases (such as a pure colloid in crystalline order) it is possible to deduce a relationship between this resistance and the potential energy between the mesoscopic elements by following the same approach described above for pure solids. In a similar way, for gels made of a network of linked polymer chains in a simple configuration, it is possible to deduce an expression for the elastic modulus as a function of the chain characteristics. Indeed each chain behaves as a small rubber with a given stiffness which is proportional to the temperature and inversely proportional to the molecule length. As a consequence the deformation of the solid leads to a deformation of the network which induces an elongation or compression of each chain, which requires some force.

Within the class of soft solids there are also all the pasty materials which behave as simple solids when they are too much deformed. These are for example muds, purées, sauces, toothpastes, mayonnaise and . . . fresh concrete. The fact that they behave as solids is readily observed when pouring them in a container: whereas for a liquid the free surface at rest is perfectly horizontal, it is generally at least slightly uneven for a pasty material and can keep the shape it has been given for the most 'pasty' among them. In general these systems contain a wide range of mesoscopic elements developing different types of interactions, as is typically the case for concrete including colloidal particles and coarse grains in a liquid. The structure is obviously also disordered and it is extremely difficult to establish a relationship between the local and the macroscopic behaviour. For some materials of simple structure such as concentrated foams or emulsions, a macroscopic deformation is intimately linked to the local deformations of each bubble or droplet, so that in the absence of other effects we find that the elastic modulus is proportional to the typical surface tension stress in such a material, i.e. the surface tension divided by the droplet or bubble radius.

1.2 Newtonian fluids

We also live in a fluid environment: we are surrounded by air, sea, we need water at any time of our life, blood flows through our veins, etc. In contrast with solids these materials (fluids) may be deformed at will and flow as soon as some stress is applied to them. Here we focus on liquids, which correspond to a dense state of matter (Fig. 1.3).

Let us for example consider a long conduit filled with water. When the conduit is horizontal the water does not move, but as soon as the conduit is inclined even slightly, the fluid starts flowing. The behaviour of many such materials is extremely simple: the velocity in the conduit is simply proportional to the applied pressure. If the conduit is long and if a large volume of fluid can be supplied from upstream, the flow will go on over a long time, implying that the deformation that the fluid can undergo without any modification of its properties is indefinite. This means that for fluids under usual conditions it is not the deformation which plays a critical role in the behaviour description but rather the flow rate.

In a simple shear experiment (Fig. 1.1), we consider the 'shear rate', i.e. the ratio of the relative velocity (V) of the solid planes to the sample thickness (H): $\dot{\gamma} = V/H$. For this simple flow the shear rate is also the time variation of the deformation: $\dot{\gamma} = (\Delta x/H\Delta t)$. Finally, for a liquid system, the constitutive equation is essentially sought in the form of the 'flow curve', i.e. the relationship between the shear stress (τ) and the shear rate ($\dot{\gamma}$) in simple shear.

Usual liquids made of small molecules such as water, oil, syrups, etc, are Newtonian materials. The shear stress is simply proportional to the shear rate:

$$\tau = \mu_0 \dot{\gamma} \tag{1.2}$$

where the coefficient of proportionality, μ_0, is the material viscosity. The flow history has no impact on such materials. In particular they flow immediately at a rate proportional to the currently applied stress, and do not store any energy as a result of the deformation.

Some other liquids (non-Newtonian) exhibit more complex behaviour which can be described by generalising the above approach, i.e. by considering the apparent viscosity (η), defined as:

$$\eta = \tau/\dot{\gamma} \qquad [1.3]$$

where η is not a constant but a function of the current shear rate.

In fact the constitutive equation of a liquid as considered above remains valid as long as the flow is not too fast. Otherwise inertia effects, namely turbulence, may take place and the apparent behaviour of the material is more complex. Turbulence occurs when the ratio of inertia effects to viscous effects is sufficiently large. This may be appreciated from the value of the Reynolds number (Re):

$$Re = \frac{\rho \dot{\gamma} H^2}{\eta} \qquad [1.4]$$

in which ρ is the fluid density and H a characteristic length scale of the flow. When Re is much larger than 1, the flow is in general 'turbulent'. When Re is much smaller than 1, the flow is 'laminar' and the constitutive equation as determined from usual rheological techniques is relevant for describing the material properties. This means that both the flow involved for the determination of the liquid properties and the flow which is to be described must be in the laminar regime.

As for solids the atoms or molecules in a simple liquid are densely packed against each other. However, in contrast with solids, the elements in a simple liquid are not fixed in specific places but are submitted to a significant thermal agitation which induces fast displacements over a distance which may be of the order of the element size (Fig. 1.3). A useful way of describing the rheophysics of such systems consists of considering that, in a liquid at rest (due to the element packing and the interactions with their neighbours), the atoms or molecules are instantaneously in potential wells from which they can exit and move in any direction with the same probability as a result of thermal agitation. When a stress is applied to the material, this makes the elements on average climb along the potential well in some specific direction so that the probability of moving out in that direction is now larger than in the opposite one. The same phenomenon takes place for all the elements, finally inducing a relative motion of two neighbouring liquid planes along this direction at a rate which is a function of the probability of jump and the value of the stress applied. This gives rise to a continuous deformation of the material, i.e. a flow, at a rate function of the stress. In this scheme the basic result is that a flow is obtained whatever the value of the shear stress applied since even an extremely low stress destabilises the system. In addition it appears that for simple liquids made of atoms or small molecules such as water, oil, honey, glycerol, alcohol, etc, the shear rate induced is simply proportional to the shear stress, i.e. we have Newtonian materials. However, our understanding of the physical origin of this result is so far only partial.

Such structure characteristics have an additional important consequence on the rheological behaviour of the material: due to the rapid agitation of the elements at our usual scale of observation the material does not retain any memory of its flow history. Its structure characteristics are thus constant and its rheological response to some stress is uniquely determined by the current stress level. This means that such fluids are not affected at all by the deformation they undergo, but this does not mean that no work is necessary to induce a flow. In fact some energy is required to push each molecule out of its potential. This energy is not stored in the material but is dissipated (in the form of a temperature increase) since, as soon as one stops the flow, the material is in the same mechanical state as before flow. This energy dissipation corresponds to the work supplied per unit time, which may be computed by integrating, over the sample volume, the product of the force applied to each plane and the relative velocity of two neighbouring planes. Finally, using the shear stress and the shear rate, the viscous dissipation per unit volume (P) can be expressed as:

$$P = \tau\dot{\gamma} \tag{1.5}$$

1.3 Suspensions

The class of Newtonian materials is much wider than the class of pure materials. Indeed many materials are made of mesoscopic objects (of a size much larger than the size of the atoms or molecules composing the liquid) dispersed in a Newtonian liquid. These are 'suspensions'. Whereas for a pure liquid the fluid layers glide over each other in simple shear, for a suspension the fluid layers follow more complex paths because of the presence of the elements. The result is a larger viscous dissipation for a larger concentration of elements but it can be shown that, as long as the distribution of elements in space remains isotropic and constant, the suspension is Newtonian.

A critical parameter for appreciating the behaviour of suspensions is the volume fraction (ϕ) of elements in suspension, i.e. the ratio of the total volume of the elements to the total sample volume. When the volume fraction is sufficiently low, i.e. in the dilute regime, the elements are far from each other so that the liquid flow around each element is perturbed at a distance smaller than the separating distance between them (Fig. 1.4). We can say that the elements do not interact hydrodynamically. In that case the suspension generally has a rheological behaviour similar to that of the interstitial liquid, possibly with slightly larger rheological parameters. For a Newtonian suspending liquid, a dilute suspension of non-colloidal grains is Newtonian with a viscosity:

$$\mu(\phi) = \mu_0\left(1 + 2.5\phi\right) \tag{1.6}$$

This expression constitutes an excellent approximation of reality for $\phi < 2\%$.

At larger solid fractions the elements interact significantly so that on the behaviour of both the behaviour of the suspending liquid and the elements. For a suspension of non-colloidal grains in a Newtonian liquid, we still have a Newtonian material in the semi-dilute regime and various empirical formulae have been proposed to describe the suspension viscosity in that range. For example one may use the following expression (Krieger–Dougherty):

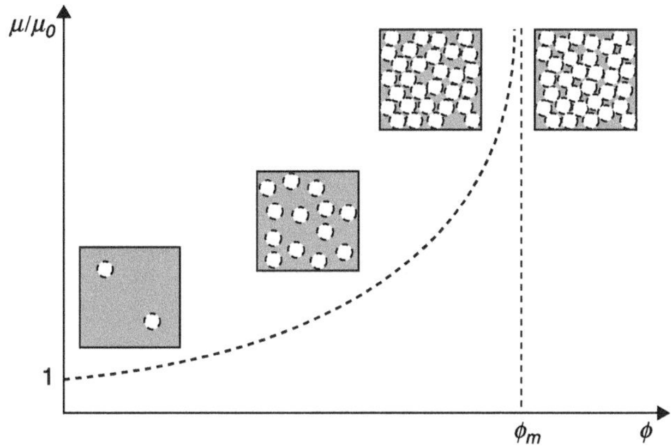

1.4 Relative viscosity of a suspension of uniform spheres in a Newtonian liquid as a function of the solid fraction. The schemes correspond to the different regimes of concentration (from left to right): dilute, semi-dilute, concentrated, compact.

$$\mu = \mu_0 \left(1 - \phi/\phi_m\right)^{-2.5\phi_m} \quad [1.7]$$

which is relevant for representing the viscosity in a wide range of solid fractions (say up to 55%). Note that a critical parameter has been introduced in this equation, namely ϕ_m, the maximum packing fraction, which is the maximum volume fraction which can be reached with the grains. When the solid fraction approaches ϕ_m we are in the concentrated regime and the viscosity tends to infinity (Fig. 1.4). It may be considered that, for a higher concentration (compact regime), the structure is jammed and the material is solid.

The maximum packing fraction is not a well-defined parameter as it depends on the spatial distribution of the elements, which may depend on the flow history. For uniform spheres it is around 55% if the particles are simply poured in a container and can be increased to 64% by a slight and long vibration. The crucial point is that the exact value to be used in Eq. 1.7 has a dramatic impact on the viscosity value in the concentrated regime, as may be seen in Fig. 1.4. Unfortunately there is not yet an appropriate procedure for determining the appropriate value of ϕ_m for use in Eq. 1.7. It seems that this value can only be determined empirically for each material from its viscosity vs. solid fraction curve: ϕ_m corresponds to the concentration at which the viscosity tends to infinity.

Although it has several empirical aspects the approach described above appears to be very general in that it works for various grain types and size distributions. Indeed Eq. 1.7 can be used to estimate the viscosity of various types of materials made of elements in suspension in a simple liquid. For that purpose it is necessary to estimate the effective maximum packing fraction for the specific elements under consideration. For example, for a suspension of particles of large aspect ratios (the extreme case being fibres), the maximum packing fraction is much smaller than for uniform spheres. Conversely, when the grain size distribution is wide the maximum packing fraction is much larger than the typical value for uniform spheres (around 60%): for example it may be up to 90% or 95% for a grain size distribution ranging from a few microns to a few millimetres.

Another problem is the fact that in the concentrated regime the particles may interact so strongly that the behaviour of the interstitial liquid becomes negligible. The rheological behaviour of the material is now highly dependent on the type of interactions between the suspended elements so that even for a suspension in a Newtonian liquid the Newtonian character may be lost. In this context, the description of the behaviour with the help of Eq. 1.7 is a rough approximation. The effective behaviour of this type of material is very complex, since depending on the flow history, the main source of viscous dissipation may vary from frictions between grains if a network of contact has formed throughout the sample (when the shear stress is more or less proportional to the normal stress) to hydrodynamic dissipations within the interstitial liquid if the grains are dispersed in such a way that they are separated by thin liquid films (when the shear stress is mainly proportional to the shear rate). The exact conditions under which the transition between these two regimes occurs are poorly identified.

Moreover in such materials density heterogeneities develop easily as some particles can migrate slightly through the sample, which tends to increase the grain concentration in the more slowly flowing regions. This effect is critical because the transition between the two regimes can occur for a slight variation of concentration or a slight change of configuration (distribution of the grains in space). Now we are dealing with a material the behaviour of which must be described by taking into account the coupling between the local flow characteristics and the possible liquid transfer through the granular phase.

For concentrated suspensions of colloidal particles, bubbles, droplets or polymers, the situation is different. The interactions between the elements play a major role in the behaviour of the system and there may be different flow regimes depending on the ratio of the intensity of these interactions to that due to simple hydrodynamic interactions (flow of the liquid). However, under usual conditions no migration of the elements leading to density heterogeneities can be expected. This is so either because we are dealing with systems in which the elements are squeezed against each other or because we are dealing with concentrated suspensions of small elements which by themselves form an equivalent porous medium of very small permeability.

1.4 Fluids with slightly non-Newtonian character

1.4.1 Shear-thinning and thixotropy

For Newtonian fluids the ratio of the shear stress to the shear rate is constant. Many other fluids have a non-Newtonian character: their apparent viscosity now varies with the shear rate and/or with the flow history. Such a character results from the fact that, in contrast with Newtonian fluids, the origin of the viscous dissipation is now modified by the flow. This is particularly the case for suspensions of asymmetrical elements able to change their orientation or their shape during flow, or objects developing mutual interactions which may vary with the flow history. For example, a long object tends to align along the flow direction: on average it occupies this type of position more often than a direction perpendicular because, in the latter case, due to shear it rapidly rotates and reaches the direction of flow. In a slightly different way polymer chains tend to stretch along the flow direction. The apparent viscosity of the system is generally lower when the asymmetrical elements are aligned along the flow direction, because in this case, the perturbation of the flow due to the presence of the elements is smaller. As a consequence we can distinguish two types of effects on the mechanical behaviour. If this alignment develops more or less instantaneously for a given shear rate and depends significantly on shear rate, we will have a 'shear-thinning' material for which the apparent viscosity decreases with shear rate (Fig. 1.5):

$$d\eta/d\dot{\gamma} < 0 \qquad [1.8]$$

If the alignment takes some time to develop we will have 'thixotropic' effects, i.e. the apparent viscosity for a given shear rate varies in time:

$$d\eta(\dot{\gamma})/dt \neq 0 \qquad [1.9]$$

From this example we see that shear-thinning and thixotropy can be confused because they may find their origin in the same physical effect. Yet they are clearly associated with different mechanical effects: variation with flow rate for shear-thinning and variation with time for thixotropy. Thixotropy is dealt with in more detail in Section 1.6.

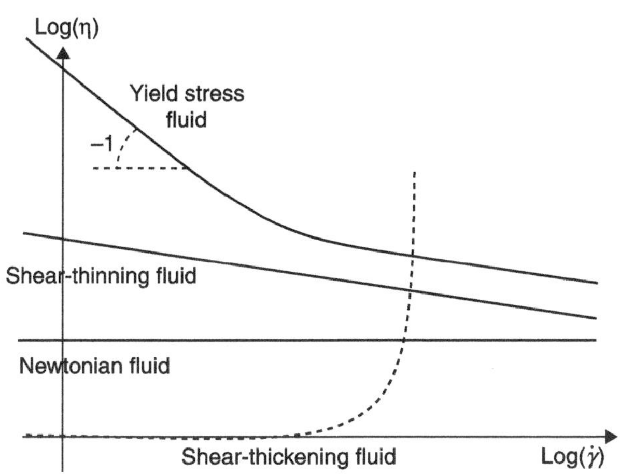

1.5 Main types of flow curves represented in terms of the apparent viscosity ($\tau/\dot{\gamma}$) as a function of the shear rate.

Another possible origin of shear-thinning is Brownian motion. Indeed, in a dilute suspension, the Brownian motion of colloidal particles leads to an average displacement of the particles from their initial position proportional to the square root of time. If the typical relative displacement of two particles induced by shear over a given time is much smaller, Brownian motion induces an additional viscous dissipation (as a result of the particle displacements through the liquid) which is much larger than that due to the mean shear flow. As a consequence the apparent viscosity at low shear rates in dilute colloidal suspensions is larger than at high shear rates. Finally the relative importance of Brownian motion and hydrodynamic dissipations may be appreciated from the Peclet number (*Pe*):

$$Pe = \frac{\eta \dot{\gamma} b^2}{k_B T} \qquad [1.10]$$

where b is the particle size, k_B the Boltzmann constant and T the temperature. Brownian motion plays a significant role if $Pe \ll 1$.

In addition, shear-thinning effects may occur in moderate or concentrated suspensions as a result of variations in colloidal interactions with shear rate. From a general point of view this effect is poorly understood. It starts to find a relatively clear explanation (transition from a jammed to a liquid state) within the frame of concentrated suspensions exhibiting a yield stress (see Section 1.5), but in that case the shear-thinning character is drastic since the apparent viscosity tends to infinity when the shear rate tends to zero.

1.4.2 Shear-thickening

A peculiar effect, namely shear-thickening, sometimes occurs with suspensions of small particles of approximately homogeneous size. This is for example observed with a suspension of cornstarch in water: when gently mixing the material with a spoon it apparently reacts as a liquid of low viscosity, but if it is mixed more rapidly, it reacts as a liquid of very high viscosity and even can break like a solid. A similar effect is observed with concentrated suspensions of uniform small spheres; in this case it was initially suggested that this was an order–disorder transition which would explain the effect. Since then other explanations have been suggested such as the formation of 'hydroclusters', namely transient clusters of particles formed by hydrodynamic interactions. However, all the explanations of shear-thickening rely on the general idea that there is some kind of jamming of the structure beyond some critical shear rate, a view which is close to the apparent macroscopic effect. This jamming is associated with the fact that the hydrodynamic force due to liquid flow between two approaching or separating particles diverges when the distance between the two solid surfaces tends to zero. There thus seems to exist a critical shear rate beyond which the structure formed by the particles in the liquid does not have enough time to relax during the macroscopic deformation: the imposed deformation tends to push the particles very close to each other, leading to very high energy dissipation associated with a diverging apparent viscosity.

Shear-thickening may be seen as an effect opposite to shear-thinning, which would be described as $d\eta/d\dot{\gamma} > 0$. In reality the trend is quite different: generally the apparent viscosity of shear-thickening materials is more or less constant up to a critical shear rate value around which it starts to dramatically increase (Fig. 1.5) and it is not clear whether or not the measurements in the latter regime are relevant. There do not seem to be known examples of 'smooth' shear-thickening, associated with a progressive increase of the apparent viscosity with shear rate. When such an effect has been observed, it could be explained by artefacts such as drying, particle migration leading to material heterogeneities or sedimentation.

1.5 Yield stress fluids

Many industrial or natural materials have a two-fold character, i.e. they behave as solids under some circumstances and as liquids otherwise. Actually these materials behave as solids when a stress smaller than a critical value is applied, so that they can keep the shape they have been given – a critical aspect for many applications. When a stress larger than the critical value is applied they apparently behave as liquids, i.e. they are able to flow more or less indefinitely as long as the stress is maintained. Examples of such materials include:

- creams, gels and clay suspensions used in cosmetics;
- various materials used in civil engineering such as ceramics slips, paints, mortars, plasters, cement pastes and fresh concrete;
- muds used for drilling or injections in the oil industry;
- foodstuffs such as marmalades, purées, soups, sauces, foams;
- food pastes used in the preparation of extruded cakes, crackers, biscuits, etc;
- sewage sludges;
- mineral waste suspensions from the mining industry;
- mud or debris flows in mountain streams.

Some of these materials exhibit a time-dependent behaviour, i.e. they are thixotropic, a behaviour illustrated by the fact that their resistance to flow increases after an increasing time at rest (see Section 1.6) – an aspect we will leave for now. Thus we are left with a material which essentially behaves as a solid, i.e. mainly elastic with possibly some viscoelastic effects, when the applied stress is smaller than the yield stress τ_c, and as a viscous liquid when the stress is larger than τ_c. In this liquid regime the apparent viscosity *a priori* has a specific aspect: in order to ensure the continuity with the solid regime it should tend towards infinity when the shear rate tends to zero, i.e. when the material stops flowing. This is indeed what is generally observed in practice: in a stress vs. shear rate diagram the data points form a horizontal plateau at low shear rates – at a level which can be considered as the yield stress level (Fig. 1.6). This implies that the apparent viscosity η is approximately $\tau_c/\dot{\gamma}$, which tends to infinity when $\dot{\gamma}$ tends to zero (Fig. 1.5).

In general the Herschel–Bulkley model, which expresses as follows in simple shear:

$$\tau > \tau_c \Rightarrow \tau = \tau_c + k\dot{\gamma}^n \quad [1.11]$$

in which k and n are material parameters, represents very well experimental data in a wide range of shear rate. In this model the three parameters depend on the material and n is generally between 0.3 and 0.5. The important point is that such a model can be very well fitted to experimental data for shear stress vs. shear rate over a range of several decades (typically four to five) of shear rates. As a consequence this model is able to represent the material behaviour from very slow flows (typically at shear rates of the order of $10^{-2}s^{-1}$) to rapid flows (at shear rates of the order of $100s^{-1}$). This contrasts with the Bingham ($n = 1$) or even the Casson models $\left(\sqrt{\tau} = \sqrt{\tau_c} + \sqrt{\mu\dot{\gamma}}\right)$ still used in some cases but which generally fit the data over a much more limited range of shear rates. As a consequence, with the two latter models a specific set of parameters may provide appropriate predictions of the flow characteristics of a given material under some specific boundary conditions, but may be quite inappropriate for describing them under other circumstances.

Although the structure of yield stress fluids relies on the arrangement of mesoscopic objects rather than on molecules, it is possible to use the micro-mechanical approaches developed for crystalline solids (see Section 1.1) to get some idea of the physical origin of the rheological parameters of such materials in the solid regime. In particular the elastic modulus may be related to the structure characteristics (interactions between elements) in some ideal cases and the yield stress value can be related to the critical stress needed to break the material structure.

In contrast, for the liquid regime no relevant physical meaning has so far been provided for the additional parameters (k and n) and the general form of the model. Thus the Herschel–Bulkley model is nothing more than an appropriate fit to data. In particular it should not be considered that two separate sources of viscous dissipation exist within the material in the liquid regime: one related to the breakage of the network of interactions and giving rise to a constant stress term (the yield stress), and another being related to the hydrodynamic viscous dissipation due to the flow of the interstitial liquid and giving rise to a power-law function of the shear rate (the second term of the Herschel–Bulkley model). In fact the sum of these two terms describes a progressive increase in the total viscous dissipation as the shear rate increases, as a result of more complex relative motions of the elements including some liquid flows, and it happens that this increase can be well represented by such a function. This remark appears even clearer for some materials for which the yield stress associated with the solid–liquid transition differs from that to be used in the Herschel–Bulkley model for

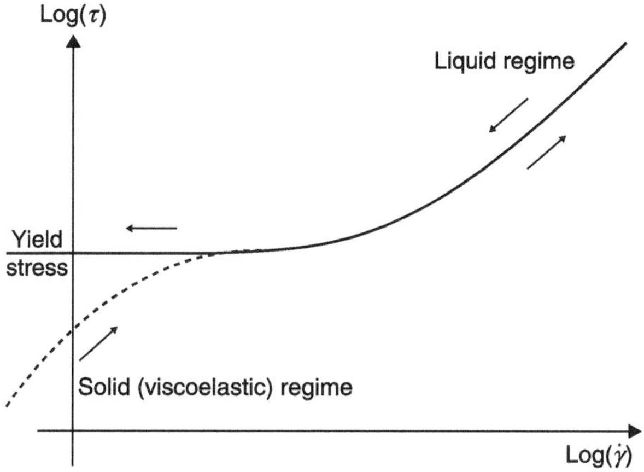

1.6 Apparent flow curve of a yield stress fluid obtained from an increasing–decreasing stress or shear rate ramp. The increase corresponds to the dotted line and the decrease to the continuous line.

the liquid regime. This is so for thixotropic materials (see Section 1.6) for which the apparent yield stress at rest varies with the flow history so that obviously no single yield value can be deduced and used to describe the liquid regime.

The apparent flow curve is usually obtained from an increasing–decreasing stress or shear rate ramp (Fig. 1.6). In this case the sample is initially at rest in its solid regime. As a consequence the response at low shear rates in the increasing stage does not generally correspond to steady state in the liquid regime. The material remains in its solid regime as long as the total deformation it has undergone is below a critical value. The measurement system nevertheless provides some shear rate values computed over some finite time, which are then reported in the rheogram. This leads to the typical aspect of the curve represented in Fig. 1.6, with an initial steep slope in the increasing curve which in fact varies with the timing of increase of the shear rate or shear stress. In contrast the flow curve in the liquid regime obtained in the decreasing stage is relatively robust, i.e. independent of the flow history.

Let us examine yield stress fluids more generally. They are usually made of a large number of elements suspended in a liquid and which develop significant mutual interactions. This is so for example for concentrated emulsions or foams in which the droplets or bubbles are squeezed against each other. When a small stress is applied the droplets or bubbles are only slightly deformed, and when the stress is released they get back to their initial state. When a large stress is applied the elements are sufficiently deformed to break the structure at least over a very short period of time. Then a similar (on average) structure reforms with the elements distributed differently. If the stress is maintained this process repeats indefinitely and the material deforms indefinitely: it flows like a liquid. Similar phenomena should occur with colloidal suspensions with particles developing mainly repulsive interactions. With attractive particles the situation is slightly different: they form a network of linked particles which may break if a sufficiently large stress is applied; then the broken material is made of flocs or elements which may be considered as individuals during some time, i.e. as long as they do not restore some link.

The materials described above are model systems. Real materials are often concentrated mixtures of a lot of element types making it not so easy to distinguish which element type, and thus which interaction type, is at the micromechanical origin of the mechanical behaviour. The general characteristics of these systems nevertheless remain:

- We are dealing with jammed, disordered structures in the solid regime.
- These structures continuously break and reform in the liquid regime.
- The disorder is similar in the liquid and the solid regimes.

1.6 Thixotropy

We already mentioned this possible character of the behaviour, i.e. the dependence of the apparent viscosity on the flow history. It is for yield stress fluids that thixotropy is most often observed and can play a significant role. Since yield stress fluids exhibit two behaviour regimes it is necessary to examine how thixotropy takes place in each.

1.6.1 Solid regime

Thixotropic effects in the solid regime can be well characterised from the increase of the elastic modulus in time as measured for small deformations. However, there is no general model for describing these evolutions and that can be considered as applicable to any material. Very roughly speaking the trends could be described by a logarithmic function of the time, i.e. $G \propto \ln t/t_0$, which at least illustrates that the rate of increase of the elastic modulus decrease with the time of rest. Another possibility for appreciating the thixotropy in the solid regime is to observe the increase of the apparent yield stress in time. A similar rough logarithmic variation is observed.

Qualitatively it is clear that the increase in strength results from the fact that the jamming increases or the interactions strengthen. For attractive colloidal systems this is related to the increase of the links in time: clear pictures of this effect have been obtained for simple colloidal systems in two dimensions, showing how aggregates progressively increase in size in time after the particles have been initially dispersed in the liquid. In this case the particles wander through the liquid as a result of Brownian motion and aggregate when they encounter other particles. However, even for such an ideal system, there is no model describing this kinetic process and there is no obvious relationship between the structure characteristics and mechanical strength.

1.6.2 Liquid regime

In the liquid regime thixotropy is observable by the fact that the apparent viscosity decreases in time while the boundary conditions are fixed: starting from the material at rest the shear stress decreases when the shear rate is imposed or the shear rate decreases when the shear stress is fixed. The importance of this decrease depends on the initial state: it is more marked for a longer time of rest. More generally a progressive decrease of the apparent viscosity in time is generally observed when one changes the level of shear rate imposed. These effects are associated with the fact that a specific steady state structure corresponds to each shear rate level, and it takes some time to reach. This phenomenon has a clear impact on the measurements with the help of the usual procedure for flow curve determination, i.e. increase then decrease of the shear rate or shear stress (Fig. 1.7). The level of the plateau (after the initial progressive increase) increases with the preliminary time of rest, then the fluid is liquefied and finally reaches the flow curve associated with a smaller time of rest.

Once again there is no general picture of the evolution of the structure state with shear rate either from a mechanical or a physical point of view. For simple suspensions of attractive particles it seems reasonable to assume that, as the imposed shear rate is increased, the network of linked particles breaks in smaller flocs, such that it might be possible to establish some correspondence between the average floc size and the shear rate. Some attempts have been made in that way

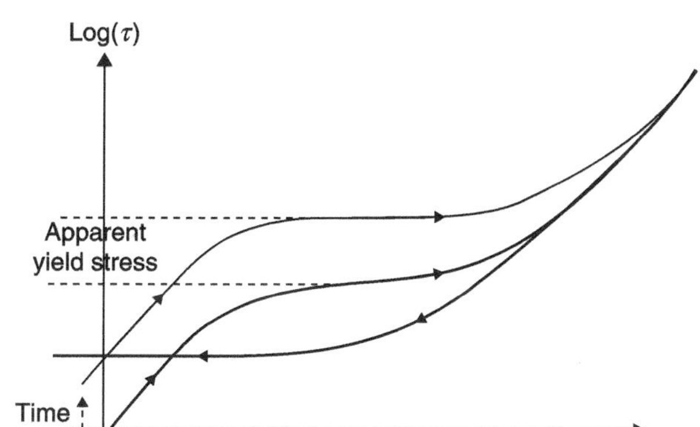

1.7 Apparent flow curve for an increasing–decreasing stress or shear rate ramp after different times of rest: the initial plateau level increases with the time of rest so that the apparent yield stress increases, but the decreasing flow curves are identical.

but there is a lack of effective knowledge of the processes that occur, so that in general fitting parameters are finally used in such models. It is worth emphasising that, although a unity of description would be useful, the physical processes at work in thixotropy processes in the solid and the liquid regimes are likely significantly different. Usually, the evolution of the state of the structure is described with the help of a single parameter, λ, and the rheological parameters are assumed to vary with this parameter. Besides, the time evolution of the state of structure of the material λ is described using a kinetic equation in which the rate of variation of λ is expressed as the difference between a rate of natural restructuring f and a rate of destructuring g:

$$\frac{d\lambda}{dt} = f(\lambda) - g(\lambda, \dot{\gamma}) \qquad [1.12]$$

However, the simple yielding behaviour described so far has some problems. First of all it is surprising to have materials behaving as solids below a critical stress and flowing homogeneously like simple liquids just beyond this critical stress. Indeed, for usual solids the flow start is associated with a breakage or strong localisation of shear in the material. Yield stress fluids would thus be special solid materials able to suddenly liquefy beyond a critical stress. Besides, the horizontal plateau observed for thixotropic yield stress fluids (Fig. 1.7) for an increasing ramp of stress or shear rate suggests that some flow instability may occurs: a material exhibiting such a plateau in its flow curve can hardly flow steadily at a shear in the range of the plateau.

Actually it seems that mainly two classes of materials exist. Physical gels, concentrated emulsions or foams seem to behave mainly as simple yield stress fluids. This is likely to be because the network of interaction associated with the yielding behaviour results from the jamming of the elements squeezed against each other and is broken during shear over successive short periods of time so that the stress values needed to induce and maintain flow are very similar. Moreover there is no significant restructuring effect at rest. Thus there is a unique yield stress and the stress needed to maintain flow in the liquid regime increases from this value as the shear rate increases. In contrast for materials containing a significant fraction of attractive colloidal particles, the strength of the material in the solid regime can be significantly larger than that in the liquid regime because the network of interactions is broken in the latter state. This explains the plateau in the increasing part of the flow curve observed after a sufficient time of rest (Fig. 1.7). For such materials, shear-banding (i.e. a strong heterogeneity in the shear rate distribution) may develop for apparent shear rates along the plateau in the flow curve. However, in practice it is not clear that this effect has a strong impact on the macroscopic characteristics of the flow. Indeed a model fitted to the apparent flow curve (including a plateau) and assuming homogeneous flow appears to provide a sufficient approximation for describing most practical cases.

1.7 Viscoelasticity

1.7.1 Solids

Until now we have assumed that when the material is submitted to a given stress the steady deformation in the solid regime or the steady state shear rate in the liquid regime are reached instantaneously if the structure is not modified. In reality there are necessarily some time effects associated with viscous dissipation in the solid regime and storage of elastic energy in the liquid regime which may need to be take into account.

Let us consider the solid case. A characteristic time is needed before the final deformation is almost reached. This finds its origin in viscous effects which are coupled or added to the elasticity of the structure. An illustrative picture of this situation is a gel made of polymers chains linked to each other and immersed in a liquid. For a given stress applied, the final deformation reached after some time depends strictly on the elastic modulus of the polymer network (which we assume here to be purely elastic). As the system is deformed the liquid is sheared so that there are also viscous effects, which may be described with the help of the apparent viscosity of the suspension of polymers in the liquid. Finally the total stress, needed to impose a deformation γ leading to an instantaneous shear rate $\dot{\gamma}$, can be written as:

$$\tau = G\gamma + \mu\dot{\gamma} \quad [1.13]$$

This equation may easily be solved for example when the stress applied is fixed to some value. It is found that the deformation increases exponentially up to its asymptotic value τ_0/G with a characteristic time μ/G. Thus, when the typical time of observation or experiment is larger than μ/G, the material essentially appears as an elastic solid. When, on the contrary starting from rest, one looks at the behaviour over timescales much smaller than μ/G, the material appears as a Newtonian material.

For other material types the origin of viscous effects may be more complex, involving not only viscous dissipations in the interstitial liquid but also in the interactions between the elements of the structure. More complex behaviour than that described above may then be found but the general principle is the same.

1.7.2 Liquids

In liquids also the steady state shear rate may not be reached immediately because initially part of the supplied energy is stored in the form of elastic energy. Let us for example consider the case of bubbles dispersed in a simple liquid. When a shear stress is applied to the system it tends to shear the liquid at a shear rate depending on its apparent viscosity, but also tends to deform the bubbles up to some deformation associated with the balance with surface tension effects. We assume that the liquid phase is a Newtonian fluid and the bubbles form a homogeneous material of elastic modulus G. Under these conditions the total deformation induced after some time can be written as:

$$\gamma = \tau/G + \int \tau/\mu dt \tag{1.14}$$

Under fixed shear rate ($\dot{\gamma}_0$) this equation is easily solved and implies that the shear stress decreases exponentially down to its asymptotic value $\mu\dot{\gamma}_0$ with a characteristic time μ/G. Thus, when the typical time of observation or experiment is larger than μ/G, the material essentially appears as a simple liquid. When, on the contrary starting from rest, one looks at the behaviour over timescales much smaller than μ/G, the material appears as an elastic solid. Note that μ/G is also the characteristic time of relaxation of the system: if one stops the steady state flow at some time the bubbles will get back to their equilibrium shape with the characteristic time μ/G.

The origin of elasticity storage in materials is the elongation of polymer chains, the deformation of flocs of colloidal particles, the deformation of droplets, etc. These effects can be modelled by more or less sophisticated models (in particular involving a large number of relaxation times), but the general principle as described above remains the same.

1.8 Conclusions

When considering the various types of possible interactions between the components of a real complex fluid and the various corresponding effects on the macroscopic behaviour, it is often very difficult to be sure of any interpretation of rheological data (macroscopic) in terms of physico-chemical interactions. In that aim it is preferable to use model systems in which the number of possible interactions and their effects on the macroscopic behaviour is limited.

For real systems it is important to keep in mind the above background for simple or model systems, which can suggest reasonable interpretation. For all systems it is critical to carry out experiments that provide relevant data concerning their rheological behaviour. This behaviour needs to identify clearly the rheological regime in which the system is (solid, liquid, transient) and to control any artefact of the measurements which would affect the material homogeneity, the material volume or the assumed velocity field in the rheometer. This is the subject of the next chapter.

2
Introduction to the rheometry of complex suspensions

G. OVARLEZ, Université Paris-Est, France

Abstract: This chapter deals with experimental methods for the measurement of the rheological properties of materials, with a focus on complex suspensions such as fresh concrete. The various behaviours studied in this chapter are described in Chapter 1. The first section provides the basis of rheometry. The second section deals with the measurements of the properties of simple yield stress fluids. The third section is devoted to the measurements of the properties of thixotropic yield stress fluids. The phenomena of shear localisation and shear banding are discussed in the fourth section, together with material heterogeneities.

Key words: rheology, rheometry, yield stress fluids, elastic modulus, yield stress, constitutive law, thixotropy, shear localisation, shear banding, migration.

2.1 Rheometry

2.1.1 Simple shear

In order to predict the behaviour of materials in any complex flow, which *a priori* involves strains in all directions of space, it is necessary to know the relationship between the deviatoric stress tensor σ_{ij} and the strain rate tensor d_{ij} (which is, for example, $\sigma_{ij} = 2\eta d_{ij}$ for a Newtonian fluid of viscosity η).[1] Such a tensional relationship is very hard to derive from experimental measurements. The objective of rheometry is to simplify the problem as much as possible by measuring the flow properties of materials in the case where there is only one non-zero component of the strain rate tensor, namely simple shear. Another simple case is extensional flow, which provides a complementary means of characterisation of materials but is not studied here (see Macosko, 1994). Note also that new methods have recently been developed to investigate the behaviour in controlled three-dimensional flows (Jop *et al.*, 2006; Ovarlez *et al.*, 2010).

Thanks to its simplicity, simple shear allows us to characterise and understand the flow behaviour of complex fluids. Moreover, such characterisation can be used directly to predict the flow behaviour in many practical situations (pipe flows, flows down a slope, etc). Nevertheless, in more complex cases, this characterisation still has to be generalised in a tensorial form.

In simple shear (Fig. 2.1), shear occurs only through the relative motion of parallel layers of the material. Noting x the direction of flow and y the direction of the spatial variations of the velocity $V_x(y)$, the only non-zero component of the

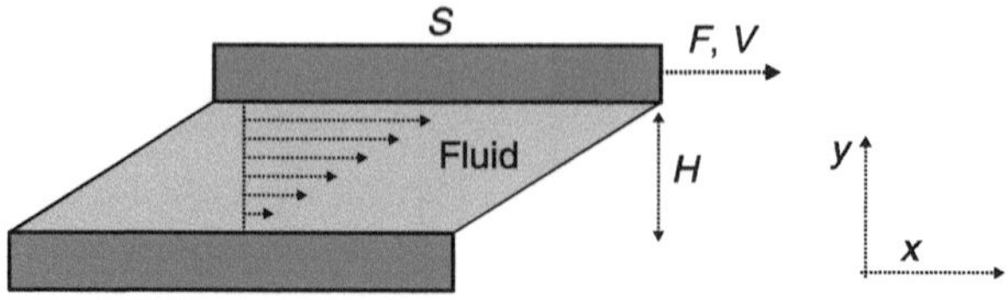

2.1 Schematic representation of a simple shear experiment.

strain rate tensor in simple shear is $d_{xy} = d_{yx}$. It is then convenient to define the shear rate as $\dot{\gamma} = 2d_{xy} = \frac{dV_x}{dy}$ which is thus the velocity gradient. If $\dot{\gamma}$ is constant along the y direction, simple shear is said to be homogeneous.

Three components of the stress tensor can be non-zero in simple shear: the shear stress $\tau = \sigma_{xy}$, and the first and second normal stress differences $N_1 = \sigma_{xx} - \sigma_{yy}$, $N_2 = \sigma_{yy} - \sigma_{zz}$. The purpose of rheometry is to characterise the dependence of the shear stress τ (and, if necessary, of the normal stress differences) on the shear strain γ and/or the shear rate $\dot{\gamma}$ in simple shear.

Homogeneous simple shear is theoretically achieved by putting a sample of the material between two infinite parallel plates separated by a distance H (named 'gap') and imposing a difference of velocity V between the two plates. The material behaviour is then characterised by measuring the relationship between two macroscopic quantities: the force $F(V)$ that has to be applied to enforce a flow characterised by the velocity V. In the ideal conditions of the homogeneous simple shear of Fig. 2.1, the shear rate and the shear stress are obtained through $\dot{\gamma} = V / H$ and $\tau = F / S$ (S being the plates' surface, supposed to tend to infinity). The macroscopic measurements $F(V)$ then provide directly the constitutive law $\tau(\dot{\gamma})$. Note also that, in some materials, a non-zero normal force F_N has to be applied to maintain a constant gap during shear; in the ideal conditions of Fig. 2.1, F_N provides a measurement of the total normal stress $\Sigma_{zz} = F_N / S$ as a function of $\dot{\gamma}$. As two normal stress differences may be non-zero, it is seen here that a single experiment in simple shear is not sufficient to obtain these two quantities. It is shown below that the use of two different geometries allows the two quantities to be measured.

In order to impose simple shear with a finite geometry, most experiments use rotational flows. The main problems of rheometry are then to:

- find conditions in which homogeneous simple shear can be achieved;
- relate in a general case (homogeneous and heterogeneous shear), the measured macroscopic quantities (rotational velocity, torque, normal force) to local quantities (shear rate and stresses) and thus to the material constitutive behaviour $\tau(\dot{\gamma})$.

It can be seen below that, in many cases, only approximate equations exist. It is thus vital to know their domain of validity and to be able to choose the equations appropriate for a given material in a given situation.

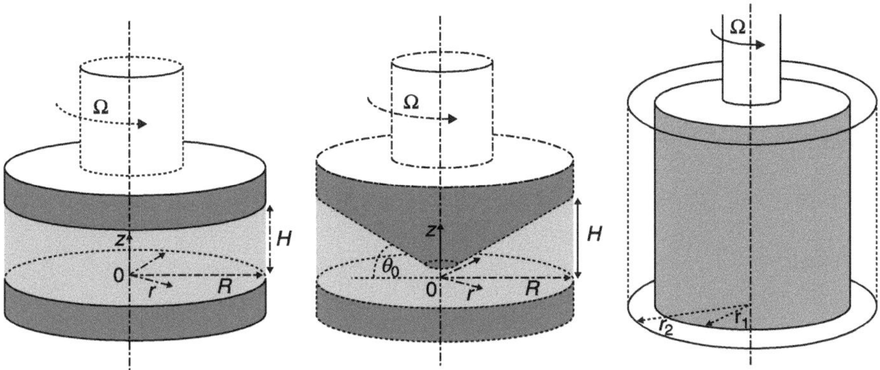

2.2 Standard geometries: parallel plates, cone and plate, Couette (Coussot, 2005).

The three main geometries that achieve simple shear flow are cone and plate, parallel plate and Couette (Fig. 2.2). Simple shear is also achieved in capillary flow and inclined plane geometry. These are not studied here but details can be found in, for example, Macosko (1994) and Coussot (2005).

Below we present the basic equations and their origin, and discuss the application to complex fluids; detailed calculations can be found in Macosko (1994).

2.1.2 Cone and plate

In a cone and plate geometry, the material is put between a cone and a disk of the same symmetry axis and the same radius R. The cone tip is truncated to avoid friction between the cone and the bottom plate. The cone virtual tip is supposed to touch the plate. The bottom plate is fixed, and shear is imposed by rotating the cone around the symmetry axis at a velocity Ω, resulting in a torque T.

In these conditions, in spherical coordinates, the only non-zero component of shear is $d_{\theta\varphi}$, which *a priori* depends on the angle only θ. From the stress equilibrium equation, it can then be shown that the shear stress $\tau = \tau_{\theta\varphi}$ (in spherical coordinates) is roughly homogeneous for small cone angles: the shear stress variation from the plate to the cone is of order of 0.5% for a cone angle θ_0=4°, which explains why cone of angles lower than 4° are always chosen. Because the shear stress τ is homogeneous and uniform in the whole gap, for materials which are characterised by a univalent relationship between stresses and strain rates (which excludes shear-banding materials), the shear rate $\dot{\gamma}$ is constant in the whole gap. With a no-slip boundary condition at the cone and plate surfaces, the shear rate can then simply be computed. The velocity at a radius r on the cone surface is Ωr; the velocity is zero on the plate surface, while the gap size at this radius is $r \tan\theta_0$

leading to $\dot{\gamma} \approx \frac{\Omega r}{r \tan\theta_0} \approx \frac{\Omega}{\theta_0}$. This equation simply relates the shear rate to the applied macroscopic rotational velocity Ω. The torque T resulting from a constant shear stress τ exerted on the cone surface is $T = \int_0^R 2\pi r^2 (\cos\theta_0)^2 \tau dr \approx \frac{2}{3}\pi R^3 \tau$. Note that the measurement of the normal force F_z exerted on the cone also provides:

$$N_1 = \tau_{\varphi\varphi} - \tau_{\theta\theta} = 2F_z/(\pi R^2) \quad [2.1]$$

To conclude, the main advantage of cone and plate geometry is that it achieves almost perfect homogeneous simple shear, provided the angle is small and that slip is avoided on the surfaces. Then, from the torque/rotational velocity relationship $T(\Omega)$, one can simply extract the constitutive behaviour $\tau(\dot{\gamma})$ thanks to reliable equations that are true whatever the material constitutive law is:

$$\tau = \frac{3}{2}\frac{T}{\pi R^3} \quad [2.2]$$

$$\dot{\gamma} = \frac{\Omega}{\theta_0} \quad [2.3]$$

This means that, together with small gap Couette geometry (see Section 2.1.4), this geometry is in general the best choice for material characterisation.

However, it should be noted that the truncature of the cone tip is usually a few tens of microns and that the gap at the edge for a 2°/40 mm diameter cone is 0.6 mm. This implies that this geometry cannot be used for many suspensions; it can be used for colloidal suspensions whose elements' size is of a few microns at most. Moreover, special care should be taken with shear-banding materials, for which two material states may coexist for a given shear stress. In this case, the approximate homogeneity of stress for small angles is still valid (it comes from stress equilibrium equation, independently of the material behaviour), but the shear rate need not be homogeneous anymore (a basic analysis of this case is found in Sections 2.3.1 and 2.4.3; details are found in Coussot, 2005; Ovarlez *et al.*, 2009).

Other problems may occur with thixotropic systems: at yield, the material may experience strong rejuvenation near the cone (where the shear stress is slightly higher) and stay solid near the plate; subsequent analysis of the properties from macroscopic measurements in such conditions are then likely to be incorrect.

2.1.3 Parallel plates

In a parallel plate geometry, the material is put between two disks of same symmetry axis and of same radius R, separated by a distance H (named the 'gap').

The bottom plate is fixed and shear is imposed by rotating the upper plate around the symmetry axis at velocity Ω, resulting in a torque T.

In these conditions, in cylindrical coordinates, the only non-zero component of shear is $d_{z\theta}$ which *a priori* depends on both r and z. From the stress equilibrium equation it can then be shown that the shear stress $\tau(r) = \tau_{z\theta}(r)$ does not depend on z. This is the only information we have about the shear stress: in contrast with a cone and plate geometry, $\tau(r)$ cannot be computed directly from the torque $T = \int_0^R 2\pi r^2 \tau(r) dr$ only. For materials characterised by a univalent relationship between stresses and strain rates, this also means that the shear rate $\dot{\gamma}(r)$ depends only on the radius r. With a no-slip boundary condition at both plate surfaces, the velocity at a radius r is Ωr on the upper plate and zero on the lower plate, leading to:

$$\dot{\gamma}(r) = \frac{\Omega r}{H} \qquad [2.4]$$

In other words, simple shear is achieved at any radius r in a parallel plate geometry, but shear is heterogeneous along the radial direction. Equation 2.4 also means that, from the knowledge of the applied macroscopic rotational velocity Ω, the local shear rate in the gap is known whatever the material constitutive law is.

To compute the shear stress from the torque measurements, there are two possibilities. First, it can be shown that the shear stress $\tau(R)$ at the edge of the geometry ($r=R$) can be computed exactly as:

$$\tau(R) = \frac{T}{2\pi R^2}\left[3 + \frac{\Omega}{T}\frac{\partial T}{\partial \Omega}\right] \qquad [2.5]$$

When associated with the shear rate $\dot{\gamma}(R) = \frac{\Omega R}{H}$ computed at the same radial position R, Eq. 2.5 provides $\tau(\dot{\gamma})$ data from $T(\Omega)$ data. However, as this expression involves the derivative of the torque with respect to the rotational velocity, many accurate $T(\Omega)$ data have to be recorded to allow for the computation of $\tau(\dot{\gamma})$ data. It is thus not very convenient to use. Another possibility consists of using expressions that are strictly valid only for a given material and to use them for any material. For example, for a viscous fluid $\tau(r) = \eta\dot{\gamma}(r)$, one gets $\tau(r) = 2\frac{T}{\pi R^3}\frac{r}{R}$. Combining this equation with $\dot{\gamma}(r) = \frac{\Omega r}{H}$ makes it possible to extract a $\tau(\dot{\gamma})$ data from a single $T(\Omega)$ measurement. These couple of data can *a priori* be computed at any radial position r. This position can be chosen so as to minimise the error made in the material characterisation when using these equations for non-Newtonian materials. It has been shown that the error is minimised at $r \approx 3R/4$ for materials of constitutive law $\tau < \dot{\gamma}^n$ with $n<1.4$; combining $\tau(3R/4)$ and $\dot{\gamma}(3R/4)$ then leads to a correct evaluation of their behaviour within 2% error (Macosko, 1994). This explains why the choice of computing the shear stress and shear rate at *3R/4* is made by default in most rheometer software.

Note that the measurement of the normal force exerted on the upper plate also provides $\tau_{\theta\theta} + \tau_{rr} - 2\tau_{zz} = N_1 - N_2$ (Macosko, 1994). If normal stress differences are important, full determination of the stress tensor in simple shear then implies a need to use both parallel plate and cone and plate geometries.

To conclude, parallel plate geometries are characterised by a heterogeneous but controlled shear rate $\dot{\gamma}(r) = \frac{\Omega r}{H}$ in most situations provided slip is avoided on the surfaces. Standard approximate equations can be used in many situations to directly relate a torque/rotational velocity measurement $T(\Omega)$ to a data point of the constitutive behaviour through:

$$\dot{\gamma} = \frac{3}{4}\frac{\Omega R}{H} \qquad [2.6]$$

$$\tau = \frac{3}{2}\frac{T}{\pi R^3} \qquad [2.7]$$

It should be noted that the shear rate computed in that way is the exact value at $3R/4$ whatever the material constitutive law, whereas the shear stress is exact only for Newtonian fluids.

Finally, note that in contrast to a cone and plate geometry, the gap is not fixed. It can be as high as a few millimetres (the main limitation being that gravity effects have to remain negligible to allow for the above equations to be derived; Coussot, 2005). Parallel plates can then be used for suspensions whose elements size is of 200 μm at most (e.g. mortars with fine particles). Shear being heterogeneous, special care should be taken with yield stress fluids at low shear rate: shear localisation is then likely to appear (Coussot, 2005) and is more difficult to take into account than in a Couette geometry (see below).

2.1.4 Couette geometry

In a Couette geometry, the material is put between two coaxial cylinders. In most cases, the outer cylinder is fixed and shear is imposed by rotating the inner cylinder of height H around the symmetry axis at a velocity Ω, resulting in a torque T. Note that the distance of the inner cylinder to the bottom of the cup is usually chosen to be much larger than the gap size to ensure that the contribution to the torque from shear at the bottom is as negligible as possible; these end effects can be corrected with appropriate procedures if necessary (Macosko, 1994).

In these conditions, in cylindrical coordinates, the only non-zero component of shear is $d_{r\theta}$ which *a priori* depends on r only. From the stress equilibrium equation, it can be shown that the shear stress $\tau(r) = \tau_{r\theta}(r)$ is known everywhere in the gap and is given by:

$$\tau(r) = \frac{T}{2\pi h r^2} \qquad [2.8]$$

i.e. the shear stress is heterogeneous and controlled. On the other hand the shear rate $\dot{\gamma}(r)$ is *a priori* unknown. This is different from the case of a parallel plate geometry, where the shear rate is heterogeneous and controlled while the shear stress is unknown.

Two cases have to be distinguished: thin gap Couette geometry and wide gap Couette geometry. A Couette geometry is said to have a thin gap when the shear stress can be considered as homogeneous within 1%. From $\tau(r) = \frac{T}{2\pi h r^2}$, this means that $\frac{R_o^2 - R_i^2}{R_i^2} < 1\%$ i.e. that the gap size $R_o - R_i$ is smaller than 0.5% of the inner radius R_i. It then implies that, for materials characterised by a univalent relationship between stresses and strain rates (which excludes shear-banding materials), the shear rate is almost constant in the whole gap. With a no-slip boundary condition at the surface of both cylinders, the shear rate can then simply be computed as:

$$\dot{\gamma} = \Omega \frac{R_i}{R_o - R_i} \qquad [2.9]$$

while the shear stress is:

$$\tau = T/2\pi h R_i^2 \qquad [2.10]$$

Other cases are referred to as wide gap Couette geometries. In this case, as for a parallel plate geometry, an exact equation exists (Coussot, 2005):

$$2T \frac{\partial \Omega}{\partial T} = \left[\dot{\gamma}(\tau(R_i)) - \dot{\gamma}(\tau(R_o)) \right] \qquad [2.11]$$

However, Eq. 2.11 is not very convenient to use as it does not provide directly the value of the shear rate at a given position, but the difference between the shear rates at the inner and outer cylinders. Nevertheless it is used in the simple case of yield stress fluids close to yielding, where shear localisation occurs (see Section 2.4.3). It then reduces to $2T \frac{\partial \Omega}{\partial T} = \dot{\gamma}(\tau(R_i))$, which automatically takes shear localisation into account (Jarny *et al.* 2008; Ovarlez *et al.*, 2008; Coussot *et al.*, 2009).

It is more usual to use approximate equations. As for a parallel plate geometry, one computes the shear rate distribution for Newtonian materials and uses the $\dot{\gamma}(r)$ equation obtained in that way for any material. To extract $\tau(\dot{\gamma})$ data from a single $T(\Omega)$ measurement, a conventional choice is to compute the shear stress $\tau(r)$ and the shear rate $\dot{\gamma}(r)$ in the middle of the gap. Standard equations used to relate a torque/rotational velocity data $T(\Omega)$ to a constitutive behaviour data point $\tau(\dot{\gamma})$ are finally, when computed in the middle of the gap $\bar{R} = (R_o + R_i)/2$:

$$\dot{\gamma} = \Omega \frac{R_i^2 R_o^2}{\bar{R}^3 (R_o - R_i)} \qquad [2.12]$$

$$\tau = T/2\pi h\bar{R}^2 \qquad [2.13]$$

Other approximate equations can be used, depending on the radius where shear stress and shear rate are computed; at the inner radius, one gets $\dot{\gamma} = \Omega \frac{R_o^2}{\bar{R}(R_o - R_i)}$ and $\tau = T/2\pi h R_i^2$. Note that flows in a Couette geometry do not provide normal stress measurements unless specific normal force probes are put on the outer cylinder surface (Deboeuf *et al.*, 2009).

The main advantage of Couette geometries for complex fluids is that the gap can be large. It is thus the preferred geometry for suspensions made of large particles such as fresh concrete. However, unless the inner cylinder radius is very large, a wide gap implies a very heterogeneous shear stress distribution which poses the problem of shear localisation and shear-induced migration – see below.

2.1.5 Vane tool

Concentrated suspensions are often studied with a vane-in-cup geometry; see Barnes and Nguyen (2001) for a review. In this geometry, the inner cylinder of the Couette geometry is replaced by a vane tool, which consists of N thin blades centred on a thin shaft (usually, $N = 4$ or 6). Compared with the coaxial cylinders Couette geometry, it *a priori* offers two main advantages. First, it allows the study of the properties of structured materials with minimal disturbance of the material structure during the insertion of the tool. Second, it is supposed to avoid wall slip (Saak *et al.*, 2001) because the material sheared in the gap of the geometry is sheared by the (same) material that is trapped between the blades. Consequently, it is widely used to study the behaviour of pasty materials containing large particles as fresh concrete (Koehler *et al.*, 2006; Estellé *et al.*, 2008; Wallevik, 2008).

The streamlines in a vane-in-cup geometry are generally not circles. This implies that the above equations for a Couette geometry cannot *a priori* be applied in this case. In practice, for a Newtonian material, the torque exerted on a vane tool is lower than that exerted on a cylinder of same inner radius driven at a same rotational velocity; it is typically 10–30% lower for standard geometries (Sherwood and Meeten, 1991; Ovarlez *et al.*, 2011). Nevertheless, a practical method, known as the Couette analogy (Bousmina *et al.*, 1999; Estellé *et al.*, 2008), can still be used to infer constitutive laws from macroscopic measurements with a vane tool. It basically consists of calibrating the geometry with linear (Hookean or Newtonian) materials. The equivalent inner radius $R_{i,eq}$ of a vane-in-cup geometry is defined as the radius of the inner cylinder of a Couette geometry that would have the same torque/shear stress and velocity/shear rate conversion factors for a linear material. For any material, all macroscopic data are then analysed as if the material was sheared in a Couette geometry of inner cylinder radius $R_{i,eq}$. The nonlinearity (which affects the flow field) is sometimes accounted for as it is in a standard Couette geometry (Estellé *et al.*, 2008). This approach may finally provide constitutive law measurements within a good approximation (Baravian *et al.*, 2002).

Note that for suspensions of coarse particles, particle depletion may be induced very rapidly by shear near the blades. It results in a kind of wall slip near the blades (Ovarlez *et al.*, 2011), whereas the vane tool is mainly used to avoid this phenomenon. This suggests that a coaxial cylinders geometry with properly roughened surfaces is preferable, when possible, when studying flows of pasty materials with large particles. A detailed analysis is provided in Section 2.4.5.

2.1.6 Geometry factors for yield stress measurements

Some of the equations derived in parallel plate and Couette geometries are only approximate: they are chosen to provide the best possible characterisation of a wide range of materials with very different nonlinear rheological properties. When studying a given kind of material, one question that arises is the accuracy of these equations for specific quantities. Below we discuss the case of yield stress measurements in yield stress fluids.

In a parallel plate geometry, at yield $\tau(r) = \tau_Y$ everywhere on the upper plate. Together with $T = \int_0^R 2\pi r^2 \tau(r) dr$ this yields:

$$\tau_Y = \frac{3}{2}\frac{T}{\pi R^3} \quad [2.14]$$

It is remarkable that Eq. 2.14, which here is exact for yield stress measurements, is the same as Eq. 2.8, which is optimal for flow properties. Equation 2.8 can then be used reliably together with Eq. 2.7 to evaluate all material properties.

In a Couette geometry, yield first occurs on the inner cylinder, where stress is maximal. This implies that at yield, $\tau(R_i) = \tau_Y$, and thus:

$$\tau_Y = T/2\pi h R_i^2 \quad [2.15]$$

This means that, in a wide gap Couette geometry, the best choice of geometry factors for yield stress measurements may not be the same as those for flow properties (see for example Eq. 2.11 and Eq. 2.12).

In a vane-in-cup geometry, cylindrical symmetry seems to be recovered in yield stress fluids near yielding (Keentok *et al.*, 1985; Yan and James, 1997; Savarmand *et al.*, 2007). This would mean that $R_{i,eq} = R_i$ can be used for yield stress measurements and one may thus compute exactly the yield stress with:

$$\tau_Y = T/2\pi h R_i^2 \quad [2.16]$$

in this specific situation. This point is discussed in more detail in Section 2.4.5.

2.1.7 Experimental problems

Most of the experimental problems found in rheometric experiments come from the main hypotheses made when computing shear stress/shear rate relationships from torque/rotational velocity measurements. First, one has to ensure that the flow and the material are homogeneous; these specific issues are dealt with in Section 2.4. Then, a no-slip boundary condition is assumed at the interfaces of both tools. Rough surfaces are usually used to ensure there is no wall slip in the experiments. Moreover, the absence of wall slip in parallel plate and Couette geometries can be checked by performing experiments with two different gaps (Coussot, 2005). If different results with two different gaps are obtained, this means that the no-slip hypothesis made in the analysis is not correct, i.e. there is wall slip.

Another important problem in rheometry arises from the fact that the applied torque is the sum of the torque applied to the material and the inertial torque applied to accelerate/decelerate the rotating tool. One thus has to be particularly careful when applying ramps and oscillations (see Section 2.2.1). Although the inertial torque can theoretically be accounted for, this may pose accuracy problems when it is dominant. This makes high frequency oscillatory measurements particularly difficult. For more details on experimental problems, see Coussot (2005).

2.2 Characterisation of simple yield stress fluids

Yield stress fluids (see Chapter 1) are materials that flow steadily only when the applied shear stress τ in simple shear is higher than a yield stress τ_Y. In addition, the behaviour of 'simple' yield stress fluids is basically independent of the flow history. Different methods are presented below devoted to characterising the flow properties of yield stress fluids when $\tau \geq \tau_Y$ and their solid properties when $\tau \leq \tau_Y$, and to determining the value of τ_Y accurately.

2.2.1 Steady flow properties

Steady flow properties in simple shear are characterised by the stationary relationship $\tau = f(\dot{\gamma})$ between the shear stress τ and the shear rate $\dot{\gamma}$. These data can be obtained in both stress controlled and shear rate controlled experiments.

Creep tests

Creep tests consist of applying a series of constant shear stresses τ_i over a long period of time, usually after a preshear (Fig. 2.3). For, $\tau_i \geq \tau_Y$ a yield stress fluid reaches steady flow at a shear rate $\dot{\gamma}_i$ after a time t_{eq} which depends on τ_i and the material. The data $\tau_i = f(\dot{\gamma}_i)$ then define the flow curve of the material. For $\tau_i < \tau_Y$,

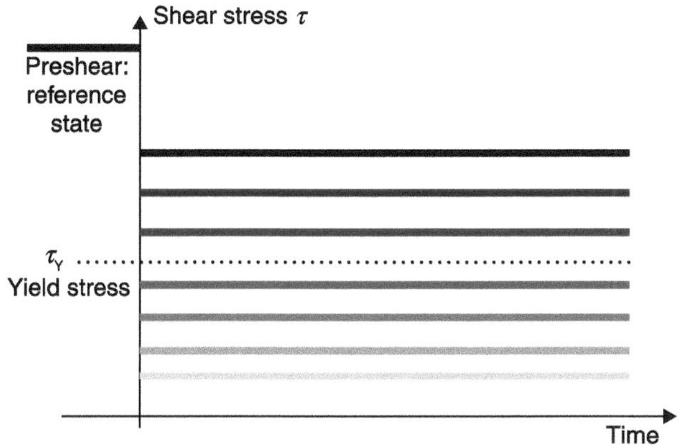

2.3 Schematic representation of a creep experiment.

one should observe unsteady flow with a vanishing shear rate $\dot{\gamma}_i(t) \xrightarrow[t\to\infty]{} 0$, which defines the yield stress τ_Y. This is illustrated in Section 2.3.1 for thixotropic yield stress fluids.

Of course, a series of constant shear rates can also be applied until a constant shear stress is reached. These last experiments are commented on in more detail in the case of thixotropic materials.

But although performing creep tests is a very accurate method for characterising materials, it is quite time consuming and does not provide a lot of data.

Ramps and sweeps

The independence of the behaviour of simple yield stress fluids on flow history usually implies that, when the applied shear stress (or shear rate) is changed abruptly, they ideally reach their new steady-state instantaneously (typically within one second). The practical consequences of this are that:

- a preshear is not necessary in principle, although a first preshear is still preferable just after loading to ensure a reproducible initial state;
- a series of applied shear stresses τ_i can be imposed in any order;
- the value of τ_i can be changed (almost) as quickly as desired to obtain a large set of $\tau_i = f(\dot{\gamma}_i)$ data.

This results in much faster experiments than creep experiments and more data. The main limitations to the rate of change of τ_i are the rheometer inertia and the material inertia.

Based on these principles, two experiments are usually performed: ramps and sweeps (Fig. 2.4). Stress ramps consist of continuously changing (upward or downward) the value of $\tau(t)$ in time and in recording $\dot{\gamma}(t)$ to give $\tau = f(\dot{\gamma})$; equivalently, shear rate ramps $\dot{\gamma}(t)$ can be applied. Two kinds of ramps are usually

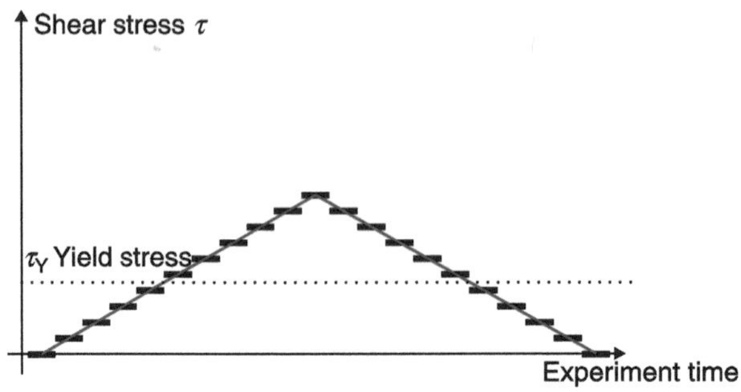

2.4 Schematic representation of stress ramp and stress sweep experiments.

distinguished: linear ramps, in which the applied shear stress or shear rate increases (or decreases) linearly with time, and logarithmic ramps in which the value of the shear stress or shear rate is multiplied by a constant value at any second (i.e. during a logarithmic shear rate ramp, the same time is spent in each decade of shear rates, and the shear rate evolution in time is $\dot{\gamma} = \beta \times 10^{\alpha t}$ with α and β free parameters of the ramp).

Usually, an up-ramp is followed by a down-ramp. This allows to show that the 'steady' behaviour of a 'simple' yield stress fluid is actually obtained by checking the reproducibility of the results obtained during the up- and down-ramps. If a hysteretic response is found, this usually means that the material is thixotropic (see Section 2.3.1). Note that ramps performed too quickly may also generate discrepancies between the results obtained during the up- and down-ramps because of the geometry and material inertia; a simple way to avoid (and test) this effect is to decrease the rate of change of shear stress/shear rate, which minimises the impact of inertia.

An alternative to ramps is sweeps, in which a series of increasing/decreasing constant shear stresses τ_i (or shear rates $\dot{\gamma}_i$) are applied during very short time periods (typically ten seconds). The advantage of sweeps is that they allow checking at each step that the behaviour is steady. The results of typical linear (up and down) stress ramps are shown in Fig. 2.5 for a model yield stress fluid (a concentrated emulsion). It is observed that the same behaviour is obtained during the up- and down-ramps (except for low shear rate). This shows that the steady behaviour $\tau(\dot{\gamma})$ is obtained and thus justifies the use of ramps.

The behaviour at low rate deserves a comment. During the first (up) ramp, the material is initially at rest and thus in its solid regime where its behaviour is basically elastic $\tau = G_{\gamma}$. This means that, when applying a linear ramp $\tau = \alpha_{\tau}$, as long as $\tau < \tau_Y$, the (elastic) material is strained at a constant rate $\dot{\gamma}(\alpha,G) = \alpha/G$. This explains the first part of the curve during the up-ramp in which a roughly constant shear rate is measured at any shear stress $\tau < \tau_Y$. Afterwards, the material flows and the $\tau(\dot{\gamma})$ data correspond to the flow behaviour. During the down-ramp,

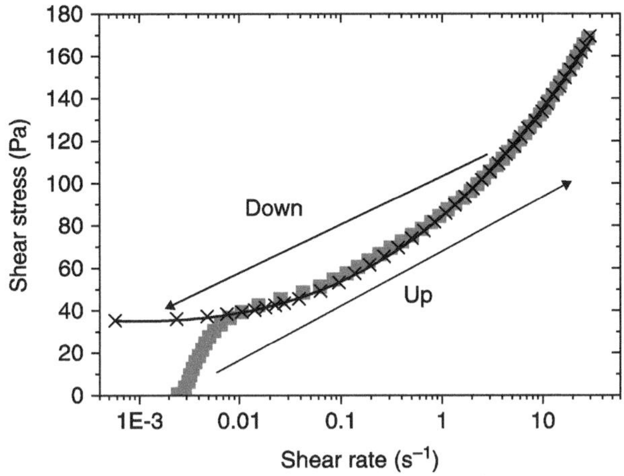

2.5 Shear stress vs. shear rate for linear with up (squares) and down (crosses) stress ramps for a model yield stress fluid (a concentrated emulsion).

the same $\tau(\dot{\gamma})$ curve is described until $\tau = \tau_Y$; when $\tau < \tau_Y$, the material progressively restores its elastic energy which results in a negative rate of strain (absent from the curve in Fig. 2.5 where the shear rate is in logarithmic coordinates). A simple way to check that data correspond to the flow behaviour is to perform the same experiment at two different ramp rates: only that part of the curve corresponding to a solid regime depends on this rate. Of course, the same features are observed when applying up/down shear rate ramps. This implies that, if only a short time is available to perform experiments, then a down-ramp should be applied, starting from a high shear stress (or shear rate), rather than an up-ramp.

Finally let us comment on the use of linear/logarithmic shear rate ramps. As yield stress fluids usually have nonlinear behaviour of the form $\tau = \tau_Y + K\dot{\gamma}^n$, if an accurate characterisation of the behaviour (in particular of the yield stress) is required, then it is better to have data for several decades of shear rates. In this case, one then has to apply a logarithmic shear rate ramp and to plot shear stress vs. shear rate in a logarithmic scale (Fig. 2.5). On the other hand, if only the behaviour in a limited range of $\dot{\gamma}$ is of interest, then a linear ramp can be applied: it will provide sufficient data for the characteristics of interest and will be faster to perform than a logarithmic ramp.

2.2.2 Elastic modulus measurements

In their solid regime, yield stress fluids usually behave as viscoelastic materials. At sufficiently low strain, their behaviour is linear and their elastic modulus is defined as $G = \tau/\gamma$ where τ is the steady shear stress that has to be applied to maintain a given steady shear strain γ. However, when strained, these materials also exhibit viscous behaviour and their elastic response is not instantaneous. In order to evaluate viscous and elastic effects separately, the classical experiment consists of applying strain oscillation:

$$\gamma(t) = \gamma_0 \sin \omega t \qquad [2.17]$$

and measuring the stress response $\tau(t)$. When the material is probed in its linear regime,[2] $\tau(t)$ is of the form:

$$\tau(t) = G'(\gamma_0 \sin \omega t) + G''(\gamma_0 \cos \omega t) \qquad [2.18]$$

As elastic effects are in phase with the strain whereas viscous effects are in phase with the strain rate (and thus in quadrature with the strain), this defines an elastic modulus G' and a viscous modulus G''. Alternatively, stress oscillations can be applied and strain oscillations measured, which does not change the definition of G' and G''.

To understand the interpretation of these coefficients, it is useful to consider the most simple model of viscoelastic solid, which is that of Kelvin–Voigt (Fig. 2.6), symbolised as a spring and a damper in parallel.

In this model, the relationship between the shear stress and the shear strain is $\tau = G\gamma + \eta\dot{\gamma}$. For the above applied strain oscillations, this yields $\tau = G\gamma_0 \sin\omega t + \eta\gamma_0\omega\cos\omega t$. For the Kelvin–Voigt model, and for that model only, G' and G' have a simple interpretation as $G'(\gamma_0,\omega) = G$, $G''(\gamma_0, \omega) = \eta\omega$. However, in general, G' and G'' have a complex dependence on both γ_0 and ω (see for example Derec *et al.*, 2003). It is thus usual to perform amplitude sweeps (where oscillations of increasing γ_0 are applied) to check in which range of strains the material has a linear behaviour: this defines a critical strain below which G' is constant (Fig. 2.7a) (Schultz and Struble, 1993; Nachbaur *et al.*, 2001). In this regime, frequency sweeps (where oscillations of increasing ω are applied) may then be performed to better understand the dissipation properties of the material.

Whereas G' and G'' depend on shear history for complex fluids (in particular on the time elapsed since the beginning of the measurement), these are constant

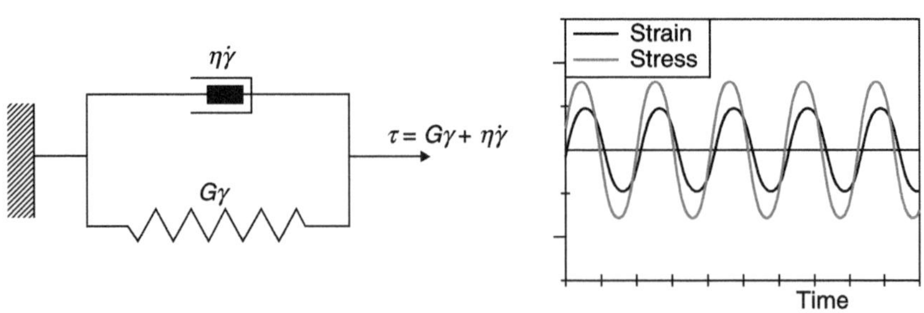

(a) (b)

2.6 (a) Kelvin–Voigt model; (b) typical oscillatory experiment.

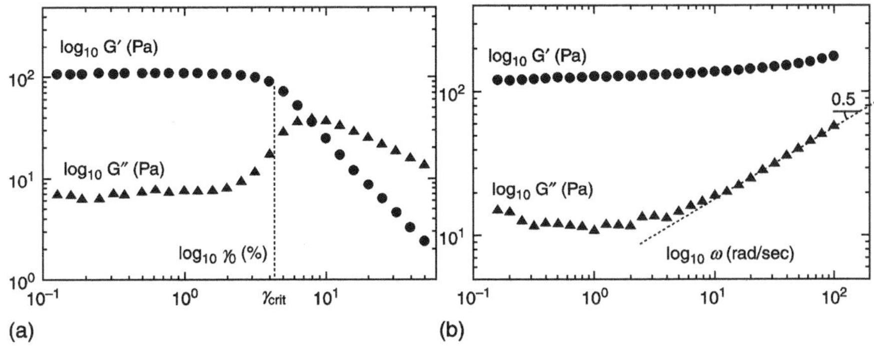

2.7 (a) G′and G″vs. strain measured in a dense colloidal suspension during a strain sweep oscillatory experiment; (b) G′and G″vs. frequency in the linear regime, measured in the same suspension during a frequency sweep oscillatory experiment. (Derec *et al.*, 2003).

material properties in simple yield stress fluids and no specific preparation is *a priori* required to perform these measurements. It is worth noting that, in yield stress fluids, G' is usually found to depend only slightly on ω (Fig. 2.7b); this implies that an experiment at low strain and at a single frequency is basically sufficient to characterise the elastic properties of yield stress fluids in their solid regime.

Note finally that the linearity of the response can be easily checked through the raw response to oscillations. If a sinusoidal response is obtained when a Sinusoidal oscillation is imposed, then the G' measurement can be trusted as the elastic modulus of the material without performing strain sweep.

2.2.3 Yield stress measurements

Simple yield stress fluids are characterised by a single yield stress τ_Y. Starting from a liquid state, flow stops when the shear stress τ is decreased below τ_Y; starting from a solid state, flow starts when τ is increased above τ_Y. Several methods exist for the determination of τ_Y.

A first method simply consists of analysing the flow curves $\tau(\dot{\gamma})$ obtained thanks to the experiments described in Section 2.2.1: the yield stress is the value τ_Y of the shear stress at vanishing shear rate $\tau(\dot{\gamma}) \xrightarrow[\dot{\gamma}\to 0]{} \tau_Y$. This means that data at a sufficiently low shear rate have to be measured; as pointed out in Section 2.2.1, the best method of measuring $\tau(\dot{\gamma})$ in these circumstances consists of applying a decreasing logarithmic shear rate ramp. A decreasing stress ramp can also be applied: in this case, it has to be slow enough because small stress changes lead to large shear rate changes at the approach of τ_Y.

When only the yield stress is of interest, it can be measured directly by applying a constant low shear rate $\dot{\gamma}_0$ to the material, starting from rest. A typical measurement is displayed in Fig. 2.8. Two regimes are observed: a roughly linear increase of the shear stress at low strain, followed by a stress plateau. Initially, the material is a solid: the first regime then corresponds to the elastic straining of the material $\tau \simeq G\gamma \simeq G\dot{\gamma}_0 t$. When the shear stress reaches the yield stress, i.e. when the strain is equal to the yield strain $\gamma_Y \simeq \tau_Y/G$, the material turns to a fluid and quickly reaches its stationary state, characterised by a resistance to flow $\tau(\dot{\gamma}_0)$. The fact that a plateau is observed is typical of simple yield stress fluids. For a sufficiently low value of $\dot{\gamma}_0$, this plateau is identified with the yield stress as $\tau(\dot{\gamma}_0) \xrightarrow[\dot{\gamma}_0 \to 0]{} \tau_Y$. 'Low' shear rate means that viscous effects are negligible, but depends on the material. For a material behaviour of the form $\tau = \tau_Y + K\dot{\gamma}^n$, this means that $\dot{\gamma}_0 << (\tau_Y/K)^{1/n}$. Typical values for standard yield stress fluids are $\dot{\gamma}_0 = 10^{-3}$ to 10^{-2} s^{-1}.

Finally, stress (or strain) amplitude sweeps in oscillatory experiments may also provide an evaluation of the yield stress. G' and G'' measured during such an experiment are shown in Fig. 2.7a. At low stress, in their linear regime, yield stress materials display dominant–constant–elastic properties with $G' >> G''$. On increasing stress, the material leaves its linear regime, G' decreases and G'' increases. At a given shear stress, which is often identified with the yield stress τ_Y, G' and G'' cross; this crossover approximately corresponds to a peak in G'', which subsequently decreases on increasing stress. In this last regime, $G'' > G'$ consistently with the interpretation that $\tau > \tau_Y$ and that the material is in its liquid regime. The decrease of G'' with increasing stress for $\tau > \tau_Y$ is a signature of the

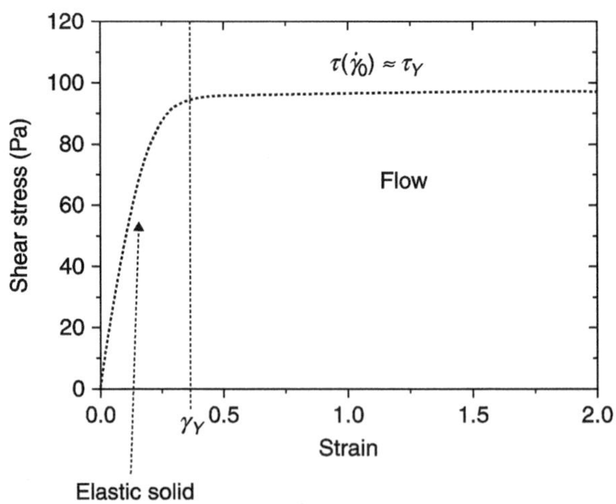

2.8 Shear stress vs. shear strain during a yield stress measurement at low constant shear rate for a model yield stress fluid (a concentrated emulsion).

shear-thinning properties of the material. Note that such determination of the yield stress is rather qualitative.

2.3 Characterisation of thixotropic yield stress fluids

Thixotropic yield stress fluids (see Chapter 1) are materials whose behaviour depends on the flow history. Their characterisation implies rigorous procedures to ensure reproducibility and to allow the prediction of the behaviour in any complex situation. Different methods are presented below devoted to characterising their flow properties, solid properties and yield stress.

2.3.1 Flow properties

This section begins by describing the most basic characterisation of thixotropy. It then explains how the most important characteristics of thixotropic materials are determined: dynamic yield stress, critical shear rate, steady-state behaviour and characteristic time to reach steady state in simple situations.

Thixotropic loop

The simplest experiment which aims at evidencing thixotropy is the so-called 'thixotropic loop' test (see for example Assaad *et al.* 2003). A thixotropic loop is the hysteretic response obtained in a shear stress vs. shear rate plot (Fig. 2.9) when a material is submitted to a cyclic shear history – typically, a series of up and down shear rate ramps, cf. Fig. 2.4. The observation of a hysteretic response shows that the material behaviour depends on shear history – in contrast with simple yield stress fluids. For a given procedure, the amplitude of the hysteresis gives an indication of the importance of the material dependence on shear history. However, it is very difficult to extract any quantitative information from such a test. Indeed, throughout the whole shear history in this test, the material structure evolves in a way which depends complexly on the parameters used (ramp rate, linear ramp vs. logarithmic ramp, etc.). It is worth noting in particular that the response to a series of up/down ramps evolves in time: it stabilises after several loops (Toorman, 1997).

To better characterise thixotropic yield stress fluids, it is better to perform creep test and shear rate steps tests. These are described below.

Creep tests

Creep tests (see Section 2.2.1) are the first important tests to perform as they allow determination of the dynamic yield stress and provide evidence of the viscosity bifurcation phenomenon specific to thixotropic materials (Coussot *et al.*, 2002; Coussot *et al.*, 2006).

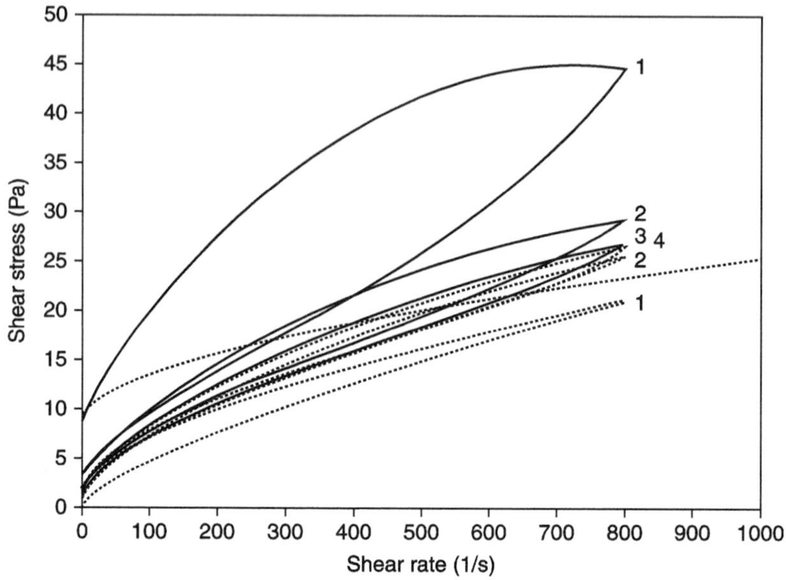

2.9 Thixotropic loops: response of a model of thixotropic material to a series of up/down ramps. Reprinted with permission from Springer Science Business Media: Toorman (1997), Figure 7a.

Before applying a given constant shear stress τ_i, it is important to preshear the material at a high shear rate for long enough to ensure a steady state is reached. This ensures that all tests always start from the same destructured state of the material. Creep tests provide the dynamic yield stress of the material, namely the value $\tau_{Y,d}$ above which a steady flow is reached. Compared with simple yield stress fluids, the specificities of the response to creep tests of thixotropic materials are that

- the time $\tau_{eq}(\tau_i)$ for which the material reaches steady-state may be long;
- the steady shear rate achieved at shear stresses τ_i close to $\tau_{Y,d}$ is usually higher than a critical value $\dot{\gamma}_c$ i.e. $\dot{\gamma}_i(\tau_i) \xrightarrow[\tau_i \to \tau_{Y,d}]{} \dot{\gamma}_c$.

This contrasts with the behaviour of simple yield stress fluids where $\dot{\gamma}_i(\tau_i) \xrightarrow[\tau_i \to \tau_{Y,d}]{} 0$. In other words, the material steady viscosity tends to a finite value when the applied stress tends to $\tau_{Y,d}$; which led this phenomenon being called 'viscosity bifurcation'. An example of the response of a thixotropic material to creep tests is shown in Fig. 2.10 and steady state flow curves obtained in creep tests experiments for various thixotropic yield stress fluids are depicted in Fig. 2.11. It is clearly seen that flow starts only above a critical shear rate $\dot{\gamma}_c$, which is a characteristic of the material.

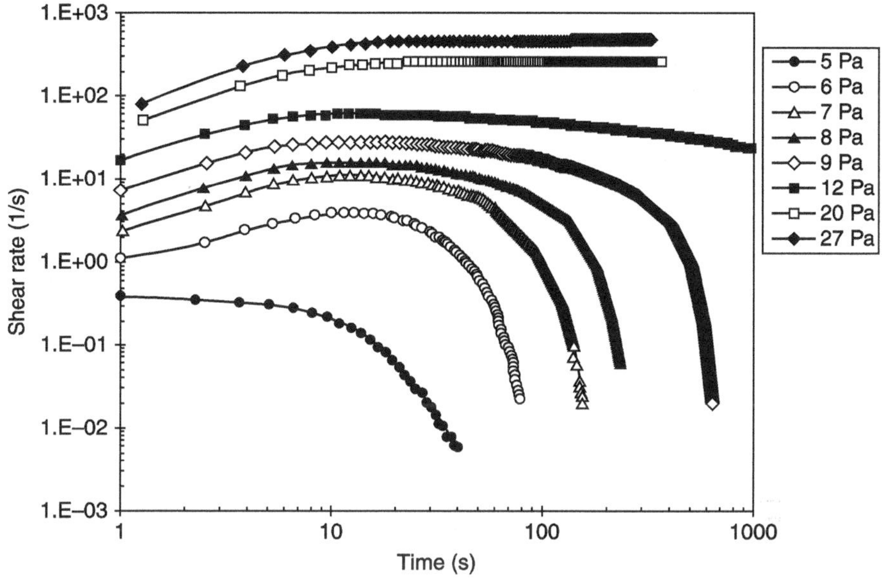

2.10 Shear rate vs. time in creep experiments for a bentonite suspension. Reprinted with permission from Coussot *et al.* (2002) copyright (2002) by the Society of Rheology.

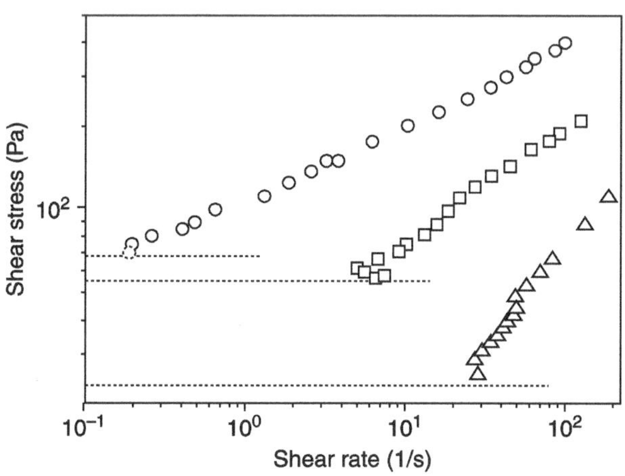

2.11 Steady state shear stress vs. shear rate in creep experiments for a gel (circles), a mustard (squares) and a colloidal suspension (triangles). For each material, no steady flow was observed below a well-defined critical shear rate. Reprinted with permission from Coussot *et al.* (2006) copyright (2006) by the Society of Rheology.

The fact that it is impossible to impose a stable flow at a shear rate lower than $\dot{\gamma}_c$ in thixotropic materials is of major importance for controlled shear rate experiments and ignorance of this feature may indeed lead to experimental errors. In a controlled shear rate mode, a macroscopic shear rate is achieved by imposing the rotational velocity of the tool, which is computed using standard equations of rheometry (see Section 2.1). These standard equations are based on the assumption of a homogeneous flow in the gap of the geometry. However, from the above observations, this macroscopic shear rate $\dot{\Gamma}$ corresponds to the true (local) shear rate of a homogeneous flow only if it is higher than $\dot{\gamma}_c$. In cases where $\dot{\Gamma}$ is lower than $\dot{\gamma}_c$, the material behaviour is not compatible with the prescribed shear rate and a phenomenon of 'shear banding' is observed: the material splits into two regions in the gap, a region flowing at $\dot{\gamma}_c$, and a non-flowing region (Coussot, 2005; Ovarlez *et al.*, 2009) (Fig. 2.12). The relative width of the two regions is such that the average shear rate equals $\dot{\Gamma}$. Characterisations obtained for macroscopic shear rates lower than $\dot{\gamma}_c$ are thus meaningless as they correspond to the response of a heterogeneous material. Moreover, if such a macroscopic shear rate is imposed once, it may affect all subsequent measurements as the material history is spatially heterogeneous.

To conclude, let us stress that only prior determination of the critical shear rate in a controlled stress mode allows the domain of validity of controlled shear rate experiments to be determined.

Shear rate steps tests

Although a first insight into the kinetics of material structuration under shear can be obtained from creep tests described above through the time necessary to reach steady state, the response to shear rate step tests appears to be more relevant. Indeed, most structural models are written as function of the shear rate $\dot{\gamma}$, as the characteristic time $1/\dot{\gamma}$ seems to fix the timescales of structuration and destructuration undershear (Mujumdar *et al.*, 2002; Roussel *et al.*, 2004; Dullaert and Mewis, 2005; Roussel, 2005).

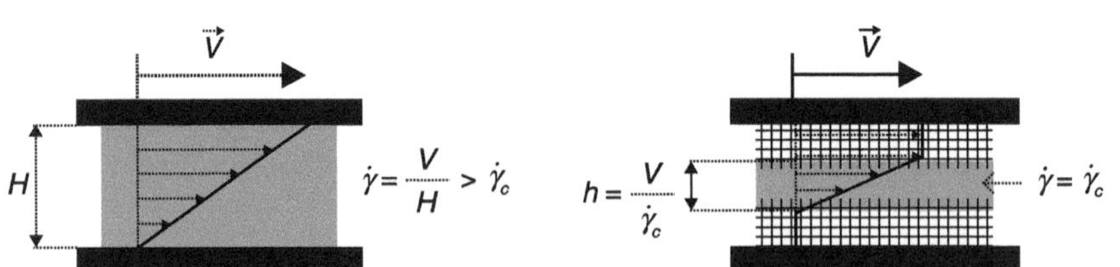

2.12 Sketch of the velocity profiles in a simple shear experiment. Left: when the macroscopic shear rate V/H is higher than the critical shear rate; right: when the macroscopic shear rate V/H is lower than the critical shear rate.

A shear rate step test is performed as follows (Fig. 2.13): starting from a steady state at a given shear rate $\dot{\gamma}_1$, the shear rate is changed abruptly to a new value $\dot{\gamma}_2$ until a new steady state is reached. The transient response is studied by plotting the shear stress (or the apparent viscosity) vs. time, the origin of time being defined as the instant of the change in shear rate.

These experiments provide several pieces of information. First, of course, the steady behaviour is obtained. Second, the instantaneous viscosities measured just after a step-change from an initial value $\dot{\gamma}_1$ characterise the intrinsic behaviour of the material whose structure is that imposed by steady shear at $\dot{\gamma}_1$ (Toorman, 1997; Ovarlez *et al.*, 2006). Finally, and most importantly, the kinetics of structuration/destructuration are determined. If $\dot{\gamma}_1 < \dot{\gamma}_2$, the transient shear stress response to the shear rate step-change in characterises destructuration (or breakdown of the structure), whereas when $\dot{\gamma}_1 > \dot{\gamma}_2$ it characterises structuration (or build-up of the structure). Analysis of these responses enables the parameters of structural models

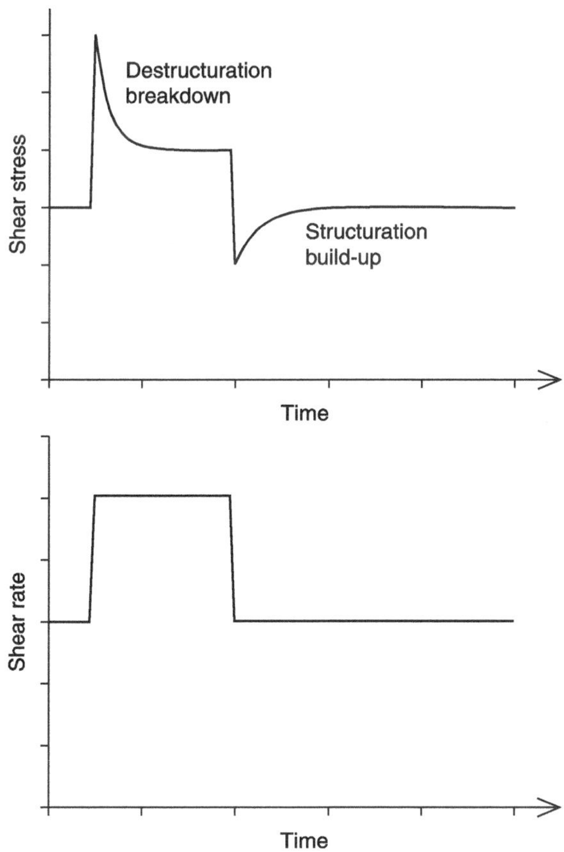

2.13 Representation of the stress response (top) to step-changes in shear rate (bottom).

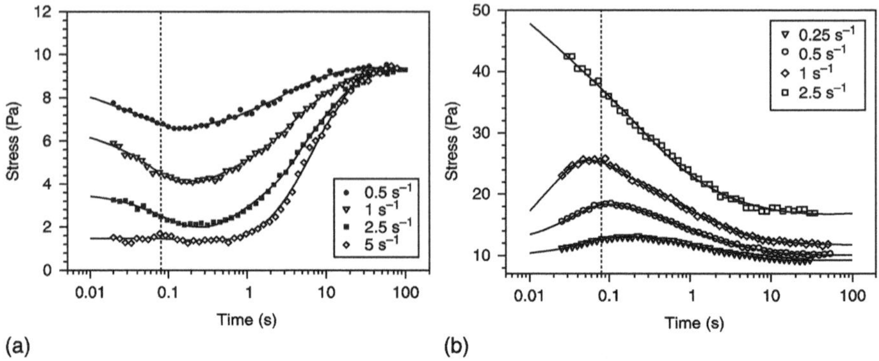

2.14 Stress response to step changes in shear rate in a thixotropic suspension: (a) when the shear rate is decreased (Dullaert and Mewis, 2005). toward from various values 0.25 s^{-1}; (b) when the shear from various values rate is increased towards 5 s^{-1}. Reprinted with permission from Dullaert and Mewis (2005). Copyright (2005) by the Society of Rheology.

to be determined. When the response can be modelled by a simple exponential, a characteristic time of structuration/destructuration is extracted (Mujumdar *et al.*, 2002; Coussot, 2005). It seems that these characteristic times mostly depend on the target shear rate $\dot{\gamma}_2$. An example of the response of a thixotropic material to shear rate step-changes in is shown in Fig. 2.14.

2.3.2 Solid properties

At rest, thixotropic materials become structured in time; they are also said to age (see Chapter 1). At a macroscopic level, this implies that their elastic modulus and static yield stress increase with the time they spend at rest. This section introduces methods devoted to characterising these evolutions.

Procedure for obtaining a well-defined aging solid state

Before characterising the structuration at rest of thixotropic materials, it is crucial to achieve a well-defined initial state. It is *a priori* possible to turn the material to a solid by starting from any liquid state and by applying any shear stress history below the dynamic yield stress. However, it has been shown that the values of the macroscopic properties of the material in its solid state depend strongly on the shear stress applied during flow stoppage (Ovarlez and Chateau, 2008) and may also depend on the shear rate history applied in the liquid state (Rueb and Zukoski, 1997).

In order to study the material structuration at rest independently of these effects (which can be studied separately), a procedure consists of:

- preshearing strongly the material in order to always start the experiments from a reproducible destructured state;

- applying a zero stress to the material (this defines a period during which the material turns from a fluid to a solid and subsequently ages).

Rest is thus here defined as a period during which a zero stress is applied (Ovarlez and Chateau, 2008). Note that it is very different to apply a zero shear rate after a preshear: if the material has the possibility to store elastic energy, its solid state is ill-defined as the applied shear stress at rest may then take any value between 0 and the dynamic yield stress.

Elastic modulus

Once the material is at rest, its structuration can be studied thanks to oscillatory experiments in the linear regime (see Section 2.2.2). From the definition of rest given above, it means that stress (and not strain) oscillations should be applied. In this case, the stress is indeed zero on average. One then has to plot the elastic modulus value vs. time. A signature of structuration is found in the continuous increase of the elastic modulus in time (Derec *et al.*, 2003; Coussot *et al.*, 2006; Ovarlez and Chateau, 2008) (Fig. 2.15).

It should be checked that it is a non-perturbative experiment, i.e. that applying these oscillations does not change the material state in time. This can be done by comparing the elastic modulus obtained after a long resting time in two situations: when oscillations are applied during all the resting period and when oscillations are applied only at the end of the resting period. It seems that as long as these measurements are performed in the linear regime of the materials, they are indeed nonperturbative (Ovarlez and Chateau, 2008); on the other hand, oscillations applied in a non-linear regime have been observed to modify the structuration kinetics (Viasnoff and Lequeux, 2002).

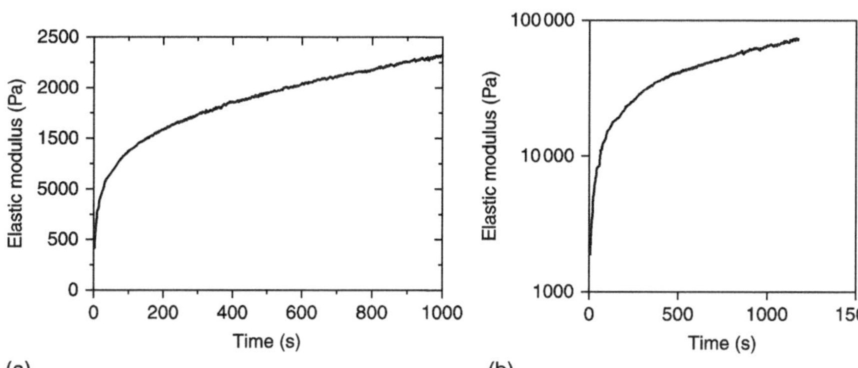

2.15 Elastic modulus vs. time spent at rest for (a) a thixotropic bentonite suspension (adapted from Ovarlez and Chateau, 2008) and (b) for a cement paste with a water to cement ratio equal to 0.35 (adapted from Roussel *et al.*, 2011).

Static yield stress

Another consequence of structuration at rest is the increase of the static yield stress with the resting time: it is harder and harder in time to restart the flow of a thixotropic yield stress fluid. The static yield stress evolution can be studied by applying the yield stress measurement procedure described in Section 2.2.3 after a preshear and various resting periods at zero stress (as stated above). In such an experiment (Fig. 2.16a), the shear stress first increases roughly linearly with strain, corresponding to the elastic straining of the material in its solid regime; in agreement with Fig. 2.15, the initial slope of the shear stress increase vs. strain (i.e. the elastic modulus) is higher for higher resting time. At some yield strain (usually of the order of 0.1 and poorly dependent of the resting time), a peak is observed and is followed by a slow decrease of the shear stress. The peak defines the static yield stress, the decrease being a signature of destructuration of the material under slow shear in the liquid regime. In this last regime, for measurements performed at a same given low shear rate, the shear stress should always tend to the same plateau value (close to the dynamic yield stress) whatever the resting period duration before the measurement. Checking that the same plateau value is reached in a series of static yield stress measurements is a good test of the reproducibility of the procedure.

In order to study the time dependence of the static yield stress, this procedure (preshear/rest/measurement) has to be repeated many times for various resting times, with each experiment providing only a single point of the static yield stress vs. time relationship. In contrast with the elastic modulus measurement method, this method is indeed perturbative as the material is destructured by shear. Static yield stress measurements in a thixotropic suspension are displayed vs. resting time in Fig. 2.16b and show a clear increase in time. Note that in many systems it is observed that the relative increase of the elastic modulus in time and that of the

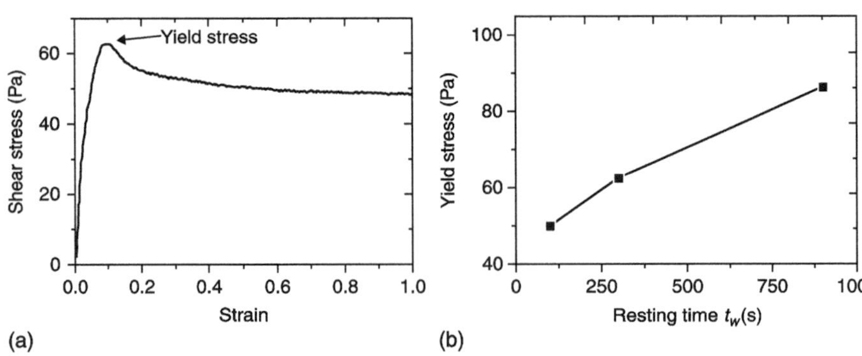

2.16 (a) Shear stress vs. strain for a yield stress measurement at low shear rate in a bentonite suspension. (b) Static yield stress vs. resting time (adapted from Ovarlez and Chateau, 2008).

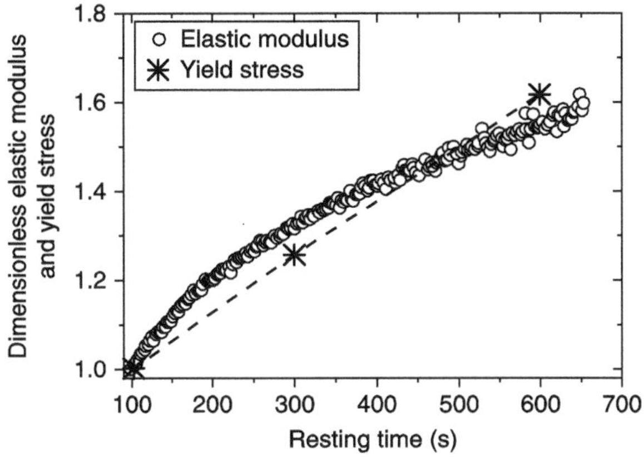

2.17 Dimensionless elastic modulus and dimensionless yield stress vs. resting time for a bentonite suspension (adapted from Ovarlez and Chateau, 2008).

static yield stress are quantitatively similar (see Fig. 2.17). In this case, the elastic modulus experiment should be preferred as a systematic method of characterisation of structuration as it is much more rapid, being nonperturbative.

However, cement pastes may be counterexamples to this generic behaviour. Recent experiments (Roussel *et al.*, 2011) tend to suggest that the linear elastic modulus and the yield stress of cement pastes may find different origins and may then have different time evolutions. Two structuration mechanisms may indeed be at work, one linked to early age hydration (the signature of which is found on the elastic modulus) and another linked to rearrangements of the colloidal network (which impacts the yield stress).

2.4 Advanced techniques for the study of local flow properties

However much care is taken, problems may occur during rheological measurements. This is indicated by the strong discrepancy between the values of the rheological parameters measured in the same materials using various concrete rheometers (Ferraris and Brower, 2001; Ferraris and Brower, 2004) (cf. Chapter 3). These discrepancies must find their origin in the violation of one or several of the main hypotheses at the basis of the conversion of the macroscopic data (torque, rotational velocity) into local data (shear stress, shear rate), namely:

- the no-slip boundary condition on the geometry walls;
- the homogeneity of the flow, and
- the homogeneity of the material.

Indeed, wall slip leads to overestimation of the shear rate in the material and thus to underestimation of its apparent viscosity. Flow heterogeneities (localisation) lead to underestimation of the true (local) shear rate and thus to overestimation of the material apparent viscosity. Material heterogeneities (migration and sedimentation) generally lead to overestimation of the material volume fraction and thus to underestimation of its apparent viscosity.

Wall slip can be mostly avoided thanks to the use of roughened surfaces. In contrast, one has *a priori* no control of the flow and material homogeneity. These characteristics of the flow cannot be checked without a means of visualisation. Below we discuss the principle of local information measurements and show how they can help to obtain access to the local and intrinsic constitutive behaviour. We then illustrate their application to the analysis of flow heterogeneities and of material heterogeneities. We finally present specific issues with a vane-in-cup geometry.

2.4.1 Experimental methods

The basic idea is to couple a well-controlled rheometric experiment, which provides macroscopic information about the flow (the velocity and the torque at the boundary), and a probe of the local material velocity and of the local material composition. Such probe has, of course, to be non-perturbative, which makes the experiments challenging.

Several methods exist:

- In optical techniques (see for example Barentin *et al.*, 2004), the motion of tracers is followed. In suspensions, the tracers are usually the constitutive elements of the material. A velocity field can be obtained either by tracking individual particles ('particle tracking' technique) or by tracking zones containing several particles thanks to image correlation ('particle image velocimetry' technique). The particle local concentration can also be measured directly. The main drawbacks of these techniques are their limitation to transparent materials and the need to seed the material with particles (which may be slightly perturbative) when the constitutive elements cannot be chosen as tracers. Note that other optical techniques exist such as those based on dynamic light scattering (Salmon *et al.*, 2003).
- In acoustic techniques (see for example Manneville *et al.*, 2004), a local velocity field can be obtained by sending ultrasonic pulses into the flowing material and correlating successive scattered signals (this provides a signature of the scatterer displacements). The scatterers are sometimes the material constitutive elements, but are more usually seeded particles, which may make the technique slightly perturbative. The main drawbacks of this technique are that the scattering properties of the material have to be compatible with it, and that the spatial attenuation of the signal does not allow measurements in wide gap geometries.
- In nuclear magnetic resonance (NMR) techniques (see for example Raynaud *et al.*, 2002; Rodts *et al.*, 2004), magnetic fields are used to interact with a

quantum mechanical property of the hydrogen nuclei in the material, namely the spin. It is a nonintrusive technique, as no seeding is required and as the spins can be handled independently of any other material property. Depending on the sequences of magnetic fields applied to the material in time, one can get information about the local hydrogen density (which provides the particle volume fraction in a suspension) or the local material velocity. The main drawback of such a technique is of course its cost and its difficulty of implementation. Its main advantage is that it can *a priori* be applied to any material, without any seeding, and that it is provides information on the bulk of the material in wide gap geometries. It is thus particularly adapted for complex suspensions such as cementitious materials.

2.4.2 Local constitutive law determination

From these techniques, the local velocity in the material $V(\vec{r})$ is known everywhere, allowing computation of the local strain rate. In order to obtain the local constitutive behaviour of the material, one then needs to know the local stress value, which implies in particular a torque measurement in rheometric experiments. As discussed in Section 2.1, in simple shear experiments, the stress distribution is known independently of the material behaviour in a Couette geometry and in a cone and plate geometry, but not in a parallel plate geometry (it is also known in capillary flows and inclined plane flows).

Below we illustrate the coupling of rheometric experiments and flow visualisation in the case of a Couette geometry. We consider a series of experiments performed at various rotational velocities Ω. We recall that the shear stress distribution $\tau(r,\Omega)$ within the gap of a Couette geometry is given by $\tau(r,\Omega) = T(\Omega)/2\pi r^2 H$, where $T(\Omega)$ is the torque exerted on the inner cylinder. The local shear rate $\dot{\gamma}(r,\Omega)$ is inferred from the velocity profiles and computed as:

$$\dot{\gamma}(r,\Omega) = \frac{V(r,\Omega)}{r} - \frac{\mathrm{d}V(r,\Omega)}{\mathrm{d}r} \qquad [2.19]$$

If the material is homogeneous, i.e. if it is the same at any position r in the gap, then all the $\tau(r,\Omega)$ and $\dot{\gamma}(r,\Omega)$ data can be combined into a series of data of the constitutive law $\tau = f(\dot{\gamma})$ measured locally. An example of such constitutive law determination is shown and commented on in Section 2.4.3 for a model yield stress fluid (a concentrated emulsion).

In the case where the material is heterogeneous, the local volume fraction $\phi(r,\Omega)$ has to be measured and taken into account, which is also possible with some of the methods described above (Ovarlez *et al.*, 2006). When the heterogeneous density profile $\phi(r)$ is independent of Ω, a change of variables can be performed between the radius r and $\phi(r)$. A local stress/strain-rate curve $\tau(\dot{\gamma},\phi)$ – at fixed and well-defined density $\phi(r)$ – is then obtained by collecting all measurements of local stress $\tau(r,\Omega)$ and shear rate $\dot{\gamma}(r,\Omega)$ for a fixed r and varying

Ω (note that the analysis is also possible but more complex when $\phi(r,\Omega)$ depends on Ω). An example of such constitutive law determination is shown and commented in Section 2.4.4 for a model mortar (sand particles in a concentrated emulsion).

Finally, it is worth noting that, in contrast with purely macroscopic experiments, this combination of techniques allows the derivation of the constitutive law of complex suspensions without any hypothesis.

2.4.3 Flow localisation

We now discuss the observation of flow heterogeneities in yield stress fluids; more details can be found in Ovarlez *et al.* (2009).

Shear localisation

In all geometries, the shear stress distribution is heterogeneous (even if the heterogeneity is slight in cone and plate and thin gap Couette geometries). In yield stress fluids, when the applied macroscopic stress is close to the yield stress τ_Y, this implies that the stress is lower than τ_Y and that the material does not flow in some regions of the gap of the geometry. This phenomenon is called 'shear localisation' as flow occurs only in some regions.

This is illustrated in Fig. 2.18a where local velocities measured in a concentrated emulsion flowing in a wide gap Couette geometry are depicted. It is observed that the material does not flow in the whole gap, the velocity being zero in a region

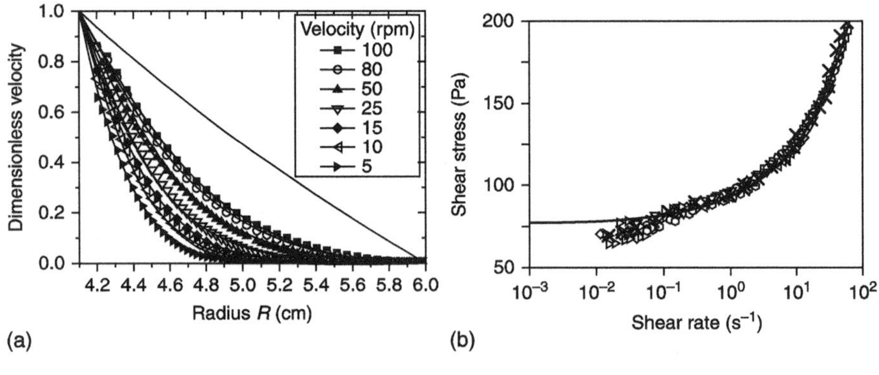

2.18 (a) Dimensionless velocity (ratio between the local azimuthal velocity and the azimuthal velocity at the inner cylinder surface) as a function of the radial position within the gap of a wide gap Couette geometry (R_i = 4.15 cm, R_o = 6 cm) for a concentrated emulsion (NMR measurements). (b) Local constitutive law inferred from the velocity profiles of Fig. 2.18a (empty symbols), and data obtained using Eq. 2.11 (crosses) (Ovarlez *et al.*, 2008).

near the outer cylinder. Moreover, the fraction of the material that is sheared decreases when the rotational velocity Ω of the inner cylinder decreases. Shear localisation is here attributed to the $1/R^2$ decrease of the shear stress $\tau(r,\Omega)$, which passes below τ_Y at the interface $R_c(\Omega)$ between the flowing and the non flowing regions. When Ω tends to 0, $R_c(\Omega)$ tends to R_i, i.e. the thickness of the flowing region tends to zero and the stress $\tau(R_i,\Omega)$ at the inner cylinder tends to τ_Y.

In standard macroscopic experiments, this phenomenon of shear localisation prevents us from calculating the true (local) shear rate within the gap with the standard equations of rheometry (see Section 2.1) and may thus lead to wrong evaluation of the rheological properties of the material. It is nevertheless possible to take this artefact into account thanks to Eq. 2.11, which is supposed to be valid without any hypothesis about the flow. When shear is localised, the shear rate at the outer cylinder is zero and this equation reduces to $2T\dfrac{\partial \Omega}{\partial T} = \dot{\gamma}(\tau(R_i))$, which is easy to handle. From a whole set of macroscopic measurements $T(\Omega)$, the material constitutive law can then be computed. Figure 2.18b shows a comparison of data obtained using this approach and data measured locally from the velocity profiles thanks to the approach presented in Section 2.4.2, in the concentrated emulsion of Fig. 2.18a. It is observed that the data obtained with Eq. 2.11 match perfectly the local behaviour, which is a strong validation of this equation (Ovarlez *et al.*, 2008; Coussot *et al.*, 2009).

In conclusion, one has to be careful when studying yield stress fluids. Shear localisation forbids the use of simple standard equations of rheometry, which lead to underestimation of the true shear rate. However, shear localisation can be properly taken into account using more elaborated equations such as Eq. 2.11 in a Couette geometry.

Shear banding

All yield stress fluids show shear localisation in a heterogeneous shear stress field. However, very different features are observed in simple and thixotropic yield stress fluids. In simple yield stress fluids, the velocity profiles tend smoothly to zero at the interface between the sheared and the unsheared regions (Fig. 2.18a). In contrast, in thixotropic yield stress fluids (such as cement pastes) flow stops abruptly: there is a discontinuity of the slope of the velocity profiles at the interface between the sheared and the unsheared regions (Fig. 2.19).

To understand these features, it should be recalled that the shear rate is $\dot{\gamma}(r) = V(r)/r - \mathrm{d}V(r)/\mathrm{d}r$ in a Couette geometry, which reduces to $\dot{\gamma}(r) = -\mathrm{d}V(r)/\mathrm{d}r$ at the interface between the sheared and the unsheared regions. A non-zero slope of the velocity profile at flow stoppage is thus the signature of the phenomenon of viscosity bifurcation (see Section 2.3.1): thixotropic materials cannot flow steadily for a shear rate below a value $\dot{\gamma}_c$. This critical shear rate $\dot{\gamma}_c$ is thus an intrinsic property of thixotropic yield stress fluids, in addition to their yield stress. In

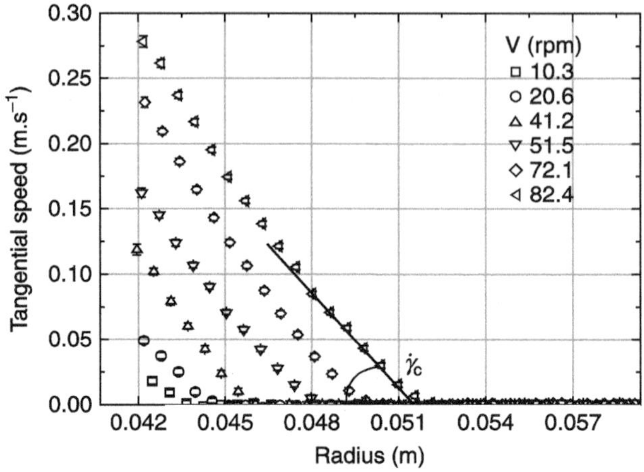

2.19 Azimuthal velocity profiles in a cement paste sheared in a wide gap Couette geometry for various velocities of the inner cylinder. Reprinted with permission from Jarny *et al.* (2005).

contrast, the smooth transition between flow and rest observed in simple yield stress fluids means that $\dot{\gamma}(\tau)\xrightarrow[\tau\to\tau_y]{}0$, consistently with standard constitutive laws such as the Bingham and Herschel–Bulkley laws. For more details on this issue, see Ovarlez *et al.* (2009).

An important consequence of this specific behaviour of thixotropic materials is that flow localisation also occurs in homogeneous stress fields for these materials, in contrast with simple yield stress fluids (see Fig. 2.18a and Section 2.3.1 for a discussion). This coexistence of two material states (here, sheared and unsheared) in a homogeneous stress field is called 'shear banding'. In any slow flow, these materials will experience shear banding, with the same consequences such distinct layers casting (Roussel, 2007) in cementitious materials.

2.4.4 Material heterogeneities

This section discusses the observations of flow-induced material heterogeneities in suspensions.

Shear-induced migration

Volume fraction measurements have led to show the development of concentration heterogeneities in noncolloidal suspensions in wide-gap Couette flows (Phillips *et al.*, 1992; Ovarlez *et al.*, 2006), parallel-plate flows (Chow *et al.*, 1994; Barentin *et al.*, 2004), pipe flows (Sinton and Chow, 1991) and extrusion experiments (Altobelli *et al.*, 1997). It has been observed in neutrally buoyant suspensions and is thus a shear-induced migration phenomenon; other phenomena occur in the presence of gravity (see below). It was observed mostly in suspensions of noncolloidal particles in Newtonian media. However, some experiments showed that the same phenomena occur when particles are suspended in viscoelastic fluids (Tehrani, 1996) and in yield stress fluids (Hafid *et al.*, 2010; Ovarlez *et al.*, 2011). In polydisperse systems, these heterogeneities can be accompanied by size segregation, which finds its origin in the particle size dependence of the kinetics of migration.

In wide-gap Couette flows, the consequence of migration is an excess of particles near the outer cylinder, i.e. particles tend to migrate from the regions of high shear to the regions of low shear (Fig. 2.20a). Migration is actually a direct consequence of shear heterogeneities. The basic mechanism is as follows: due to their interactions, particles experience shear-induced diffusion (Leighton and Acrivos, 1987). The shear-induced diffusion coefficient being an increasing function of the shear rate, shear heterogeneities naturally lead to a flux of particles towards the zones of low shear. For more details and theoretical developments, the reader is referred to Phillips *et al.* (1992), Morris and Boulay (1999), Ovarlez *et al.* (2006) and Lhuillier (2009).

When dealing with macroscopic rheometric measurements, the main problem is to avoid this phenomenon. If migration occurs during an experiment, then macroscopic data cannot be analysed as the response of a well-defined system. Data obtained on a heterogeneous system generally lead to underestimation of the apparent viscosity of the material (Ovarlez *et al.*, 2006). It is often considered in the literature on complex suspensions (such as cementitious materials) that migration is a slow phenomenon and that it is therefore possible to perform rheological measurements on a homogeneous material before the development of any significant heterogeneity. It was indeed observed in Newtonian suspensions

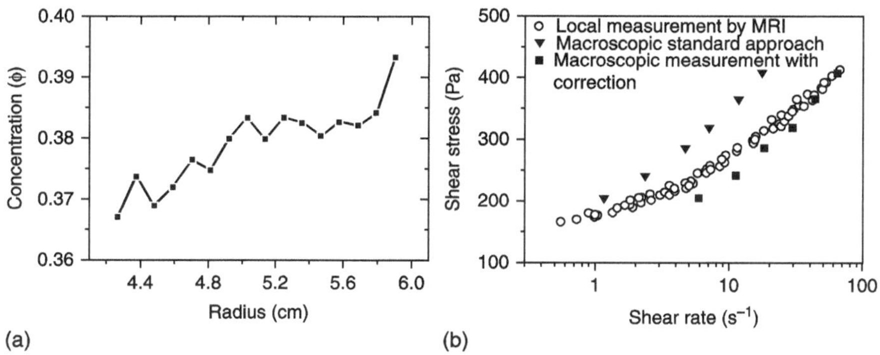

2.20 (a) Radial concentration profiles in a Couette cell for a suspension containing 38% of sand (sand size 160–200 μm) after an up-ramp in rotational velocity. (b) Macroscopic and local flow curves. Reprinted with kind permission from Springer Science + Business Media: Hafid *et al.* (2010), Figures 1 and 4.

that it is usually a slow diffusive phenomenon. The typical strainscale for migration to occur is expected to be rate independent and to scale as $\propto$ gap$^2/a^2$ where a is the particle size, i.e. the use of large gap to particle size ratio is likely to prevent migration. This would, for example, lead to an expected strain of the order of 10 000 for migration to occur when the gap is 100 particles wide.

However, it has been recently shown that, at high particle volume fraction, migration may be very rapid in some cases (depending on the flow characteristics) and thus unavoidable: the strain to induce migration can be smaller than 100 with 40 μm particles in a 2 cm gap, which is more than 500 times faster than expected (Fall *et al.*, 2011). It was also observed in a concentrated model mortar (sand in a model yield stress fluid): the volume fraction profiles depicted in Fig. 2.20a were obtained after a very rapid up-ramp in velocity (Hafid *et al.*, 2010). Consequently, this phenomenon can basically not be avoided when performing rheological measurements on such systems and it may occur in most fresh concrete rheology studies.

The impact of migration on the rheological measurements can be quantified by comparing the true (local) behaviour $\tau(\dot{\gamma},\phi_0)$ of the material at a well-defined constant volume fraction ϕ_0, obtained thanks to the technique detailed in Section 2.4.2, and the behaviour obtained from macroscopic experiments by assuming (wrongly) that the material is homogeneous. It is clearly observed that the macroscopic measurements tend to underestimate the apparent viscosity of the material at any rate and to underestimate its yield stress. This is a consequence of the volume fraction decrease near the outer cylinder. Note that the discrepancy between local/macroscopic measurements increases with the volume fraction (Ovarlez *et al.*, 2006; Hafid *et al.*, 2010). This phenomenon is most likely the main source of error in cementitious materials rheology (Hafid *et al.*, 2010).

Migration is part of the behaviour of the material and cannot be prevented. The only thing one can do in a rheological experiment is to evaluate the possible occurrence of migration and its impact. As migration is an irreversible process, the classical experiment consists of applying exactly the same macroscopic shear rate (or shear stress) at the beginning and at the end of a series of experiments: if the same response is obtained in both cases, it should mean that the material – supposed to be homogeneous just after loading – has remained homogeneous and that all measurements can be trusted. However, much care has to be taken with this technique: remember that migration can be very fast. Preshear thus has to be avoided; if a steady heterogeneous state is reached during the preshear, then the above procedure may lead one to wrongly conclude that the material is homogeneous. Ideally, the response at the very beginning of shear (typically, for a strain lower than ten) should be taken as a reference. It thus seems that clear conclusions can be reached from such approach in non-thixotropic systems only. In thixotropic materials, the time needed to reach steady state during the first measurement can be sufficient to generate steady heterogeneities, in which case the above experiment may again lead one to wrongly conclude that the material is homogeneous.

Shear-induced settling

Particle volume fraction measurements coupled with rheometric experiments also allow the coupling between flow and settling to be studied. It has indeed been shown that particles that are stable at rest in a yield stress fluid tend to settle when the material flows (Ovarlez *et al.*, 2010). To study this phenomenon properly, the effect of shear has to be decoupled as much as possible from the effect of gravity. A Couette geometry is then the best choice that can be made as it imposes shear in the plane perpendicular to sedimentation.

This phenomenon is illustrated in Fig. 2.21 for a suspension of noncolloidal particles in a yield stress fluid. After introduction in the Couette cell, the sample is homogeneous. When left at rest, the sample remains homogeneous and the particles do not settle. When the sample is sheared, particles move downwards. This is evidenced by the volume fraction profiles, which show classical features of sedimentation. The upper part is at 0% concentration; the bottom part remains at the initial 5% concentration (note that the particles tend to accumulate out of the measurement window, below the inner cylinder, which explains why the concentration in the lower part does not change). The transition zone between these two parts defines a sedimentation front that moves regularly towards the bottom as the flow duration increases. This sedimentation in a suspension that is stable at rest means that, under shear in the parallel plane, there is no more yield resistance to flow in the vertical direction. More details and a quantitative analysis can be found in Ovarlez *et al.* (2010). The main result is that the sedimentation velocity of particles in a yield stress fluid sheared at a shear rate $\dot{\gamma}$ is conveniently approximated by the Stokes velocity $V \propto \Delta\rho g R^2/\eta_{app}$ in a fluid of viscosity equal

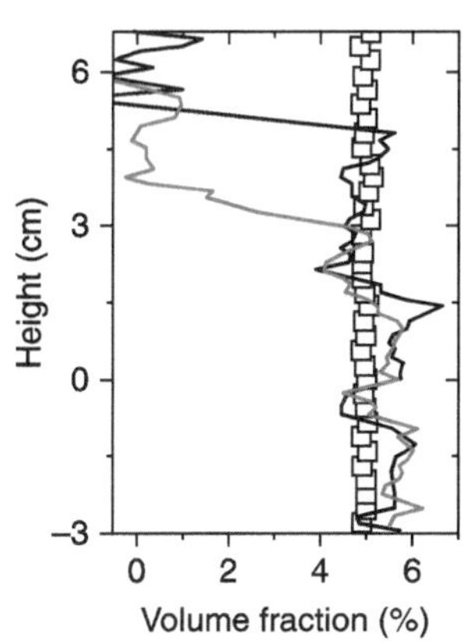

2.21 Vertical volume fraction profiles observed in the gap of a Couette geometry in a suspension of 5% of glass beads in a concentrated emulsion, after a 24 h rest (squares) and after 15 min (black line) and 25 min (grey line) of shear at 4 s^{-1}. Reprinted with permission from Ovarlez *et al.* (2010).

to the apparent viscosity $\eta_{app} = \tau(\dot{\gamma})/\dot{\gamma}$. This simple equation allows evaluation of the sedimentation velocity and thus of the possible occurrence of this phenomenon when studying suspensions of particles in yield stress fluids such as mortars and fresh concrete. Note that this settling velocity decreases when the particle volume fraction increases.

Finally, let us also mention the phenomenon of shear-induced resuspension, which was evidenced using the same methods (Acrivos *et al.*, 1993): in a Couette cell, particles that tend to settle at rest are resuspended by shear. The origin of this phenomenon is basically the same as that of shear-induced migration. The efficiency of this phenomenon is enhanced at high shear rate and high volume fraction.

2.4.5 Flow in a vane-in-cup geometry

Flow visualisations are also crucial for studying flows in geometries for which no theory exists, as a complement to simulations. For example, for a vane-in-cup geometry, theoretical equations exist only for linear materials in an infinite cup. Below we illustrate the contribution of flow visualisation to the study of flows with a vane tool with two examples. For more details, the reader is referred to Ovarlez *et al.* (2011).

Flows near the yield stress

A practical method used to study the flow properties of complex suspensions with a vane tool, known as the Couette analogy (see Section 2.1.5), consists of analysing macroscopic data as if the material was sheared in a Couette geometry of inner cylinder radius $R_{i,eq}$. While the streamlines go into the virtual cylinder delimited by the blades in Newtonian media (Baravian *et al.*, 2002), yielding a $R_{i,eq}$ value lower than the vane radius (Sherwood and Meeten, 1991), it was found that the streamlines are nearly cylindrical (Barnes and Carnali, 1990; Savarmand *et al.*, 2007) and thus that $R_{i,eq} = R_i$ for shear-thinning fluids of index n lower than 0.5. For yield stress fluids, several works (Keentok *et al.*, 1985; Yan and James, 1997; Savarmand *et al.*, 2007) show that at yield (i.e. at low shear rates), the material flows uniformly in a thin cylindrical layer near the blades. This is now widely accepted (Barnes and Nguyen, 2001) and used to perform a Couette analogy with $R_{i,eq} = R_i$; the yield stress τ_Y is then simply extracted using Eq. 2.16 (Nguyen and Boger, 1992).

Recent velocity measurements in a Herschel–Bulkley fluid of index $n=0.5$ show that the picture may be more complex. Figure 2.22a shows ten azimuthal velocity profiles measured between two adjacent blades of a six-bladed vane rotated at low velocity (0.1 rpm). There is a strong θ-dependence of the velocity profiles. The velocity profile that starts near from a blade tip (at $\theta=0°$) has a much steeper slope than the profile measured midway between adjacent blades (at

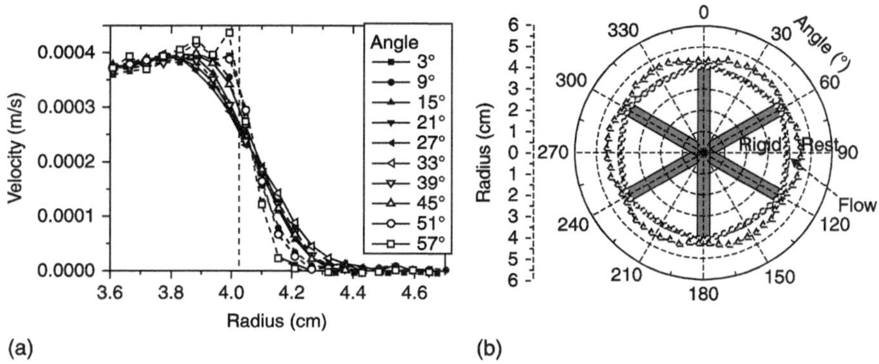

2.22 (a) Azimuthal velocity profile of a yield stress fluid (concentrated emulsion) sheared at 0.1 rpm in a six-bladed vane-in-cup geometry, for various angular positions θ between two adjacent blades (from $\theta=0°$ to $\theta=60°$). The vertical dashed line shows the radius of the vane. (b) Two-dimensional plot of the limit between rigid motion and shear (circles) and between shear and rest (triangles). The grey rectangles correspond to the blades. Reprinted with permission from Ovarlez *et al.* (2011). Copyright (2011) by Society of Rheology.

$\theta=30°$), which shows that the blade tip neighbourhoods are regions of high shear. Meanwhile the flow stops at a radius R_c which is larger at $\theta=30°$ than at $\theta=0°$. Figure 2.22b shows a two-dimensional map of the flow field, indicating both the boundary between the region of rigid body rotation (between the blades) and the sheared region, and the boundary between the sheared region and the outer region of fluid at rest (i.e. the position where the yield criterion is satisfied). Flow is found to occur in a layer of complex shape which is far from being cylindrical even at this very low velocity.

These observations contradict the usually accepted picture for yield stress fluid flows at low rates, namely that the material flows uniformly in a thin cylindrical layer near the blades. This means that further investigation on the exact conditions under which this symmetry can be recovered is needed. It is particularly striking and counterintuitive that R_c is largest at the angular position ($\theta = 30°$) where shear at R_i is smallest. It thus seems that the link between the yield stress τ_Y and the torque T measured at yield with a vane-in-cup geometry is still an open question, although the classical formula given in Eq. 2.16 probably provides a sufficiently accurate determination of τ_Y in practice.

Slip with a vane tool

Figure 2.23 shows an image of a suspension of large particles in a yield stress fluid after a flow in a vane-in-cup geometry. In this image obtained by magnetic resonance imaging (MRI), brighter zones contain less particles than the bulk

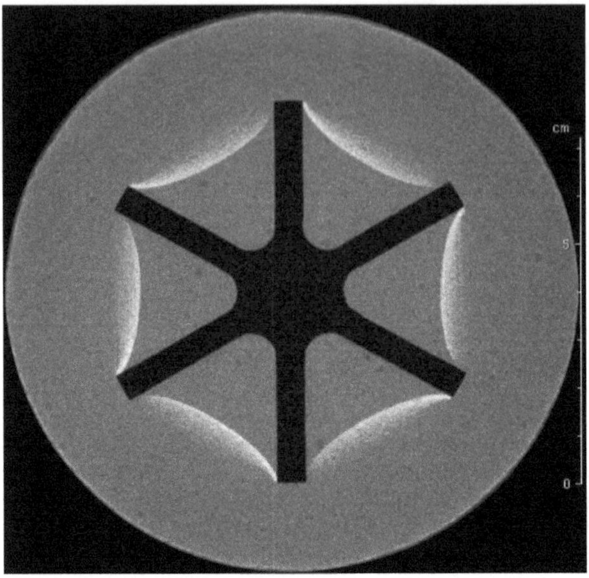

2.23 Two-dimensional magnetic resonance image of a suspension of particles in a yield stress fluid after shear in a vane-in-cup geometry. Reprinted with permission from Ovarlez *et al.* (2011). Copyright (2011) by the Society of Rheology.

material. Quantitative analysis (Ovarlez *et al.*, 2011) shows that there are actually no more particles in a thin layer (a few particle diameters wide), the position of which marks the transition between the unsheared material (which rotates as a rigid body between the blades) and the sheared material.

This observation of particle depletion near the blades when the yield stress fluid contains noncolloidal particles has important consequences. It has to be noted that the phenomenon is very rapid (Ovarlez *et al.*, 2011) and irreversible, and thus probably unavoidable when studying flows of suspensions of large particles. It likely occurs when studying polydisperse pastes such as mortars and fresh concrete. It results in the creation of a pure interstitial yield stress fluid layer and thus in a kind of wall slip near the blades. It contrasts with the classical assumption that is made in the field of concentrated suspension rheology where the vane tool is mainly used to avoid this phenomenon (Saak *et al.*, 2001). As the flow of the suspension is localised near the inner tool at low shear rate, only the pure interstitial fluid created by migration near the blades remains in the sheared layer at a sufficiently low rotational velocity of the vane tool. This explains why the same apparent value of the yield stress can be found in the suspension as in the pure interstitial fluid (Mahaut *et al.*, 2008a; Ovarlez *et al.*, 2011). Consequently, in the case of pasty materials with large particles, the vane tool may finally be suitable only for the study of the solid (elastic) properties of materials and for the static yield stress measurements (Alderman *et al.*, 1991, Mahaut *et al.*, 2008a; Mahaut *et al.*, 2008b). As the yield stress measurement may induce irreversible particle depletion near the blades, any new measurement then requires a new sample preparation. In order to study accurately the flows of pasty materials with large particles, these results suggest that when possible a coaxial cylinders geometry with properly roughened surfaces is preferable. If the use of a vane tool cannot be avoided, one should bear in mind these observations in order to carefully interpret any result.

2.5 Notes

1 For the sake of simplicity, the discussion is mostly based on the case of simple constitutive laws of the form $\sigma_{ij} = f(d_{ij})$. This can be easily generalised to cases where the behaviour also depends on the strain tensor and the strain history.

2 When the behaviour is nonlinear, the response is not sinusoidal, though one can still define G' and G'' through Fourier analysis: they are the coefficients of the first harmonic in the Fourier series decomposition of the response $\tau(t)$ (for details see Wilhelm *et al.*, 1998; Ewoldt *et al.*, 2008). In this case, while G'' still encodes all the dissipative properties of the material, G' has in general no simple quantitative meaning.

2.6 References

Acrivos A, Mauri R and Fan X (1993), 'Shear-induced resuspension in a Couette device', *Int J Multiphase Flow*, 19, 797–802.

Alderman N J, Meeten G H and Sherwood J D (1991), 'Vane rheometry of bentonite gels', *J Non-Newtonian Fluid Mech*, 39, 291–310.

Altobelli S A, Fukushima E and Mondy L A (1997), 'Nuclear magnetic resonance imaging of particle migration in suspensions undergoing extrusion', *J Rheol*, 41, 1105–1115.

Assaad J, Khayat K and Mesbah H (2003), 'Assessment of thixotropy of flowable and self-consolidating concrete', *ACI Mater J* 100, 99–107.

Baravian C, Lalante A and Parker A (2002), 'Vane rheometry with a large, finite gap', *Appl Rheol*, 12, 81–87.

Barentin C, Azanza E and Pouligny B (2004), 'Flow and segregation in sheared granular slurries', *Europhys Lett*, 66, 139–145.

Barnes H A and Carnali J O (1990), 'The vane-in-cup as a novel rheometer geometry for shear thinning and thixotropic materials', *J Rheol*, 34, 841–866.

Barnes H A and Nguyen Q D (2001), 'Rotating vane rheometry – a review', *J Non-Newtonian Fluid Mech*, 98, 1–14.

Bousmina M, Aït-Kadi A and Faisant J B (1999), 'Determination of shear rate and viscosity from batch mixer data', *J Rheol*, 43, 415–433.

Chow A W, Sinton S W, Iwamiya J H and Stephens T S (1994), 'Shear-induced particle migration in Couette and parallel-plate viscometers: NMR imaging and stress measurements', *Phys Fluids*, 6, 2561–2576.

Coussot P (2005), *Rheometry of Pastes, Suspensions and Granular Materials*, New York, John Wiley & Sons.

Coussot P, Nguyen Q D, Huynh H T and Bonn D (2002), 'Viscosity bifurcation in thixotropic, yielding fluids', *J Rheol*, 46, 573–589.

Coussot P, Tabuteau H, Chateau X, Tocquer L and Ovarlez G (2006), 'Aging and solid or liquid behavior in pastes', *J Rheol*, 50, 975–994.

Coussot P, Tocquer L, Lanos C and Ovarlez G (2009), Macroscopic vs. local rheology of yield stress fluids, *J Non-Newtonian Fluid Mech*, 158, 85–90.

Deboeuf A, Gauthier G, Martin J, Yurkovetsky Y and Morris J F (2009), 'Particle pressure in a sheared suspension: a bridge from osmosis to granular dilatancy', *Phys Rev Lett*, 102, 108301.

Derec C, Ducouret G, Ajdari A and Lequeux F (2003), 'Aging and nonlinear rheology in suspensions of polyethylene oxide-protected silica particles', *Phys Rev E*, 67, 061403.

Dullaert K and Mewis J (2005), 'Thixotropy: Build-up and breakdown curves during flow', *J Rheol*, 49, 1213–1230.

Estellé P, Lanos C, Perrot A and Amziane S (2008), 'Processing the vane shear flow data from Couette analogy', *Appl Rheol*, 18, 34037–34481.

Ewoldt R H, Hosoi A E and McKinley G H (2008), 'New measures for characterizing nonlinear viscoelasticity in large amplitude oscillatory shear', *J Rheol*, 52, 1427–1458.

Fall A, Lemaître A, Bertrand F, Bonn D and Ovarlez G (2010), 'Shear thickening and migration in granular suspensions', *Phys Rev Lett*, 105, 268303.

Ferraris C F and Brower L E (2001), *Comparison of concrete rheometers: International tests at LCPC (Nantes, France) in October, 2000*, National Institute of Standards and Technology Interagency Report (NISTIR) 6819.

Ferraris C F and Brower L E (2004), *Comparison of concrete rheometers: International tests at MB (Cleveland OH, USA) in May, 2003*, National Institute of Standards and Technology Interagency Report (NISTIR) 7154.

Hafid H, Ovarlez G, Toussaint F, Jezequel P H and Roussel N (2010), 'Estimating measurement artifacts in concrete rheometers from MRI measurement on model materials', in Khayat K H and Feys D (ed.), *Design, Production and Placement of Self-Consolidating Concrete*, RILEM Book Series, 1, pp. 127–137.

Jarny S, Roussel N, Rodts S, Le Roy R and Coussot P (2005), 'Rheological behavior of cement pastes from MRI velocimetry', *Cem Concr Res*, 35, 1873–1881.

Jarny S, Roussel N, Le Roy R and Coussot P (2008), 'Modelling thixotropic behavior of fresh cement pastes from MRI measurements', *Cem Concr Res*, 38, 616–623.

Jop P, Forterre Y and Pouliquen O (2006), 'A constitutive law for dense granular flows', *Nature*, 441, 727–730.

Keentok M., Milthorpe J F and O'Donovan E (1985), 'On the shearing zone around rotating vanes in plastic liquids: theory and experiment', *J Non-Newtonian Fluid Mech*, 17, 23–35.

Koehler E P, Fowler D W, Ferraris C F and Amziane S A (2006), 'New, portable rheometer for fresh self-consolidating concrete', *ACI Mater J*, 233, 97–116.

Leighton D and Acrivos A (1987), 'Measurement of shear-induced self-diffusion in concentrated suspensions of spheres', *J Fluid Mech*, 177, 109–131.

Lhuillier D (2009), 'Migration of rigid particles in non-Brownian viscous suspensions', *Phys Fluids*, 21, 023302.

Macosko C W (1994), *Rheology: principles, measurements, and applications*, New York, Wiley-VCH.

Mahaut F, Chateau X, Coussot P and Ovarlez G (2008a), 'Yield stress and elastic modulus of suspensions of noncolloidal particles in yield stress fluids', *J Rheol*, 52, 287–313.

Mahaut F, Mokéddem S, Chateau X, Roussel N and Ovarlez G (2008b), 'Effect of coarse particle volume fraction on the yield stress and thixotropy of cementitious materials', *Cem Concr Res*, 38, 1276–1285.

Manneville S, Becu L and Colin A (2004), 'High-frequency ultrasonic speckle velocimetry in sheared complex fluids', *Eur Phys J Appl Phys*, 28, 361–373.

Morris J F and Boulay F (1999), 'Curvilinear flows of noncolloidal suspensions: The role of normal stresses', *J Rheol*, 43, 1213–1237.

Mujumdar A, Beris A N and Metzner A B (2002), 'Transient phenomena in thixotropic systems', *J Non-Newtonian Fluid Mech*, 102, 157–178.

Nachbaur L, Mutin J C, Nonat A and Choplin L (2001), 'Dynamic mode rheology of cement and tricalcium silicate pastes from mixing to setting', *Cem Concr Res*, 31, 183–192.

Nguyen Q D and Boger D V (1992), 'Measuring the flow properties of yield stress fluids', *Annu Rev Fluid Mech*, 24, 47–88.

Ovarlez G and Chateau X (2008), 'Influence of shear stress applied during flow stoppage and rest period on the mechanical properties of thixotropic suspensions', *Phys Rev E*, 77, 061403.

Ovarlez G, Bertrand F and Rodts S (2006), 'Local determination of the constitutive law of a dense suspension of noncolloidal particles through magnetic resonance imaging', *J Rheol*, 50, 259–292.

Ovarlez G, Rodts S, Ragouilliaux A, Coussot P, Goyon J and Colin A (2008), 'Wide-gap Couette flows of dense emulsions: Local concentration measurements, and comparison between macroscopic and local constitutive law measurements through magnetic resonance imaging', *Phys Rev E*, 78, 036307.

Ovarlez G, Rodts S, Chateau X and Coussot P (2009), 'Phenomenology and physical origin of shear localization and shear banding in complex fluids', *Rheol Acta*, 48, 831–844.

Ovarlez G, Barral Q and Coussot P (2010), 'Three-dimensional jamming and flows of soft glassy materials', *Nat Mater*, 9, 115–119.

Ovarlez G, Mahaut F, Bertrand F and Chateau X (2011), 'Flows and heterogeneities with a vane tool: MRI measurements', *J Rheol*, 55, 197–223.

Phillips R J, Armstrong R C, Brown R A, Graham A L and Abbott J R (1992), 'A constitutive equation for concentrated suspensions that accounts for shear-induced particle migration', *Phys Fluids*, 4, 30–40.

Raynaud J S, Moucheront P, Baudez J C, Bertrand F, Guilbaud J P and Coussot P (2002), 'Direct determination by NMR of the thixotropic and yielding behavior of suspensions', *J Rheol*, 46, 709–732.

Rodts S, Bertrand F, Jarny S, Poullain P and Moucheront P (2004), 'Développements récents dans l'application de l'IRM à la rhéologie et à la mécanique des fluides', *C R Chim*, 7, 275–282.

Roussel N (2005), 'Steady and transient flow behaviour of fresh cement pastes', *Cem Concr Res*, 35, 1656–1664.

Roussel N (2007), 'Rheology of fresh concrete: from measurements to predictions of casting processes', *Mater Struct*, 40, 1001–1012.

Roussel N, Le Roy R and Coussot P (2004), 'Thixotropy modelling at local and macroscopic scales', *J Non-Newtonian Fluid Mech*, 117, 85–95.

Roussel N, Ovarlez G, Garrault S and Brumaud C (2011), 'The origins of thixotropy of fresh cement pastes', accepted to *Cement and Concrete Research*.

Rueb C J and Zukoski C F (1997), 'Viscoelastic properties of colloidal gels', *J Rheol*, 41, 197–218.

Saak A W, Jennings H M and Shah S P (2001), 'The influence of wall slip on yield stress and viscoelastic measurements of cement paste', *Cem Concr Res*, 31, 205–212.

Salmon J B, Manneville S, Colin A and Pouligny B (2003), 'An optical fiber based interferometer to measure velocity profiles in sheared complex fluids', *Eur Phys J Appl Phys*, 22, 1436154.

Savarmand S, Heniche M, Bechard V, Bertrand F and Carreau P J (2007), 'Analysis of the vane rheometer using 3D finite element simulation', *J Rheol*, 51, 161–177.

Schultz M A and Struble L (1993), 'Use of oscillatory shear to study flow behavior of fresh cement paste', *Cem Concr Res*, 23, 273–282.

Sherwood J D and Meeten G H (1991), 'The use of the vane to measure the shear modulus of linear elastic solids', *J Non-Newtonian Fluid Mech*, 41, 101–118.

Sinton S W and Chow A W (1991), 'NMR flow imaging of fluids and solid suspensions in Poiseuille flow', *J Rheol*, 35, 735–772.

Tehrani M A (1996), 'An experimental study of particle migration in pipe flow of viscoelastic fluids', *J Rheol*, 40, 1057–1077.

Toorman E A (1997), 'Modelling the thixotropic behaviour of dense cohesive sediment suspensions', *Rheol Acta*, 36, 56–65.

Viasnoff V and Lequeux F (2002), 'Rejuvenation and overaging in a colloidal glass under shear', *Phys Rev Lett*, 89, 065701.

Wallevik J E (2008), 'Minimizing end-effects in the coaxial cylinders viscometer: Viscoplastic flow inside the ConTec BML Viscometer 3', *J Non-Newtonian Fluid Mech*, 155, 116–123.

Wilhelm M, Maring D and Spiess H W (1998), 'Fourier-transform rheology', *Rheol Acta*, 37, 399–405.

Yan J and James A E (1997), 'The yield surface of viscoelastic and plastic fluids in a vane viscometer', *J Non-Newtonian Fluid Mech*, 70, 237–253.

3
Concrete rheometers

C.F. FERRARIS and N.S. MARTYS,
National Institute of Standards and Technology, USA

Abstract: This chapter introduces current technology for measuring the rheological properties of concrete. The need to evaluate or assess the flow properties of concrete has resulted in the development and, in a few cases, standardization of almost a 100 test methods. These can be divided into empirical tests that try to simulate field conditions and tests that attempt to quantify direct rheological property measurements. In the past 20 years, there has been a shift from empirical tests toward scientifically based methods that produce results in fundamental units. Numerical models have also been developed to help interpret experimental results. These models can account for the influence of particle shape and concentration on the flow of concrete in concrete rheometers.

Key words: concrete rheometers, multiscale approach, concrete flow simulation.

3.1 Introduction

Over the past decade, around 100 tests have been developed to measure the flow of concrete. Often, the aim of these tests has been to evaluate the quality of delivered concrete at the construction site or to design the best mixture for a specific application. These tests are often based on an empirical approach instead of rheological principles. An American Concrete Institute (ACI) report (ACI 238) gives the most extensive list and description of these tests.[1]

Empirical tests served the industry well for decades by providing information on the flow under a specific shear stress. One very widely used example is the slump cone,[2] developed over a century ago, which provides the decrease in height of a cone of concrete under gravitational force (self-weight). Tattersall revolutionized the field by applying rheological principles to concrete. He developed the Two-Point Test (see Section 3.2.3), which bears his name from the fact that it measures at least two data points instead of one. This test measures the shear stress under at least two shear rates while allowing for the calculation of rheological parameters such as yield stress and plastic viscosity.

The categorization by the US National Institute of Standards and Technology (NIST) of all tests methods into four classes[3] was further refined to a table by report ACI 238[1] as shown in Table 3.1. This classification facilitates selection of the best tests for a given application and an easier interpretation of what the tests are really measuring in terms of rheological parameters. A more detailed discussion of each test is beyond the scope of this report; instead the main focus of this paper is on

Table 3.1 NIST categorization of concrete rheology test methods [1,3]

Category	Definition
Confined flow tests	The material flows under its own weight or under applied pressure through a narrow orifice.
Free flow tests	The material either flows under its own weight, without any confinement, or an object penetrates the material by gravitational settling.
Vibration tests	The material flows under the influence of applied vibration. The vibration is applied using a vibrating table, dropping the base supporting the material, an external vibrator, or an internal vibrator.
Rotational rheometers	The material is sheared between two surfaces, one or both of which are rotating.

Source: with permission from the American Concrete Institute.

commercially available rotational rheometers. Note that commercial equipment, instruments and materials mentioned in this report are identified to foster understanding. Such identification does not imply recommendation or endorsement by the National Institute of Standards and Technology, nor does it imply that the materials or equipment identified are necessarily the best available for the purpose.

In a typical rotational rheometer, the material is sheared between two surfaces, at least one of which is rotating. During operation, both the rotational speed and the torque are measured. The rheometers can be designed to control either: stress-controlled implies that the torque is imposed and controlled while the rotational speed is measured; rate-controlled implies that the rotational speed is imposed and controlled while the torque is measured. If the geometry is simple and meets certain criteria, the rotation speed can be converted into a shear rate while torque is converted into a shear stress.[1,3]

A concrete rotational rheometer is similar to rheometers used for fluids, such as polymers or oils. Concrete rotational rheometers differ in that they need to accommodate a wide range of aggregates sizes, e.g. 2–20 mm diameter. The size of the aggregates poses a challenge because a fluid rheometer usually has gaps between the shearing planes of a few millimeters or less. To measure concrete, the gap needs to be at least five times the size of largest aggregates, i.e. 50–100 mm. Over the last 20 years, various concrete rotational rheometers have been developed having a different geometries and configurations because no one approach was standardized.

3.2 Rotational rheometers for concrete

Rotational rheometers can be divided into three main categories: parallel plates, coaxial, and vane (more details on the geometries can be found in Chapter 2). In a parallel plate rheometer, the two shearing surfaces are disks positioned parallel to each other; one disk rotates while the other remains stationary. In a coaxial rheometer, the material is placed in a cylindrical cup and a smaller cylinder, called a bob, is immersed in the material. Operationally, either the bob or the container is rotated. The vane rotational rheometer consists of a cylindrical container and a rotating vane immersed into the container. The shape of the vane is important and various designs have been employed.

All commercially available concrete rotational rheometers are rate controlled. One shear surface rotates at a controlled speed and the resistance of the material is measured by a torque often measured on the same rotating surface. The measurement regime consists of increasing and then decreasing the speed (in predetermined discrete steps) and measuring the torque at each speed. Torque versus speed measurements are plotted and the yield stress and plastic viscosity determined via regression analysis. The challenge is relating speed and torque to the shear rate and shear stress so as to be able to calculate plastic viscosity and yield stress in fundamental units. In the rest of this section, commercially available concrete rotational rheometers designed according to each category are described further.

3.2.1 Parallel plate rheometer

A parallel plate rheometer consists of two parallel discs, one of which rotates. Usually the torque is measured on the rotating disc. In a parallel plate rheometer for a fluid, such as oils or polymers, the material is confined between the two shearing plates by capillary forces and does not flow out. The gap between the disks is commonly less than 1 mm. This is not possible for concrete or even for mortar due to the presence of particles, sand and coarse aggregates. For cementitious materials, the gap between the shearing plates must be 5–10 times the maximum particle size in the cementitious mix or, equivalently, between 50 and 100 mm. For this size gap, the material will not remain between the plates and, therefore, a confinement system is necessary. The most obvious solution is a container with a plate on top that rotates, shearing the material between the top plate and the bottom of the container. The challenge here is that the addition of the container results in added resistance of the material to be sheared and the analytical solution to determine the shear rate and shear stress no longer applies. The only commercially available parallel plate rheometer, the BTRHEOM, resolves this problem by simulating the flow of the concrete and assuming slippage of the concrete at the wall of the container. The reasoning is that if the concrete slips at the container surface, its influence can be neglected. The extensive research by Hu and colleagues provided an analytical solution to calculate the Bingham parameters in fundamental units.[4,5]

The BTRHEOM was developed under the direction of de Larrard *et al.* at the Laboratoire des Ponts et Chaussées (LCPC) in Paris (since 2011 the Institut Français des Sciences et Technologies des Transports, de l'Aménagement et des Réseaux,

IFSTTAR).[6–9] The BTRHEOM requires about 7 L of concrete that has at least 100 mm of slump. This rheometer was designed to be portable so that it could be used at a construction site. Therefore, the motor is located on the base and an axis at the center of the container connects the motor to the rotating disc. The torque resulting from the resistance of the concrete to be sheared is measured through the upper blades. The rotating disc is designed to avoid slippage of the concrete with openings as shown in Fig. 3.1. Another identical disc is placed at the bottom, again to avoid slippage. The dimensions of the rheometer are: internal axis diameter (connection between motor and upper disc) of 20 mm, internal container diameter of 120 mm, and a gap between the discs of 100 mm. Vibration can be applied to the concrete either to consolidate the specimen in the container or during measurements. It should be noted that rheological measurements during vibration have not been extensively investigated or reported. Special software controls the rotation speed and vibration, and also measures the torque and calculates the rheological parameters. The rotation speed range is 0.63 rad/s (0.1 rev/s) to 6.3 rad/s (1 rev/s). The maximum measurable torque is about 14 N·m.

A modification of the BTRHEOM was developed by other researchers,[10,11] but not commercialized. The main modification consisted of suspending the top plate from a shaft held above the concrete eliminating the center axis inside the concrete container.

3.1 BTRHEOM rheometer showing the blades at the top of the bucket containing the concrete (by permission from the American Concrete Institute).

3.2.2 Coaxial rheometers

In a typical coaxial rheometer, either the container or the bob rotates and the torque is measured on the bob. For fluids, the ratio of the container radius to that of the bob is typical of the order of 1.1 with a gap of less than 1 mm in order to impose a uniform shear within the gap and to avoid the material flowing out from between the two plates. For concrete, the gap should be 100 mm, leading to a radius of the container of over 1 m. Such a large coaxial rheometer was built for mud testing and it is called CEMAGREF;[12] it has a bob of 760 mm, a container radius of 1200 mm and a capacity of 500 L of concrete. It is not commercially available and would be impractical for routine tests due to the large amount of concrete required.

The only commercially available coaxial rheometer is the BML (Fig. 3.2). Wallevik and colleagues developed this rheometer based on the Tattersall Two-Point Test instrument.[13–15] The instrument is available in different models, with containers and bob of various diameters depending on the size of the aggregates in the concrete or mortar. A typical dimension used by the authors has a bob diameter of 100 mm and a container with a diameter of 145 mm, resulting in a capacity of 17 L of concrete. It is designed for concrete with slumps greater than 120 mm. In this rheometer, the outer cylinder rotates while the inner bob registers the torque and remains stationary.[15]

To avoid slippage of the concrete both on the bob and the container, the bob consists of a series of vertical blades and the container has vertical ribs. As the shear rate of the concrete at the bottom of the container is not uniform, the torque is measured only on the central portion of the bob. This is achieved by segmenting each blade of the bob into three parts with only the central portion measuring the torque. A computer controls the rotational speed of the container, records the torque, and calculates the yield stress value and plastic viscosity values. These

3.2 BML viscometer (left) with an enlarged view of inner blades (right) (by permission from the American Concrete Institute).

values are related to the Bingham parameters by careful calibration of the rheometer using standard oils.

3.2.3 Vane rheometers

Vane rheometers are the most common rotational rheometer due to their simplicity. An impeller, having a wide variety of shapes, is immersed in a container and rotated at various speeds; the resistance of the material to the rotation is measured on the impeller. For Newtonian fluids the ratio between the resistance or torque and the speed is directly related to viscosity. Tattersall and colleagues were the first to apply this concept to concrete.[16,17] They called their instrument the 'Two-Point Test'. Their device revolutionized the field of concrete flow measurement. It introduced the concept that to really characterize the flow of concrete, which is a non-Newtonian material, at least two points are needed – thus the name of the instrument. Since then, at least two others rheometers have been developed and commercialized – the IBB[18] and the ICAR[19] devices. A short description of each follows.

Two-point rheometer

The design of the impeller was developed by Tattersall to avoid sedimentation during testing. The impeller has four angled blades set in a helical pattern around a central shaft, which imparts a stirring and mixing action to the concrete (named the MH system). The impeller is driven by a variable speed hydraulic drive unit motor through a gearbox. Torque is measured indirectly through the oil pressure in the drive unit. The linear relationship between the oil pressure and torque was obtained by prior calibration with a plummer block, radius arm and spring balance system.[16] This rheometer is designed for concrete with a slump higher than 100 mm. Dimensions of the impeller and bowl are given in Fig. 3.3. In the most common model, the speed is controlled manually while it is measured by a tachometer connected to a computer. The oil pressure is captured by the computer using a pressure transducer. The instrument is calibrated using an oil to obtain values in fundamental units.

IBB

This instrument was designed by Beaupré.[18] The most used configuration is that the vane motion is planetary and not axial. The impeller motion, however, can be made axial very easily. The impeller has the shape of an H. Its rotation leaves a gap of 50 mm between the impeller and the container. The last major modification is that the instrument is fully computer controlled. The speed is programmable and the torque and speed are recorded automatically. The software calculates the slope and intercept of a linear regression of speed vs. torque. This analysis

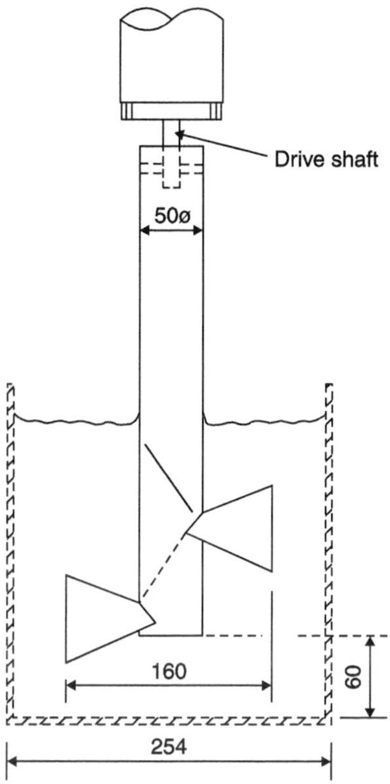

3.3 Impeller and bowl dimensions (in mm).[24]

provides rheological values related to yield stress (in N·m) and plastic viscosity (in N·m·s). The flow pattern of the concrete in this rheometer is not conductive to an analytical solution and the rheometer was not calibrated using a standard oil. Therefore, it is not possible to extract the shear rate and shear stress or the rheological parameters in fundamental units (s^{-1} and Pa·s respectively).

Figure 3.4 shows a picture of the instrument as used with a bowl capacity of about 21 L of concrete. Other bowl sizes are available to be used for mortar. It is designed for concrete with slumps between 20 and 300 mm.

RHM-3000 ICAR rheometer

The most recently developed vane rheometer is the RHM-3000 ICAR rheometer. Its configuration can be likened to a power drill with a vane attached that is immersed in a concrete container. It was developed at the International Center for Aggregates Research (ICAR) at the University of Texas at Austin by Koehler and colleagues.[19,20] The main difference from the two-point device is the shape of the vane and the location of the controls for speed and torque in the drill head. This

3.4 IBB rheometer.[24]

part is connected to a computer for programming the test sequence, and recording and analyzing the results. Most of the data published so far are with a four-blade vane. The size of the container and the impeller is determined by the size of the aggregates of the concrete to be tested. This choice is so that the gap between the blade and the container is not too small for the aggregates selected. The ICAR device is designed for concrete with slump ranging from 75 mm to self-consolidating concrete (SCC). A picture of the rheometer is shown in Fig. 3.5.

Fresh concrete tester 101 (FCT 101)

This rheometer, FCT 101,[21,22] differs in two ways from the instruments described so far. The first is that the vane is directly immersed in concrete *in situ* (e.g. in a wheelbarrow or any other container). The vane is made of two small hemispheres attached to a shaft. The shaft is connected to a drill-like device that rotates the shaft at one speed while measuring torque. The second reason is that this device can impart only one speed to the vane. Therefore, it cannot generate the Bingham rheological parameters since a curve of rotational speed vs. torque cannot be generated. It is mainly used for quality control to correlate the result with slump. This operation requires calibration with the specific concrete mixtures.

Convi Visco probe and ball measuring device

The last two rotational rheometers described in this chapter operate under the same principle: inserted obstruction. An object, in this case a ball, is inserted in the concrete and the resistance force acting on the object or obstruction is measured while the fresh concrete is mixed. If the speed of the concrete is varied, either by

3.5 RHM-3000 ICAR rheometer (by permission from Claus Germann Petersen – Germann Instruments).

changing the mixing pattern or by placing the obstruction in various locations in the mixer, the Bingham parameters can be extracted.

The Convi Visco is designed to be used inside a stationary concrete mixer, while the ball measuring device is an off-line test. The Convi Visco obstruction is a ball attached to a rod that is attached to a moving part of the mixer. The ball measuring device is a self contained system composed of a container in which the concrete is placed and an off-center rotating ball. The rotation and torque are measured and controlled by a computer.

For both instruments, the position of the ball in the container and the size of the ball can be changed. The various positions allow measurements at different speeds, and varying the size of the ball can accommodate concrete with various maximum size aggregates.

3.2.4 Summary of the concrete rheometer descriptions

All the rheometers described above have one principle in common: the resistance of the material to an applied shear rate is measured. The design approaches are varied and, in most cases, an analytical solution to calculate the material properties into fundamental units does not exist. This situation makes the direct comparison of data obtained using different rheometers impossible. A specifier must request that a concrete have a certain Bingham parameters along with the rheometer used to make the measurement.

Two solutions have been discussed by ACI Committee 238 over the past ten years. The first was to conduct a round-robin to compare the results obtained from the various rheometers on the same concrete. The second was to develop a reference material suitable for concrete rheometers, i.e. particles suspended in a fluid (granular) and Bingham. The next section summarizes the results from two round-robins that highlight the progress made in developing a reference material.

3.3 Comparison of concrete rheometers

3.3.1 International round-robins

In 2000 and 2003, two international round-robins were organized under the leadership and sponsorship of ACI Committee 238 (formerly ACI 236A). The first round-robin, in October 2000, was held at the Laboratoire des Ponts et Chaussées (LCPC) in Nantes, France.[23] The second round-robin, in May 2003, was held in the laboratories of Master Builders (MB) (now BASF) in Cleveland OH, USA.[24] The round-robins included the following rheometers: BTRHEOM, IBB, Two-Point Test, and BML. In the trial in France, the CEMAGREF was also used. The other rheometers were either not yet invented (e.g. RHM-3000 ICAR) or not available.

In the trial at LCPC, 12 concrete mixtures were produced in a mixing study station with a 1 m^3 pan mixer, and had slumps ranging from 90 to 235 mm, with a wide range of combinations of yield stress and plastic viscosity. It was found that the rheometers gave different values for the Bingham constants of yield stress and plastic viscosity, even for instruments that purportedly generated outputs in fundamental units. All of the rheometers, however, ranked the mixtures in the same order for both yield stress and for plastic viscosity. The degree of correlation of both yield stress and plastic viscosity measurements between any pair of rheometers was close to one – an encouraging result. All the rheometers used are therefore able to characterize the rheological and flow properties of fresh concrete in the same order. It was also confirmed that the slump test correlates well with the yield stress as measured with any of the rheometers. However, the report provided correlation equations between the various rheometers based on only 12 mixtures.[23] For these correlations to be generally adopted, a larger statistical sample of concrete mixtures would need to be tested.

In the trial at MB, 23 mixtures were prepared in a 170 L drum mixer and tested by each of the various rheometers. Seventeen mixtures were concrete and five were mortars (no coarse aggregates added). The mixtures had slumps ranging from 121 to 248 mm, with some of the concrete being SCC. As in the first trial, the mixtures had a wide range of yield stresses and plastic viscosities. Some tests using oil, as the first attempt to develop a reference material, were also performed. The conclusions obtained from the trial at LCPC were substantiated, but it was also found that the correlation functions between any pair of rheometers, especially for viscosity, obtained at LCPC (Phase I) and at MB (Phase II) were not high. This obviously makes it difficult, if not impossible, to compare data between laboratories. The concept of relative viscosity first described by Ferraris and Martys could be used to compare results from rheometers that do not provide results using the same units, as the relative viscosity is independent of the geometry used.[25] The relative viscosity is defined as the ratio of the concrete viscosity and the mortar viscosity. The mortar is designed as a concrete without the coarse aggregates. It is not possible to define in the same way a relative yield stress.

3.3.2 Reference material

During the trial at MB, all rheometers with the exception of the BTRHEOM used an oil as the fluid material. The BTRHEOM rheometer is not designed to measure a Newtonian fluid, because it assumes slip at the container surface, while oil usually does not slip at the solid interface. The proportionality factor between devices calculated varied over a wide range, 0.7–3 (1 would imply complete proportionality), making the oil not suitable as a reference material.[24] Another attempt to use oil for calibration of mortar rheometers was described by Ferraris and colleagues.[26] The approach relies on a combination of models and measurements leading to correction factors for the mortar rheometers. This approach is not easy to scale up to a concrete rheometer since various gaps are necessary to validate the correction factor. In most concrete rheometers, the gaps are not easily changed.

ACI Committee 238 has had extensive discussions on how to develop a reference material specifically designed for concrete rheometers. The first idea was to use an oil and add aggregates of various sizes. However, the idea was abandoned when no suitable and cost-effective aggregates were found. The next idea discussed was to use a multiscale approach. Here a reference cement paste would first be characterized using a fluid rheometer. Then a mortar reference material would be developed by adding sand and subsequently a concrete reference material would be obtained by adding coarse aggregates to the mortar. The rheological properties of the mortar would be determined using a model based on the paste measurements and then the concrete properties would be determined from the mortar.[27]

Laboratories around the world are using several reference materials to mimic concrete fresh properties. NIST, for instance, has prepared a reference material based on corn syrup and limestone powder as the first reference material for cement paste. Later, sand and coarse aggregates will be added to the paste to produce mortar and concrete. Models will link the rheological values from cement paste to concrete. The models relating cement paste to mortar and then to concrete are in advanced development and are discussed in the next section.

3.4 Modeling of concrete rheometers

Concrete rheometer measurements establish a quantitative relationship between the applied torque versus the resulting angular velocity of the propeller. The values of torque vs. angular velocity are compared for different mixtures in order to rank them in relative fashion. It is difficult to obtain actual rheological parameters (i.e. viscosity and yield stress) in fundamental units with such measurements because the flow fields are complex due to the geometry of the rheometer container or propeller and hence not amendable to exact analytic solutions of the Navier–Stokes equations. Indeed, the non-Newtonian nature of fresh concrete can make it difficult to simply scale measurements using rheometers calibrated with a Newtonian fluid using, say, a simple correction factor. There are further complications in that concrete is a very complex granular fluid (basically particles of different shapes and size embedded in a non-Newtonian fluid matrix) making continuum approximations of the fluid properties difficult.

In an attempt to lessen this complexity and to extract rheological parameters in fundamental units from the experimental data, models of the flow of a suspension sheared between two surfaces can be developed. The idea is to have as input the properties of the embedding matrix fluid (paste), obtained from accurate experimental measurements, along with information about the size and shape distribution of the aggregates. The model output would be the rheological properties of the mixture. Obviously, the model has to be validated by experimental tests. The power of this approach is that it allows for detailed study of the general flow patterns in a rheometer, the interaction of the granular fluid with the impeller or the container, and the degree of sedimentation or segregation of aggregates within the rheometer. In the future, these models could be used to visualize or virtually determine the flow of concrete in a complicated geometry building structure to optimize the placement and consolidation of the concrete. Modeling could also provide general insights into the physical mechanisms that control the flow of concrete with the goal of improving its overall performance.

To be useful, these models need to represent the aggregates in a realistic fashion, so solely using mono-sized spherical particles is not acceptable. The shape of the aggregates can be determined using tomographic methods.[28] The particle sized distribution should also be considered. Figure 3.6[29–31] shows, as an example, a virtual piece of concrete with crushed aggregates that was used in a simulation carried out at NIST. The simulation results showed that the use of crushed aggregates had a tendency to increase the overall viscosity compared with using rounded aggregates.

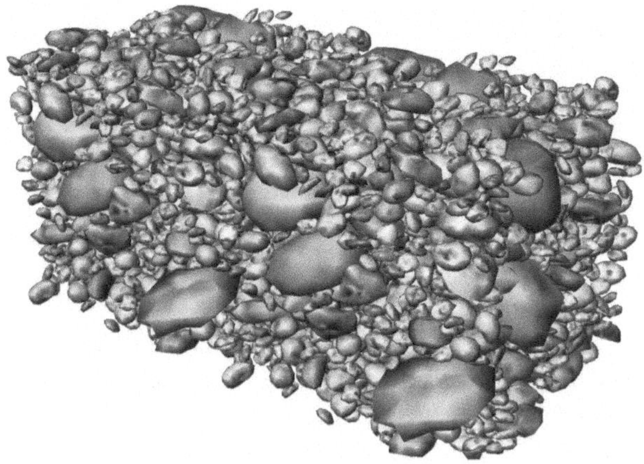

3.6 Simulation of a concrete composed of crushed aggregates.

3.4.1 Modeling rheometers

Modeling rheometers has been a topic of research for some time. Two main approaches have been used:

1. continuum model where concrete is considered a homogeneous medium;
2. discrete models where the concrete is considered as aggregates embedded in a fluid matrix, i.e. the motion of aggregates are modeled in detail (see Chapter 10 for more details).

Early models of concrete rheometers were largely based on continuum approaches. The continuum model developed by Hu *et al.* was based on a finite element method to study the BTRHEOM.[4] The authors were able to simulate slippage or not at the walls on the rheometer container, as well as determine the uncertainty of the results. They validated the model by testing concrete using another rheometer beside the BTRHEOM.

Wallevik[32] developed another continuum model to interpret the results of the BML rheometer. Because concrete is a non-Newtonian fluid, an effort was made to account for many of its rheological properties from a fundamental point of view. For example, the constitutive relation that describes stress in concrete was modified to account for collisions of aggregates, yield stress and thixotropy. This model was applied to model flow in cone and parallel plate geometries.

Two-dimensional (2D) simulations of a four-blade vane rheometer were developed by Zhu *et al.* to gain insight into how stress is distributed as a function of the material yield stress.[33] Again this approach was based on a continuum model. Figures 3.7 and 3.8 show the stress field distribution with the vane rotating at the same speed for two materials with two different yield stresses.

The following observations could be made:

- All the material is not sheared in the fluid with the high yield stress (regions marked 'A' in Fig. 3.7 and Fig. 3.8). This result is well known.
- There is a localized region of high stress at the edge of the blades.
- The iso-stress contours on Fig. 3.7 and Fig. 3.8 (e.g. lines where stress is the same) are roughly linear between the blades and therefore the overall shape is not quite a cylinder as usually assumed to analytically calculate rheological parameters. A higher number of blades will more closely approximate the cylinder.

From this 2D model, it was also found that a Newtonian fluid measured with a vane results in a viscosity value about 25% lower than the actual value as determined by a coaxial rheometer (also shown in ref. 34).

Several discrete models[35,36] have been developed that account for the granular nature of concrete in order to fully describe its flow properties in a variety of flow geometries.

An example of such approach for modeling flow in a vane rheometer is the current NIST model that can fully account for the three-dimensional (3D) geometry of the vane rheometer and allows for the embedding fluid to be Newtonian or Non-Newtonian. As an illustration, Fig. 3.9 shows the initial configuration of spheres that make up a dense suspension (50% by volume) in a four-blade vane rheometer.

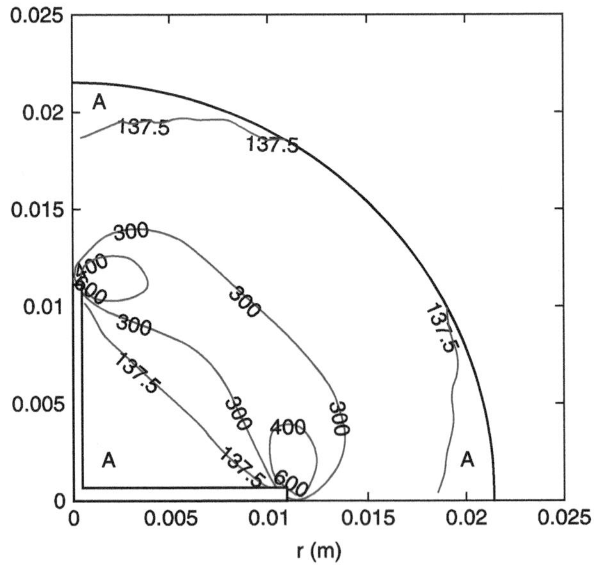

3.7 Contour plot of the stress field (Lagrangian reference frame): flow of Bingham fluid with a yield stress of 137.5 Pa in vane rheometer (angular velocity: 7.5 rad/s (1.20 rps)). The regions that are marked 'A' correspond to areas where there is little or no flow.[34]

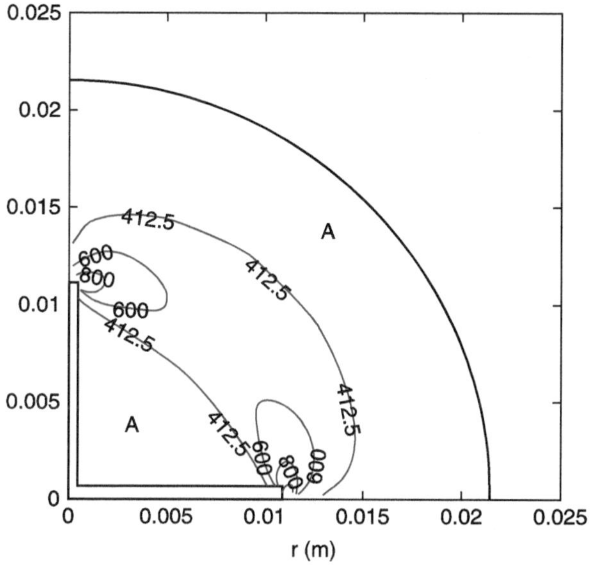

3.8 Contour plot of the stress field (Lagrangian reference frame): flow of Bingham fluid with a yield stress of 412.5 Pa in vane rheometer (angular velocity: 7.5 rad/s (1.20 rps)). The regions that are marked 'A' correspond to areas where there is little or no flow.[34]

3.9 View of vane imbedded in sphere suspension. The coloring of the spheres corresponds to their initial placement in each octant of the rheometer container. Spheres located in one of the near octants have been made transparent to reveal the blade.

3.10 Cross-section view of suspension subject to rotation by vane blade. Initially the spheres were located in each quadrant of the figure.

Spheres are a useful test case but are not acceptable for final results. The material in the container was divided in octants each in a different color for visualization purposes. The spheres in the most forward octant have been made transparent to reveal the vane propeller. Figure 3.10 shows a view from just under the vane as it begins to turn. In this case the suspending fluid is Newtonian (i.e. without a yield stress); thus the spheres may freely mix in the blades and are not restrained to remain within a single blade quadrant. The embedding fluid may also be made non-Newtonian to represent the cement paste and the suspended particles could be the sand. This model would be used for instance with the reference material described previously.[37]

3.4.2 Multiscale modeling

Regardless of the quality or sophistication of a computational model, it is impossible to model concrete in its entirety. Even when ignoring the effects of hydration, the size distribution of the particles in concrete is too wide ranging to be accounted for in a single model, even on the most powerful supercomputers. As a result, it is necessary to adopt a multiscale approach similar to the concept for developing the reference material. In a multiscale approach for suspensions, one starts at the smallest scales, defining a suspension by a matrix or embedding fluid that contains solid inclusions. This suspension now becomes the matrix fluid for larger solid inclusions (typically ten times the size of the particles in the matrix fluid). This process can again be repeated over and over until the final macroscopic scale of interest is reached.

Consider a simple multiscale model of concrete. Starting with the smallest scale, combine water (the embedding fluid) and cement particles of the order of ten micrometers in size (the solid inclusion) to make a new fluid called cement paste. If sand (of order of millimeters in size) is added to the cement paste, fluid mortar is formed. Treating the mortar as a fluid matrix, coarse aggregates (centimeter size in scale) are added to finally arrive at a model of concrete. Note that the matrix fluid properties are very different at each length scale. Water is Newtonian, while the cement paste typically shear thins, is thixotropic and changes with time to become viscoelastic as it forms a gel and undergoes hydration. Mortar is also non-Newtonian and typically shear thins. The concrete flow properties will depend strongly on the mortar's rheological properties and coarse aggregate size and shape distribution. One major application of a multiscale approach is the development of a reference material, where the properties of the concrete are calculated from the results of the simulation of mortar based on the experimental measurements of the cement paste. Models, such as the one developed at NIST, will provide the rheological properties of concrete using as input the paste properties and the aggregates (fine and coarse) shape and particle size distribution.

3.5 Conclusions

While major progress has been made in the rheological science of concrete since Tattersall first developed his rheometer in the 1980s, in practice this approach is still underutilized due to missing links such as the measurement of rheological parameters in fundamental units independently from the rheometer used, or the correlation between placement effectiveness and rheological parameters. Key to further advancement is the merging of experimental measurements and modeling for determination of the concrete rheological properties in fundamental units. Future research challenges include the modeling of the viscoelastic properties of concrete as it begins to set. Clearly it is an important challenge to link modeling and rheological measurements to develop a complete picture of cement or concrete flow pattern in any geometry, not only limited to rheometers but any formwork shape or empirical tests such as the J-box or the slump. The perfect solution needs to combine experimental results with modeling.

3.6 Acknowledgments

We thank NASA and DOE for providing supercomputer time for the simulations. We express deep gratitude to the many past and present members of the Virtual Cement and Concrete Testing Laboratory (VCCTL) consortium for intellectual, moral, and financial support. The authors also would like to thank the participants of the international round-robins for the concrete reports and NIST for supporting the advancement of measurement science for the flow of concrete. We also thank Dr Edward Garboczi (NIST), Dr Kenneth Snyder (NIST) and Dr Nicholas Dagalakis (NIST) for their constructive review.

3.7 References

1. ACI 238-1R-08, *Report on Measurements of Workability and Rheology of Fresh Concrete*, American Concrete Institute, Farmington Hills, MI, 2008.
2. ASTM C143/C143M, *Standard Test Method for Slump of Hydraulic-Cement Concrete*, ASTM International, West Conshohocken, PA, 2010.
3. Hackley, V. A and Ferraris, C. F., *The Use of Nomenclature in Dispersion Science and Technology*, NIST Recommended Practice Guide, SP 960–3, 2001 (http://www.nist.gov/public_affairs/practiceguides/SP960–3.pdf).
4. Hu, C., de Larrard, F., Sedran, T., Bonlag, C., Bose, F. and Deflorenne, F., 'Validation of BTRHEOM, the New Rheometer for Soft-to-Fluid Concrete', *Materials and Structures*, 1996, Vol. 29, No. 194, pp. 620–631.
5. Hu, C., *Rhéologie des bétons fluides*, Etudes et Recherches des Laboratoires des Ponts et Chaussées, Report no. LPC-ER-OA-95–16, Paris, 1995.
6. Sedran, R., *Rhéologie et rhéométrie des bétons. Application aux bétons autonivelants* [Rheometry and Rheology of Concrete. An Application to Self-Compacting Concrete], PhD Thesis, Ecole nationale des ponts et chaussées, Paris, 1999.
7. de Larrard, F., Hu, C., Sedran, T., Szitkar, J. C., Joly, M., Claux, F. and Derkx, F., 'A New Rheometer for Soft to-Fluid Fresh Concrete', *ACI Materials Journal*, 1997, Vol. 94, No. 3, pp. 234–243.
8. de Larrard, F., Szitkar J.-C., Hu, C. and Joly, M., 'Design of a Rheometer for Fluid Concretes', RILEM Workshop Special Concretes – Workability and Mixing, 201–208, 1993.
9. de Larrard, F., Sedran, T., Hu, C., Szitkar, J. C., Joly, M. and Derkx, F., 'Evolution of the Workability of Superplasticized Concretes: Assessment with BTRHEOM Rheometer', RILEM International Conference on Production Methods and Workability of Concrete, RILEM Proceedings 32, Glasgow, Scotland, 3–5 June, pp. 377–388, 1996.
10. Szecsy, R. S., *Concrete Rheology*, PhD Dissertation, University of Illinois at Urbana-Champaign, Urbana, IL, 1997.
11. Struble, L., Puri, U. and Ji, X., 'Concrete Rheometer', *Advances in Cement Research*, 2001, Vol. 13, No. 2, pp. 53–63.
12. Coussot, P., '*Rheologie des boues et laves torrentielles. Etudes de dispersions et suspensions concentrées*, Thèse de doctorat de l'Institut National Polytechnique de Grenoble et Etudes du Cemagref, série Montagne n°5, 1993.
13. Wallevik, O. H., 'Rheology – A Scientific Approach to Develop Self-Compacting Concrete', Third International Symposium on SCC, RILEM, Reykjavik, Iceland, August 2003, pp. 23–31, 2003.
14. Wallevik, O. H. and Gjørv, O. E., 'Development of Coaxial Cylinder Viscometer for Fresh Concrete', Proceedings of the RILEM Colloquium: Properties of Fresh Concrete, 1990, pp. 213–224, 1990.
15. Wallevik, O. H., Hjartarson, B. and Palsson, O. P., 'Rheometer-4SCC: A Portable Rheometer for Self Compacting Concrete', Nord Design, Reykjavik, Iceland, August 2006.

16. Tattersall, G. H. and Banfill, P. F. G., *The Rheology of Fresh Concrete*, Pitman, London, 1983.
17. Tattersall, G. H. and Bloomer, S. J., 'Further Development of the Two-Point Test for Workability and Extension of its Range', *Magazine of Concrete Research*, 1979, Vol. 31, pp. 202–210.
18. Beaupré, D., *Rheology of High Performance Shotcrete*, PhD Thesis, University of British Columbia, Canada, 1994.
19. Koehler, E. P., *Development of a Portable Rheometer for Fresh Portland Cement Concrete*, MS Thesis, University of Texas at Austin, Austin, TX, 2004.
20. Koehler, E. P., Fowler, D. W., Ferraris, C. F. and Amziane, S., 'A New Portable Rheometer for Fresh Self-Consolidating Concrete', Proceedings of ACI session in New York: 'Workability of SCC: Roles of its Constituents and Measurements Techniques' SP-233, pp. 97–116, 2005.
21. Steiner, T., 1996, 'Continuous Control of Fresh Concrete Using the FCT 101 Tester', Production Methods and Workability of Concrete, Proceedings of the Conference, Bartos, P. J. M., Marrs, C. L. and Cleland, D. J. (eds.), RILEM, E&FN Spon, pp. 365–367.
22. Wong, G. S., Alexander, A. M., Haskins, R., Poole, T. S., Malone, P. G. and Wakeley, L., *Portland-Cement Concrete Rheology and Workability: Final Report*, FHWARD-00–025, US Federal Highway Administration, McLean, VA, 2000.
23. Ferraris, C. and Brower, L. (eds.), 'Comparison of Concrete Rheometers: International Tests at LCPC (Nantes, France) in October 2000', NISTIR 6819, 2001 (http://fire.nist.gov/bfrlpubs/build01/PDF/b01074.pdf).
24. Ferraris, C. and Brower, L. (eds.), 'Comparison of Concrete Rheometers: International Tests at MB (Cleveland OH, USA) in May 2003', NISTIR 7154, 2004 (http://ciks.cbt.nist.gov/~ferraris/PDF/DraftRheo2003V11.4.pdf).
25. Ferraris, C. F. and Martys, N. S., 'Relating Fresh Concrete Viscosity Measurements from Different Rheometers', *Journal of Research of the National Institute of Standards and Technology*, 2003, Vol. 108, No. 3, pp. 229–234.
26. Ferraris, C. F., Geiker, M., Martys, N. S. and Muzzatti, N., 'Parallel-plate Rheometer Calibration Using Oil and Lattice Boltzmann simulation', *Journal of Advanced Concrete Technology*, 2007, Vol. 5, No. 3, pp. 363–371.
27. Ferraris, C. F., Li, Z., Mohseni, M. and Franson, N., 'Reference Materials for Rheometers and the ASTM Flow Table', 3rd International RILEM Symposium on Rheology of Cement Suspensions such as Fresh Concrete, Wallevik, O., Kubens, S. and Oesterheld, S. (eds.), Reykjavik, Iceland, August 2009, pp 257–264, 2009.
28. Garboczi, E. J., 'Three-dimensional Mathematical Analysis of Particle Shape using X-ray Tomography and Spherical Harmonics: Application to Aggregates used in Concrete', *Cement and Concrete Research*, 2002, Vol. 32, No. 10, pp. 1621–1638.
29. Flatt, R. J., Martys, N. S. and Bergstrom, L., 'The Rheology of Cementitious Materials', *MRS Bulletin*, 2004, Vol. 29, No. 5, pp. 314–318.
30. Erdogan, S. T., Martys, N. S., Ferraris, C. F. and Fowler, D. W., 'Influence of the Shape and Roughness of Inclusions on the Rheological Properties of a Cementitious Suspension', *Cement & Concrete Composites*, 2008, Vol. 30, No. 5, pp. 393–402.
31. Martys, N. S., Ferraris, C. F., Gupta, V., Cheung, J. H., Hagedorn, J. G., *et al.*, 'Computational Model Predictions of Suspension Rheology: Comparison to Experiment', 12th International Congress of the Chemistry of Cement, Montreal (Canada), paper # W6–06.3, July 2007.
32. Wallevik, J. E., *Rheology of Particle Suspensions*, Thesis, Norwegian University of Science and Technology, 2003.

33. Zhu, H., Martys, M. S., Ferraris, C. F. and De Kee, D., 'A Numerical Study of the Flow of Bingham-like Fluids in Two-dimensional Vane and Cylinder Rheometers using a Smoothed Particle Hydrodynamics (SPH) Based Method', *Journal of Non-Newtonian Fluid Mechanics*, 2010, Vol. 165, pp. 362–375.
34. Atkinson, C. and Sherwood, J. D., 'The Torque on a Rotating N-bladed Vane in a Newtonian Fluid or Linear Elastic Medium', *Proceedings of the Royal Society of London, Series A Mathematical Physical and Engineering Sciences*, 1992, Vol. 438, pp. 183–196.
35. Martys, N. S., 'Study of a Dissipative Particle Dynamics-Based Approach for Modeling Suspensions', *Journal of Rheology*, 2005, Vol. 49 No. 2, pp. 401–424.
36. Martys, N. S., George, W. L., Chun, B.-W. and Lootens D., 'A Smoothed Particle Hydrodynamics-Based Fluid Model with a Spatially Dependent Viscosity: Application to Flow of a Suspension with a Non-Newtonian Fluid Matrix', *Rheologica Acta*, 2010, Vol. 49, No. 10, pp. 1059–1069.
37. Bullard, J. W., Garboczi, E. J., George, W. L., Martys, N. S., Satterfield, S. G. and Terrill, J. E., 'Advancing the Materials Science of Concrete with Supercomputers', *Concrete International*, 2011, Vol. 33, No. 1, pp. 24–29.

4
From industrial testing to rheological parameters for concrete

N. ROUSSEL, Université Paris-Est, IFSTTAR, France

Abstract: This chapter starts by presenting correlations between slump test, slump flow test and mini cone test and yield stress along with their limits. It is demonstrated that, in the case of self-compacting concrete (SCC) and slump flow test, the measured spread diameter can not be correlated to the yield stress of the SCC tested. An alternative way to measure the yield stress of a given SCC is then presented. Considering the experimental validation obtained in the case of fine suspensions, this test seems to be able to provide a cheap, simple and rather precise measurement of the yield stress of any SCC.

Key words: yield stress, concrete, slump, slump flow.

4.1 Introduction

As discussed in the previous chapters, the behaviour of fresh concrete in steady state is often approximated by a yield stress model of the following general form (Tattersall and Banfill, 1983; de Larrard and Hu, 1996).

$$\dot{\gamma} = 0 \rightarrow \tau < \tau_0 \quad [4.1a]$$

$$\dot{\gamma} \neq 0 \rightarrow \tau = \tau_0 + f(\dot{\gamma}) \quad [4.1b]$$

where τ_0 is the yield stress, $\dot{\gamma}$ the shear rate and f a positive continuously increasing function of the shear rate with $f(0) = 0$. In the particular case of a Bingham fluid with a plastic viscosity μ_p, $f(\dot{\gamma}) = \mu_p\dot{\gamma}$.

From a practical point of view, yield stress may be associated with filling capacity and more generally to whether or not concrete will flow or stop flowing under an applied stress, whereas plastic viscosity may be associated to the velocity at which a given concrete will flow once flow is initiated.

Although measurements of plastic viscosity have several practical applications (pumping, casting rates etc), yield stress is often considered the most important parameter for formwork filling in practice (Roussel, 2007a). If we consider for instance the casting of a wall, a purely viscous fluid (i.e. no yield stress) would self level under the effect of gravity. Gravity would indeed induce a pressure gradient in the fluid if the upper surface of the material is not horizontal. This pressure gradient would generate a shear stress in the material that creates a shear rate and forces the material to flow until the upper surface becomes horizontal and until the pressure gradient at the origin of flow has disappeared. The viscosity of the material will only play a role in the time needed to obtain a horizontal surface. In the case of a yield stress fluid such as concrete, gravity and the pressure gradient also generate a shear stress. But if this shear stress, which is a complex function of the formwork thickness and the local density of steel reinforcements, becomes lower than the yield stress of the concrete, flow stops before the concrete self levels or before the formwork is entirely filled.

In practice, when engineers wants to quantify or qualify the ability of a given cementitious material to be cast, fast, cheap and simple industrial tests are often preferred to the viscometers described in Chapter 2 or the concrete rheometers described in Chapter 3, although these devices allow for the measurement of both yield stress and plastic viscosity. Indeed, as described in the previous chapter, even if these rheometers give the same ranking of a given set of tested concretes, they do not give the same absolute values of the measured rheological parameters. As a consequence, no matter how interesting the recent developments in concrete rheometers and how much progress has been achieved in the last 30 years, civil engineers still do not have simple tools to measure the rheological parameters of a given cementitious material both quickly and easily. That is probably why industrial tests such as the Abrams cone for traditional concrete or slump flow test for self-compacting concrete (SCC) are still the favourite everyday tests to quantify the workability of cementitious materials at the building site.

This chapter tries to make a link between the measurements of yield stress of fresh cementitious materials and industrial tests such as slump or slump flow. In the case of many modern cementitious materials, however, the knowledge of the yield stress at the end of the mixing phase is not sufficient to describe the observed behaviour from the concrete factory to the casting phase as an evolution of the material rheological behaviour is often noted during this time period. This evolution is mainly due to the thixotropic behaviour of cementitious materials (Roussel, 2005, 2006a). At rest, the material builds up a structure and its apparent yield stress and viscosity increase. When flowing at high shear rates, the material structure decreases back to its initial state. The material often reaches its most destructured state during the mixing phase and, according to its flow history between the mixing plant and the formwork (mixing truck, rest period, casting phase, etc), its behaviour continuously evolves. More details on these aspects can be found in chapters 11 and 12.

This chapter looks mainly at tests allowing for the measurement of yield stress. These tests mostly focus on the way the tested material stops flowing. As measurement of a stress is complex, geometrical quantities such as a length or height are often measured.

The following general procedure can be considered. A mould or reservoir of a given standard shape or volume is filled with the cementitious material to be tested. The material is poured, a mould is lifted or a gate is opened, and flow occurs if the shear stress generated by gravity inside the sample is higher than the yield stress of the material. As the material is flowing, the thickness of the sample and, therefore, the stress generated by gravity decrease. If the stress in the tested sample becomes smaller than the yield stress, flow stops. The shape at stoppage is therefore directly linked to the yield stress of the tested material if inertia or surface tension effects can be neglected (Roussel and Coussot, 2005).

The first part of the chapter considers slump and the slump flow test. Correlations between yield stress and result of these tests are proposed. It is also shown that, in the case of the slump flow test for SCC, the test result can not be correlated to the rheological parameters of the SCC tested. Indeed, although most of the conditions for the correlation to be valid (negligible surface tension effects and inertia effects) are fulfilled, the thickness of the sample at flow stoppage is of the same order as the largest particle. The impact of this phenomenon on a practical and empirical use of the slump flow is analysed.

The second part of the chapter presents an alternative way to measure the yield stress of a given SCC. This approach has been rigorously validated in the case of limestone powder suspensions but, due to the discrepancy between concrete rheometers, it has not been compared with other experimental measurements in the case of concretes. But given the excellent agreement obtained in the case of fine suspensions and given that all theoretical conditions are fulfilled for the proposed analysis to be valid, the author believes that this test is able to give a cheap, simple and rather precise measurement of the yield stress of any SCC.

4.2 The slump test family and its limits

4.2.1 Description of the test

Typically, in a slump test, a mould of a given conical shape (Fig. 4.1) is filled with the material to be tested. The mould is then lifted and flow occurs. From a practical point of view, two geometrical quantities can be measured as the result of the test: the slump (i.e. the difference between the initial height of the sample and its final height) or the spread diameter, also called slump flow. As an example, in the case of the Abrams cone test, the height H_0 is 30 cm, the initial diameters D_{max} and D_{min} are 20 and 10 cm respectively. When testing fluid concretes such as SCC, the final

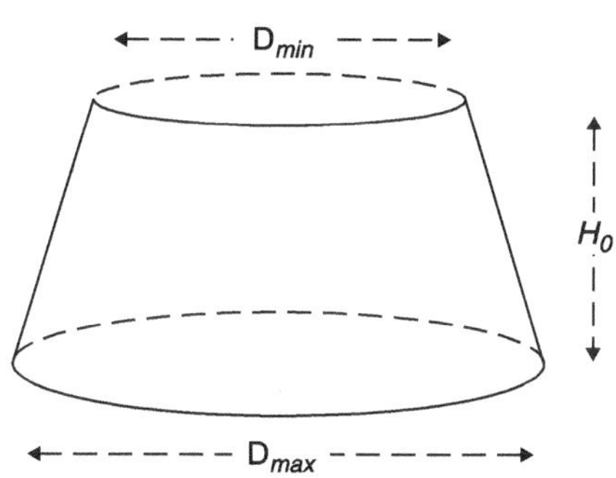

4.1 Initial cone shape.

spread diameter or slump flow is contained between 600 and 900 mm. When testing traditional concretes, the slump is between 5 and 20 cm.

4.2.2 Analysis of the flow

Several attempts to relate slump to yield stress can be found in the literature. Murata (1984), Pashias *et al.* (1996), Schowalter and Christensen (1998), Clayton *et al.* (2003) and Saak *et al.* (2004) all proposed correlations between slump and yield stress by assuming that the cone could be divided into two parts. In the upper part of the cone, the shear stress does not reach the yield stress and no flow occurs. In the lower part, the shear stress induced by the tested material own weight is higher than the yield stress and flow occurs. The height of the flowing lower part decreases until the shear stress in this zone is equal to the yield stress and flow stops.

However, as discrepancies still existed between predictions and experimental results, Roussel and Coussot (2005) recently revisited the analysis by studying the flow from two asymptotic regimes, namely $H >> R$ and $R << H$, in which H and R are the final height and radius of the sample respectively. In these two regimes, it is possible to write exact analytical correlations between the measured geometrical quantity and yield stress.

In the 'slump' regime, when $H >> R$, this correlation takes the following form:

$$\frac{\rho g(H_0 - S)}{\sqrt{3}} = \tau_0 \qquad [4.2]$$

where S is the slump, τ_0 the yield stress, ρ the density of the concrete and H_0 the initial height of the cone. In the case of the Abrams cone, from a theoretical point of view, this relation holds only for very low slumps. When the value of the slump increases, the sample height and sample diameter become of the same order of magnitude and Eq. 4.2 does not hold any more.

In the 'spread' regime, the flow thickness is in general much smaller than the radial extent of the sample, so that the radial velocity can be expected to be much larger than the vertical velocity. An analytical solution suitable for describing the slump flow of SCC or the mini cone spread flow of cement paste was obtained by Roussel and Coussot (2005), and experimentally and numerically validated for this asymptotic case. In this case, the shape at stoppage is described by:

$$h(r) = \left(\frac{2\tau_0 (R - r)}{\rho g} \right)^{1/2} \qquad [4.3]$$

where r is the radial coordinate in the sample, ρ is the density in kg/m^3, h is the thickness of the tested material and R is the final spread radius. Roussel and Coussot (2005) also calculated the expression for R as a function of the yield stress and sample volume Ω:

$$\tau_0 = \frac{225\rho g\Omega^2}{128\pi^2 R^5} \qquad [4.4]$$

where Ω is the volume of the tested sample.

In the case of the Abrams cone, in the intermediate regime (i.e. slumps of 5–25 cm), when the height of the sample is of the same order as the radius, there is no exact analytical solution. Numerical simulations of the slump test were therefore carried out (Roussel, 2006b). In this intermediate regime, the obtained numerical correlation between slump and yield stress is written as:

$$S = 25.5 - 17.6\frac{\tau_0}{\rho} \qquad [4.5]$$

where S is the measured slump and ρ is the density of the tested concrete.

4.2.3 Validity of correlations in the case of the mini slump test for cement pastes

The mini cone test is a smaller version of the Abrams cone for cement pastes and grouts. The height H_0 is 35 mm and the initial diameters D_{max} and D_{min} are 100 and 70 mm respectively. As this test is mostly used for fluid cement pastes, a spread diameter is often measured.

From a general point of view, in order for a test to be representative of the rheological behaviour of the tested cementitious material and for a direct correlation between yield stress and the measured geometrical quantity to exist, several conditions must be fulfilled (Roussel and Coussot, 2005):

- Surface tension effects must be negligible. The flow must stop only because of the yield stress of the material tested.
- Inertia effects must be negligible. If this is not the case, the kinetic energy could force the material to flow past the final shape computed from the value of the yield stress.
- The tested volume must be representative of the mixture.
- The thickness of the sample must be at least five times the size of the largest particle (Coussot and Ancey, 1999) for the equations of fluid mechanics needed to establish Equations 4.2 to 4.5 to be applicable. Otherwise, the material can not be considered as a homogeneous fluid.

Surface tension effects

Roussel *et al.* (2005) proposed a semi-empirical quantification of the surface tension effect on the spread radius. Subsequently, Roussel and Coussot (2005) demonstrated that the surface tension effects were negligible when the following condition was fulfilled:

$$\tau_0 >> \frac{3(1-\cos\theta)\gamma_{LG}}{R} \quad [4.6]$$

where θ is the wetting angle and γ_{LG} is the liquid–gas interfacial tension. As a first approximation, it can be considered as typically equal to the surface tension of water with air, i.e. 0.07 Pa.m. The thin solid line on the left in Fig. 4.2 shows this limit in the case of the mini cone and is of the order of a couple of pascals (Pa). As long as the yield stress of the tested cement paste is higher than this value, the surface tension contribution can be neglected.

Inertia effects

Roussel and Coussot (2005) roughly quantified the effect of inertia. When it is negligible, this in particular means that the slump is not affected by the lifting velocity of the mould, the flow inertia before stoppage or the 'plastic' viscosity. Inertia effects might indeed affect the final shape by inducing flow with a fluid thickness smaller than the one that can be computed from Equations 4.2 to 4.5. The typical inertia stress ($I = \rho V^2$) is of the order of a couple pascals as the flow duration is about two seconds for a flow distance of the order of 5 cm. This value is smaller than most yield stresses tested with this device, which means that inertia effects can most often be neglected.

Tested volume

The tested volume is equal to 0.287 L, approximately 10^{11} times the volume of an average cement particle. The tested volume is thus representative of the tested concrete.

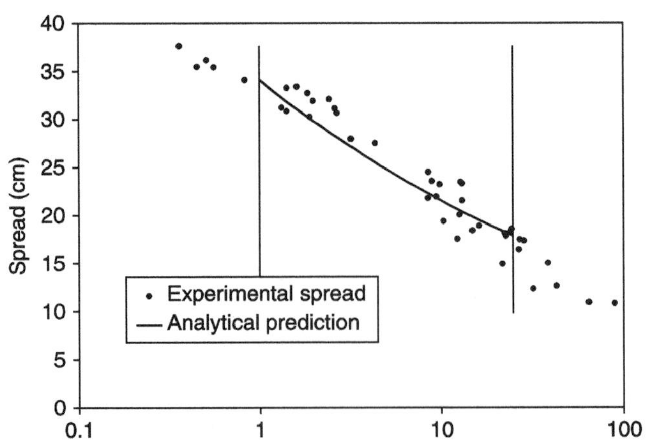

4.2 Spread of various cement pastes in terms of the ratio yield stress/ specific gravity for the mini cone test.

Minimum thickness

The thickness of the sample at flow stoppage should be at least five times the diameter of the largest particles to be able to consider the flow of a homogeneous mixture (Coussot and Ancey, 1999). As the largest particles in a cement paste are of the order of 100 μm, this means that, as long as the thickness of the flow remains more than half a mm, the material can be considered as homogeneous.

Experimental validation

Equation 4.4 was experimentally validated in the case of cement pastes and mini slump test (Fig. 4.2). The yield stress of the materials tested was measured independently using the vane test procedure described by Nguyen and Boger (1985) and a HAAKE ViscoTester® VT550 with a rotation speed of 0.4 rad. min^{-1}; see Roussel and Coussot (2005) for more details of the experiments. The analytical relationship of Eq. 4.4 only holds when H/R is far larger than 1. We consider here a limit value of $H/R = 5$. This limit of the correlation is plotted in Fig. 4.2. The lower limit of the correlation is either due to surface tension or inertia effects as described above.

There is good agreement between experimental and theoretical spread flow values but, when the yield value tends towards zero, the correlation between the model and the measurements decreases. The predicted spread tends towards infinity for a fluid with no yield stress, whereas the experimental spread flow for water is between 45 and 55 cm. Surface tension effects for low or no yield stress fluids can not be neglected in the stress equilibrium.

4.2.4 Validity of the correlation in the case of the self-compacting concrete (SCC) slump flow

Surface tension effects

In the case of the most fluid SCC (the ones that could be affected by surface tension effects), slump flow is of the order of 900 mm. If a wetting angle of the order of 30° for the interstitial fluid is assumed, it is possible to conclude from Eq. 4.6 that surface tension has no effect as long as the yield stress of the tested SCC is higher than 1 Pa, which is always the case.

Inertia effects

Let us compare the typical inertia stress ($I = \rho V^2$) to the material yield stress in the case of the SCC slump flow. The flow duration is about five seconds for a slump of the order of 20 cm. We thus have $I \cong 3$ Pa, a value much smaller than the material yield stress in that case (typically of the order of several tenths of a pascal).

Tested volume

The tested volume is equal to 6 L, which is approximately a thousand times the volume of the largest particle. The tested volume is thus representative of the tested concrete.

Minimum thickness

The thickness of the sample at flow stoppage should be at least five times the diameter of the largest particles to be able to consider the flow of an homogeneous mixture (Coussot and Ancey, 1999). The SCC slump flow, because of its final thickness/particle ratio, can therefore not be considered as a homogeneous flow and, as such, should be considered with care. For example, a higher slump flow can not be automatically associated to a lower yield stress and to an easier casting.

Let us imagine the case of two stable SCC with the same yield stress: the first one is prepared with a very fluid cement paste and a high volume fraction of rather small inclusions and the second one with a less fluid cement paste and a lower volume fraction of large inclusions. As yield stress of SCC is a function of both yield stress of the cement paste and volume fraction of the aggregates, both may have the same yield stress (Yammine *et al.*, 2008). Although both SCC are stable, the slump flow value of the first SCC will be higher than the slump flow value of the second. A rough biphasic explanation is as follows: for the first SCC, the contribution of the granular skeleton to the yield stress dominates the contribution of the cement paste while, for the second SCC, it is the opposite. If the flow could be considered as the flow of a homogeneous and continuous material, the origin of the yield stress of the mixture would not play any role in the flow stoppage. However, this is not the case as the thickness is of the same order as the size of the inclusions.

This does not mean that slump flow can not be used as a reception test. For a given SCC with a chosen granular skeleton, the slump flow test is a handy tool to spot, for example, a water to cement ratio variation during production. But the measured spread will not be correlated with the yield stress of the SCC tested. Some experimental results obtained at Laboratoires des Ponts et chaussées (LCPC) on some SCC with various granular skeletons, cement powder and additives having all roughly the same slump flow (700±50 mm) are gathered in Fig. 4.3 (Roussel, 2007b). Their yield stresses were measured using a BTRHEOM. Although the measured absolute values of the yield stress could be discussed (see Chapter 3), the relative values are meaningful. It has to be noted that all the SCC prepared were stable (sieve stability test < 15%). We can see in Fig. 4.3 that, as already stated, there is no direct correlation between yield stress and slump flow whereas this correlation was obvious in the case of the mini cone (cf. Fig. 4.2).

The thickness at stoppage should be around 60 mm to be representative of the rheological behaviour of standard SCC considering a particle maximum diameter of 12 mm. In the case of more conventional concretes (larger particles), higher

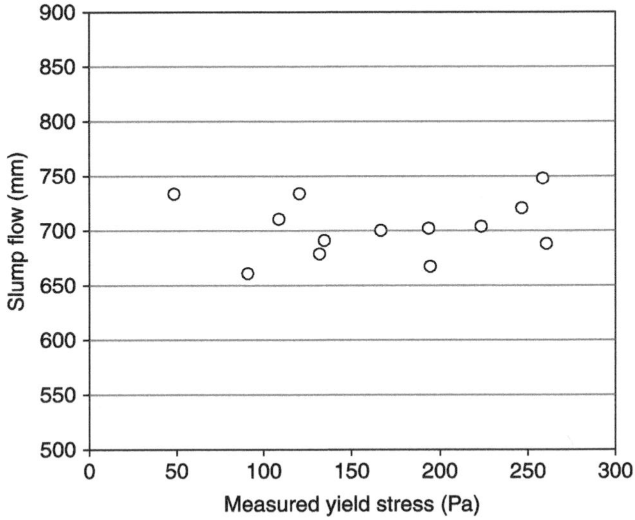

4.3 Yield stress measured using BTRHEOM for various SCC all displaying a 700±50 mm slump flow.

thickness values could be needed. Using Eq. 4.3 and Eq. 4.4, the minimum volume of the SCC to be tested to fulfil this assumption can be computed: it should be at least 300 L. This is neither realistic at a building site nor in the laboratory. In the next section, an alternative method, the LCPC BOX, which requires the same amount of SCC as the slump flow test but fulfils the minimum thickness condition, is described.

4.3 The LCPC BOX test

4.3.1 Test concept

This test is inspired by the fact that, in an axi-symmetrical test, the information on the shape of the sample at stoppage is redundant. Indeed, on each diameter, if the testing surface is horizontal, the same spread is measured. The idea behind this test is thus to study only one of these diameters by studying a channel flow (cf. Fig. 4.4). The width of the channel is l_0 = 20 cm and the length is 120 cm. The studied volume of SCC is the same as the one used in the slump flow (6 L) as it is sufficient to be representative of the tested concrete. As the flow is nearly unidirectional, the thickness of the sample at stoppage for the same sample volume is greater than in the slump flow test. It was checked that emptying a 6 L bucket was generating a flow slow enough for any inertia effects to be negligible. From an experimental point of view, the results of the test were the same in a wood or sand paper covered box for both cement pastes and SCC. In the analysis, the flow was therefore assumed to be sticking to the walls of the box.

4.4 LCPC BOX for SCC yield stress measurements.

The analysis of this flow and its stoppage is given by Nguyen *et al.* (2006). The analysis takes into account the shear stress at the lateral walls and bottom wall to predict the shape at stoppage of a given volume of a yield stress fluid flowing slowly enough for any inertia effects to be negligible. If we pose $A = 2\tau_0/\rho g l_0$ with l_0 the width of the channel, the spread length L at stoppage is then:

$$L = \frac{h_0}{A} + \frac{l_0}{2A} LN\left(\frac{l_0}{l_0 + 2h_0}\right) \qquad [4.7]$$

where h_0 is the thickness of the sample at the extremity of the box where the concrete is poured. The tested volume V is equal to:

$$V = l_0 \int_0^{h_0} x dh = \frac{l_0^3}{4A}\left(LN(1+u_0) + \frac{u_0(u_0-2)}{2}\right) \qquad [4.8]$$

Equation 4.8 allows the computation of h_0 for a given tested volume V. The ability of Eq. 4.7 and Eq. 4.8 to predict the shape of the material at flow stoppage was demonstrated by experimental measurements on limestone powder suspensions (Roussel, 2007b).

The test was used to study three different SCC (numbered 1–3). Their respective largest particle diameters were 12, 8 and 12.5 mm. They were prepared with various additives (superplasticizer and viscosity agent) and their respective water to (cement and filler) ratios were 0.30, 0.40 and 0.37. Their slump flow values

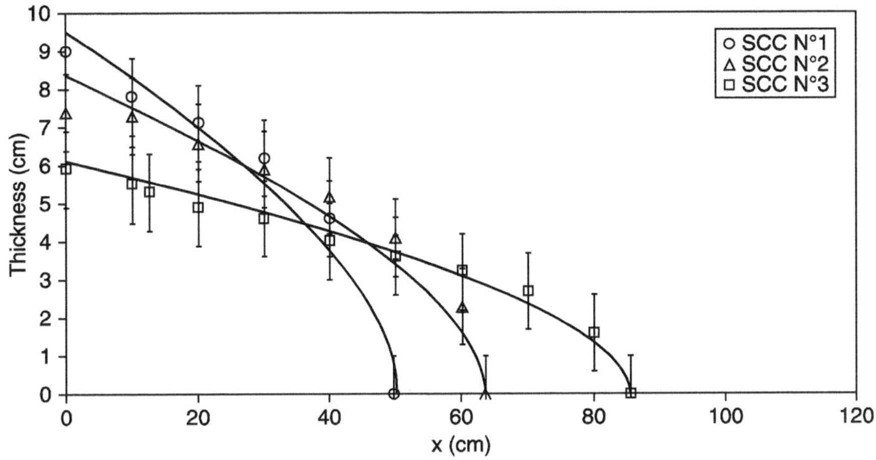

4.5 Shape of the material at stoppage for three SCC.

were 640, 730 and 690 mm respectively. The measured shapes of the sample when flow stopped are shown in Fig. 4.5. The yield stresses of the tested concretes were then calculated using Eq. 4.6 and Eq. 4.7 and were equal to 130 Pa for SCC 1, 65 Pa for SCC 2 and 35 Pa for SCC 3. As there is no concrete rheometer that gives the absolute value of the yield stress of a given concrete, there is no way of independently checking the validity of this calculation in the specific case of concrete. But as all the conditions needed to apply the equations of fluid mechanics (sufficient ratio between fluid characteristic thickness and constitutive particles size, negligible effect of inertia, negligible effect of surface tension) are fulfilled and as the boundary conditions have been checked (sticking flow at the walls), the test and its analysis can be used in the case of SCC in the opinion of the authors (Roussel, 2007b).

Test procedure

1. Fill a calibrated pre-wetted bucket with 6 L of the SCC to be tested.
2. Slowly pour the SCC at one extreme end of the BOX (the emptying of the bucket should take about 30 s); see Fig. 4.4.
3. Wait until flow stops (two minutes max.).
4. Measure the spread length *L* in the BOX (average of maximum spread length and lateral wall spread length).
5. Read the ratio of yield stress to specific gravity on Fig. 4.6.

The amount of SCC left in the bucket may be quantified by weighing the bucket before and after the test. The tested volume may then be corrected in Eq. 4.8 but, most of the time, this correction is negligible.

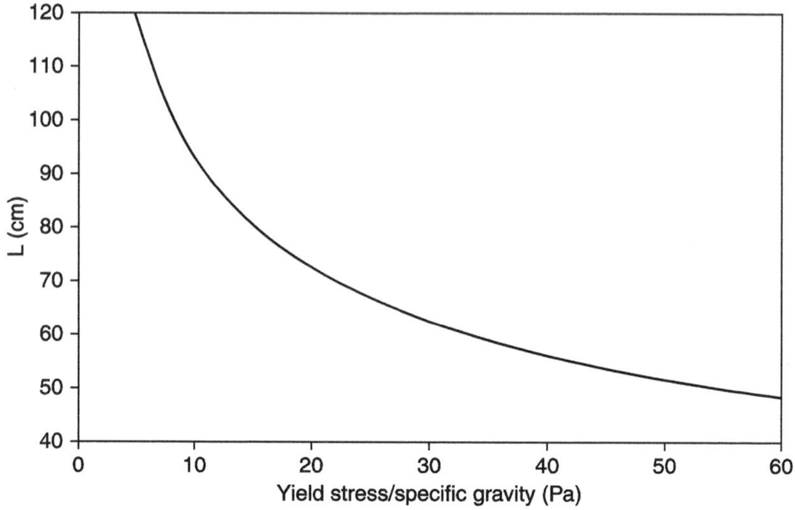

4.6 Correlation between the spread length L in the LCPC BOX and the ratio yield stress/specific gravity for the SCC tested.

4.4 Conclusions

This chapter gives the correlations between slump test, slump flow test and mini cone test and yield stress, and examines the limits of these correlations. It has been demonstrated that, in the case of SCC and the slump flow test, the measured spread diameter can not be correlated to the yield stress of the SCC tested. Indeed, although most of the conditions for the correlation to be valid (negligible surface tension effects and inertia effects) are fulfilled, the thickness of the sample at flow stoppage is of the same order as the largest particles. However, this does not prevent the slump flow test from being a good quality control and reception test as it is able to spot a variation in mix proportions, but it can not be used to measure a value for the yield stress.

The second part of the chapter presents an alternative way to measure the yield stress of a given SCC. This method is based on the fact that, in an axi-symmetrical flow, most of the sample volume is used to give redundant information as the shape of the material at stoppage is the same on every diameter. A more 'rational' test should therefore be based on a channel flow. Because of the discrepancy between concrete rheometers, it is not possible at present to confirm the validity of the method in the case of concrete. However, given the excellent agreement obtained in the case of fine suspensions and given that all the theoretical conditions are fulfilled for the proposed analysis to be valid in the case of SCC, this test seems to be able to give a cheap, simple and rather precise measurement of the yield stress of any SCC.

4.5 References

Clayton, S., Grice, T.G. and Boger, D.V. (2003), 'Analysis of the slump test for on-site yield stress measurement of mineral suspensions', *Int. J. Miner. Process.*, 70, 53–21.

Coussot, P. and Ancey, C. (1999), *Rhéophysique des pâtes et des suspensions*, Paris, EDP Sciences [in French].

De Larrard, F. and Hu, C. (1996), 'The rheology of fresh high-performance concrete', *Cem. Concr. Res.*, 26(2), 283–294.

Murata, J. (1984), 'Flow and deformation of fresh concrete', *Mater. Struct.*, 98, 117–129.

Nguyen, Q.D. and Boger, D.V. (1985), 'Direct yield stress measurement with the vane method', *J. Rheol.*, 29, 335–347.

Nguyen, T.L.H., Roussel, N. and Coussot, P. (2006), 'Correlation between L-box test and rheological parameters of a homogeneous yield stress fluid', *Cem. Concr. Res.*, 36(10), 1789–1796.

Pashias, N., Boger, D.V., Summers, J. and Glenister, D.J. (1996), 'A fifty cent rheometer for yield stress measurements', *J. Rheol.*, 40(6), 1179–1189.

Roussel, N. (2005), 'Steady and transient flow behaviour of fresh cement pastes', *Cem. Concr. Res.*, 35(9), 1656–1664.

Roussel, N. (2006a), 'A thixotropy model for fresh fluid concretes: Theory, validation and applications', *Cem. Concr. Res.*, 36(10), 1797–1806.

Roussel, N. (2006b), 'Correlation between yield stress and slump: Comparison between numerical simulations and concrete rheometers results', *Mater. Struct.*, 39(4), 501–509.

Roussel, N. (2007a), 'Rheology of fresh concrete: From measurements to predictions of casting processes', *Mater. Struct.*, 40(10), 1001–1012.

Roussel, N. (2007b), 'The LCPC BOX: a cheap and simple technique for yield stress measurements of SCC', *Mater. Struct.*, 40(9), 889–896.

Roussel, N. and Coussot, P. (2005), 'Fifty-cent rheometer' for yield stress measurements: from slump to spreading flow', *J. Rheol.*, 49(3), 705–718.

Roussel, N., Stefani, C. and Leroy, R. (2005), 'From mini cone test to Abrams cone test: Measurement of cement based materials yield stress using slump tests', *Cem. Concr. Res.*, 35(5), 817–822.

Saak, A.W., Jennings, H.M. and Shah, S.P. (2004), 'A generalized approach for the determination of yield stress by slump and slump flow', *Cem. Concr. Res.*, 34, 363–371.

Schowalter, W.R. and Christensen, G. (1998), 'Toward a rationalization of the slump test for fresh concrete: comparisons of calculations and experiments', *J. Rheol.*, 42(4), 865–870.

Tattersall, G.H. and Banfill, P.G.F. (1983), *The Rheology of Fresh Concrete*, Pitman, London.

Yammine, J., Chaouche, M., Guerinet, M., Moranville, M. and Roussel, N. (2008), 'From ordinary rheology concrete to self compacting concrete: a transition between frictional and hydrodynamic interactions', *Cem. Concr. Res.*, 38, 890–896.

5

The rheology of cement during setting

S. GAUFFINET-GARRAULT,
University of Bourgogne, France

Abstract: The evolution of cement paste from a soft material to a hard solid occurs in several stages. Dynamic mode rheometry is a valuable tool for the non-destructive study of the evolution of the structure of cement pastes and the forces between particles from mixing to setting. Dynamic mode rheometry gives access, through the measurement of the elastic modulus, to the evolution of cohesion over time, at early age. This chapter describes the chemical processes and highlights the correlation between hydration and cohesion evolution. Through a modification of the hydration process, some parameters can influence the evolution of cohesion of the paste and the 'mechanical efficiency of hydrates'.

Key words: hydration, rheology, setting, calcium silicate hydrates.

5.1 Hydration: chemical reactions and kinetics

When cement is mixed with water, each phase dissolves at least partially, leading to the establishment of a supersaturated solution with respect to different hydrates, which can precipitate. Hydration is the word used to describe all the reactions involved. Like all chemical reactions, they obey the laws of thermodynamics and kinetics.

Cement is a polyphasic material composed of calcium silicate and calcium aluminate phases (Taylor, 1997). The hydration products of silicate phases are calcium hydrosilicate – known as CSH (C for CaO, S for SiO_2 and H for H_2O) in cement nomenclature – and calcium hydroxide, $Ca(OH)_2$. Aluminate phases exhibit a high reactivity in water, causing a rapid stiffening of the paste and low mechanical strength, which related to the precipitation of calcium hydroaluminate. To avoid this phenomenon of rapid setting, calcium sulfate is added to clinker to produce cement. This addition prevents the formation of calcium hydroaluminate by promoting the precipitation of calcium trisulfoaluminate (ettringite).

5.1.1 Chemical stages

Tricalcium silicate (Ca_3SiO_5 or C_3S) is the main phase of Portland cement clinker. It is called 'alite' in clinker because it is not pure tricalcium silicate and contains a number of impurities substituted in its crystal lattice. Once in contact with water, Ca_3SiO_5 is hydroxylated and the surface dissolves according to Eq. 5.1 (Barret *et al.*, 1983; Barret and Bertrandie, 1986):

$$Ca_3SiO_5 + 3\,H_2O \rightarrow 3\,Ca^{2+} + 4\,OH^- + H_2SiO_4^{2-} \qquad [5.1]$$

The solution quickly becomes supersaturated with respect to calcium hydrosilicate (CSH), which is less soluble than tricalcium silicate. When the maximum degree of supersaturation is reached, CSH precipitates according to Eq. 5.2:

$$C/S\,Ca^{2+} + 2\,(C/S-1)\,OH^- + H_2SiO_4^{2-} \rightarrow CSH \qquad [5.2]$$

where C/S is the Ca/Si ratio of the CSH formed and C/S < 2. The system never reaches the solubility of tricalcium silicate.

Because all the calcium ions are not consumed by the precipitation of CSH, increasing concentrations of calcium and hydroxide ions in solution lead to the maximum supersaturation of the liquid phase with respect to calcium hydroxide (this corresponds to $Ca\,(OH)_2$ = 30–36 mmol/l at 25°C), which then precipitates as Portlandite according to the reaction in Eq. 5.3:

$$Ca^{2+} + 2\,OH^- \rightarrow Ca(OH)_2 \qquad [5.3]$$

From that moment, the three reactions shown in Eq. 5.1 to 5.3 become simultaneous. The concentration of calcium hydroxide is always greater than its solubility (22 mmol/l at 25°C) and the system is still supersaturated with respect to CSH.

The progress of these chemical processes (i.e. hydration) can be described by a sigmoidal evolution with time (Fig. 5.1). We observe an initial part where progress remains low, which depends on the number of initial germs of CSH precipitating on the surface of the grain (Garrault and Nonat, 2001). Acceleration is linked to the free growth of CSH on this surface. During this stage, hydration is controlled

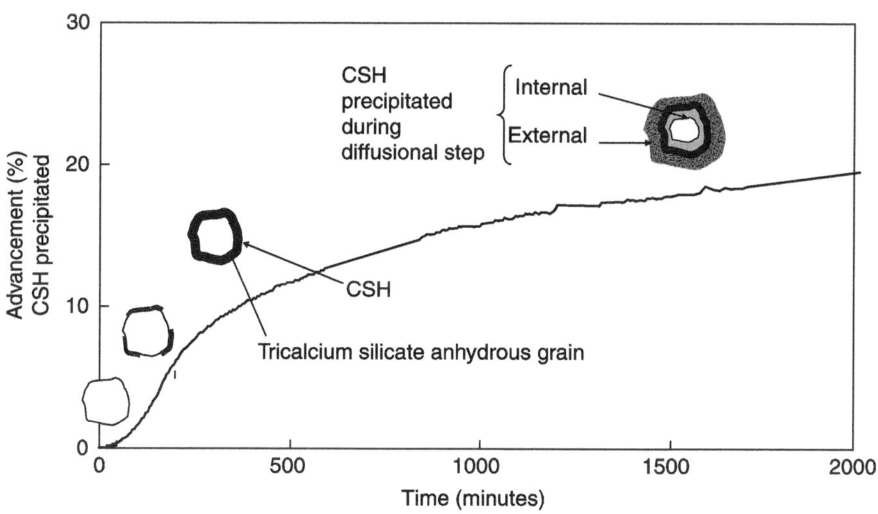

5.1 Evolution of the degree of hydration of tricalcium silicate vs. time.

by the rate of growth of CSH. The covering of anhydrous grains by CSH slows down the reaction, which becomes limited by the diffusion of reactants through the continuous layer of hydrate around the grains. Then there is growth of CSH instead of dissolving tricalcium silicate between the anhydrous grains and first CSH layer and outwardly on the solution–CSH layer interface (Tennis and Jennings, 2000; Jennings, 2008). The reaction becomes very slow and may take several months or years. These chemical reactions start very slowly, allowing for the safe casting of concrete before setting.

5.1.2 Details of calcium hydrosilicate (CSH)

Nucleation of CSH during the hydration of calcium silicates is a heterogeneous phenomenon that takes place on the cement particles (Garrault and Nonat, 1999). Growth of CSH can be seen as an aggregation process of anisotropic particles of 60 × 30 × 5 nm (Fig. 5.2) around the initial seed (Gauffinet *et al.*, 1998). This aggregation of particles is oriented in the direction of the maximum development of nanoscale particles until the surface is completely covered by the layer of CSH.

Electron microscopy observations (Scrivener *et al.*, 1985; Richardson, 1999; Famy *et al.*, 2002; Gallucci *et al.*, 2007) often show a dense layer of CSH around the anhydrous grain (Fig. 5.3). This dense layer corresponds to the hydrate layer formed during the first stage of hydration. The CSH which then develops in the pores of the paste (CSH external) over the period during which the hydration reaction is limited by diffusion has a much more fibrillar aspect. Internal CSH, which occupy the space vacated by the grain that dissolves, has a more compact morphology.

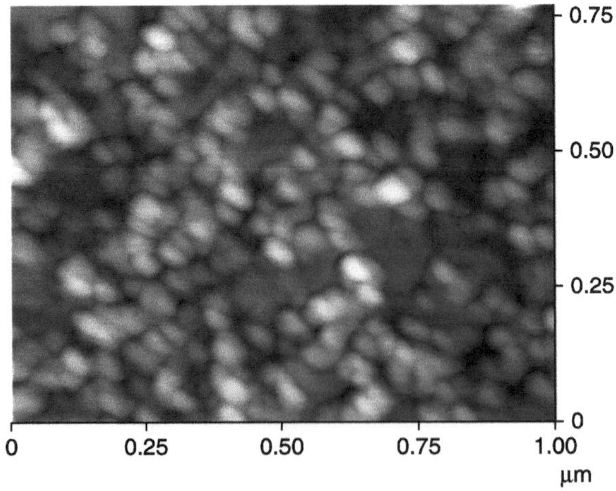

5.2 Direct observation by atomic force microscopy of CSH growth through aggregation of nanoparticles.

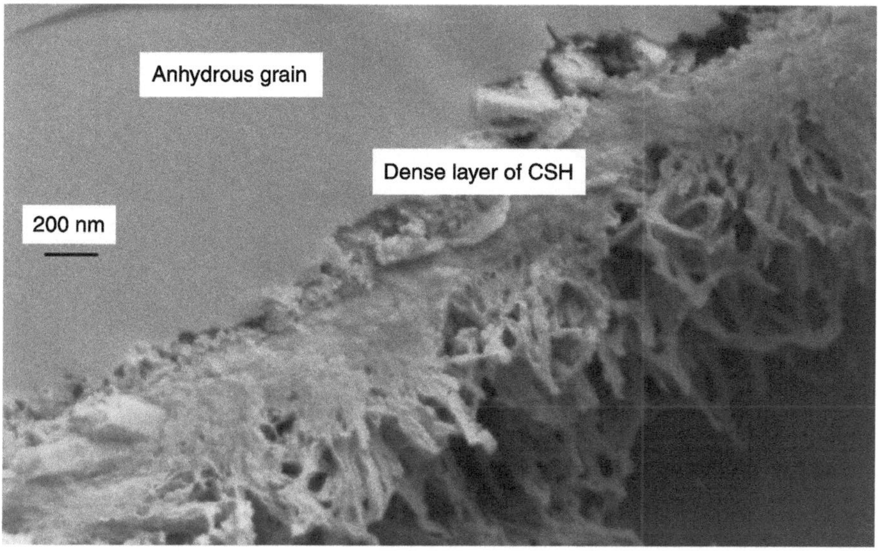

5.3 Scanning electron microscopy image of tricalcium silicate hydrated grain (Nicoleau, BASF).

From a crystallographic point of view, CSH exhibits a lamellar structure consisting of two levels of calcium ions coordinated by the oxygen of tetrahedra of silicates (Nonat, 2004). The silicates are in the form of parallel dimers, some of which can be bridged by a third tetrahedron forming pentamers. The interlayer contains water molecules and some calcium ions. The number of bridging tetrahedra and calcium interlayers depends on the pore solution concentration of calcium hydroxide. It is this dependence that is at the origin of the variation in the Ca/Si ratio in CSH.

5.1.3 Hydration of tricalcium aluminate

Unlike tricalcium silicate hydration, hydration of $Ca_3Al_2O_6$ (or C_3A) is very fast (Fig. 5.4). Although the hydration rate is continuously decreasing, it is controlled only by the dissolution rate of $Ca_3Al_2O_6$. The same type of process as in the case of calcium silicates occurs when tricalcium aluminate is in contact with water. Its hydrolysis leads to its dissolution:

$$Ca_3Al_2O_6 + 6\,H_2O \rightarrow 3\,Ca^{2+} + 2\,Al^{3+} + 12\,OH^- \qquad [5.4]$$

Contrary to the case of calcium silicates, there are several phases of hydrated calcium aluminates less soluble than the anhydrous:

- AFm phases, $Ca_2Al_2(OH)_{10}.3H_2O(C_2AH_8)$ and $Ca_4Al_2(OH)_{14}{\cdot}6H_2O$ (C_4AH_{13}), which are in fact the boundaries of a solid solution in which

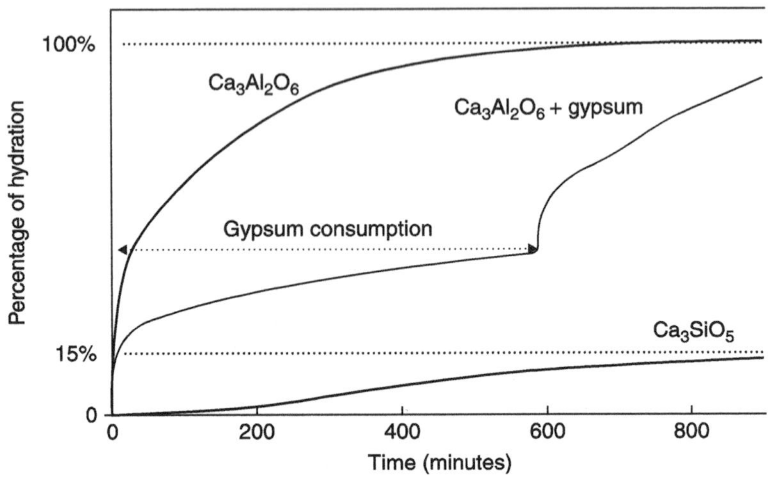

5.4 Evolution vs. time of the degree of hydration of tricalcium silicate and tricalcium aluminate with and without calcium sulphate.

the Ca/Al ratio varies between two and four according to the concentration of calcium hydroxide in solution.

- Mixed hydroxide $Ca_3Al_2(OH)_{12}(C_3AH_6)$. This phase is thermodynamically more stable (less soluble) and is therefore the end state towards which the system '$Ca_3Al_2O_6$ water' tends in the absence of any other component. The AFM phases, $Ca_2Al_2(OH)_{10}.3H_2O(C_2AH_8)$ and $Ca_4Al_2(OH)_{14}\cdot 6H_2O$ (C_4AH_{13}), dissolve and C_3AH_6 precipitates.

The first rapid formation of calcium hydroaluminate leads to premature stiffening of the paste (flash set). To prevent this stiffening, calcium sulfate is added to clinker. The calcium sulfate is added as gypsum ($CaSO_4\cdot 2H_2O$), hemi-hydrate ($CaSO_4\cdot 0.5H_2O$) or anhydrite ($CaSO_4$). Under these conditions, the less soluble phase is calcium trisulfoaluminate hydrate, $Ca_6Al_2(SO_4)_3(OH)_{12}$,$26H_2O$ or ettringite. This phase is formed as the sulfate concentration in solution is sufficient (Fig. 5.5, period 1) according to Eq. 5.5 (written here with gypsum as the only source of sulfate):

$$Ca_3Al_2O_6 + 3CaSO_4.2H_2O + 30H_2O \rightarrow Ca_6Al_2\left(SO_4\right)_3\left(OH\right)_{12}.26H_2O \qquad [5.5]$$

When all sulfates are consumed, calcium hydroaluminate then precipitates as previously without sulfate (Fig. 5.5, period 2).

During the first period of hydration, the rate of formation of ettringite, which is a much slower process than that of AFM, controls the rate of hydration (Minard *et al.*, 2007). Once all calcium sulfate has been consumed to form ettringite, calcium hydroaluminate is formed again rapidly, at a rate controlled by the dissolution rate of $Ca_3Al_2O_6$ corresponding to the second peak on the heat flow curve (Fig. 5.5). Under these conditions, ettringite is not stable and it partially

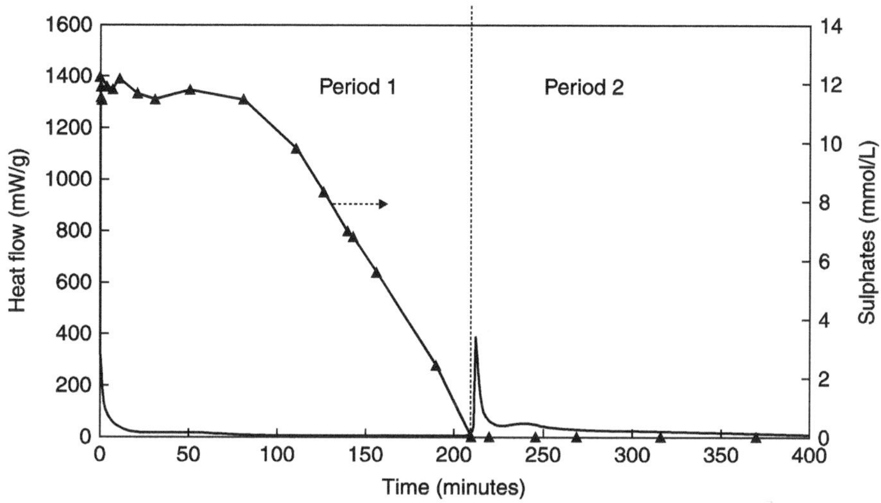

5.5 Heat flow and sulphate concentration evolutions vs. time during tricalcium aluminate hydration in presence of gypsum.

dissolves, with the released ions contributing to the precipitation of calcium monosulfoaluminate.

Cement is composed of many different phases that dissolve and bring various ions to the pore solution of cement paste. It is important to take into account the fact that concentrations of different species evolve over time and are dependent on each other. Indeed, the hydration reaction combining dissolution of anhydrous and hydrate precipitation imposes relationships between the different ion concentrations. In addition, the composition of the liquid phase can modify the solubility of the formed products. For all these reasons, hydration of the mixture of the phases constituting cement does not behave exactly as the simple superposition of the hydration of each phase considered alone.

Thus, a precise kinetic study of the hydration of cement is always difficult and especially since cement often incorporates more or less reactive additions. However, a first approximation could be that the hydration mechanisms of the different phases are not substantially altered, especially regarding the hydration of tricalcium silicate, which is the major phase of Portland cement. A portion of CSH may however precipitate on other solid supports (fillers, other carbohydrates, etc.) which alter the percentage for which the hydration reaction becomes limited by a continuous layer of hydrate.

5.2 Rheology of cement pastes

Initially dispersed grains of clinker or tricalcium silicate come into contact and form a connected structure or stable aggregates respectively in paste or dilute suspension (Jiang *et al.*, 1995, 1996). This occurs very rapidly in the first tens of seconds after mixing. As tricalcium silicate dissolves and CSH precipitates on its surface, the coagulated structure is reinforced: in dilute suspension it becomes difficult to break the aggregates and paste 'set' (Nonat and Mutin, 1992). The stiffening is not due to larger forces but to the multiplication of contacts between particles. CSH are nanocrystallites (the surface area of CSH is about 250 m^2/g) and increase dramatically the number of links of the interconnected system.

5.2.1 Experimental method

The evolution of the viscoelastic properties of the pastes can be measured by dynamic mode rheology using controlled strain rheometers (see Chapter 3). To perform the experiments, the cement paste sample is submitted to a sinusoidal shear strain defined at constant frequency (ω) by:

$$\gamma = \gamma_o \, e^{i\omega t} \qquad [5.6]$$

The resulting stress is measured by the intermediary of the torque and is also sinusoidal with a δ phase lag with respect to the applied strain:

$$\tau = \tau_0 e^{i\omega t + \delta} \qquad [5.7]$$

Within the linear viscoelasticity range and using oscillatory mode, the relation between shear strain and stress is given by:

$$\tau = G^*(\omega)\gamma \qquad [5.8]$$

where G^*, the complex shear modulus, is expressed as the sum of the shear storage modulus (G') and the shear loss modulus (G'') as:

$$G^* = G' + iG'' \qquad [5.9]$$

The shear strain value is chosen within the linear viscoelasticity range of the cement paste so as to avoid destructive measurements. Indeed, above a particular value of the shear strain called the critical strain γ_c, the destructuration of the material would occur. In case of a cement paste, the limit of the linear viscoelastic domain has been shown to occur when the strain reaches about 0.03% (Schultz and Stuble, 1993; Nachbaur *et al.*, 2001).

Nachbaur *et al.* (2001) used two different measurement tools: a classic parallel plates geometry (Fig. 5.6) and a mixer type tool (Fig. 5.7). These tools were designed to maintain a saturated water atmosphere within the cell containing the sample, thus preventing the cement paste drying out. The surface of the plates had also been roughened to avoid wall slippage (Louis-Tisserand, 1995). The sample was a disk of paste with a diameter of 6 mm and a thickness of 2 mm. With these dimensions and the characteristics of the rheometer, the lowest strain possible is 0.001%.

For the measurements made with the non-conventional geometry, the paste was prepared by mixing the powder (cement or tricalcium silicate) and water by hand

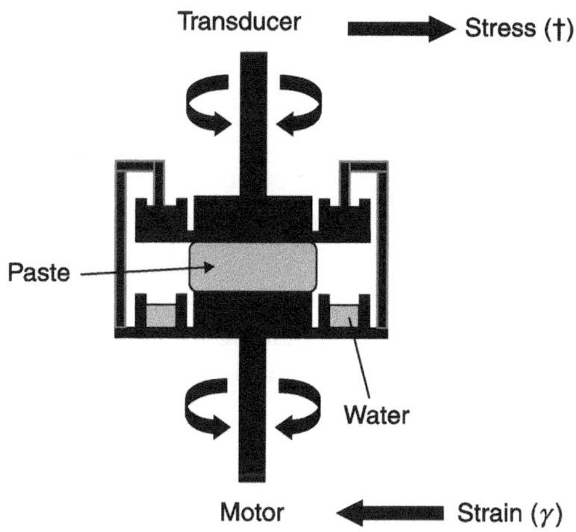

5.6 Parallel plates geometry.

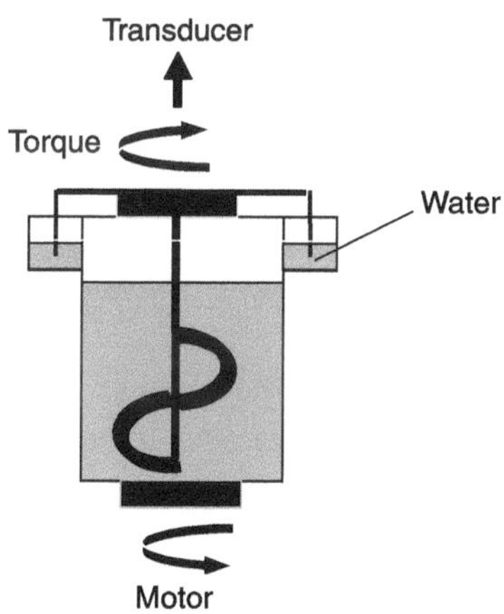

5.7 Non-conventional geometry: helicoidal ribbon.

in a beaker before transferring it to the measurement tool. This resulted in a delay (about two minutes) between the end of mixing and the beginning of the dynamic mode measurement. Moreover, despite the care taken, additional stress can be added to the sample during this transfer; the measurements would then not be made on the paste in its state reached immediately after mixing.

The second tool used was a non-conventional one and consisted of a helicoidal ribbon in a cylindrical cell. With this geometry, the mixing and the measurements in dynamic mode can be performed using the same tool without transfer. This suppresses the problems of delay and additional stress mentioned above. This geometry provides the same rheological information as the conventional parallel plates geometry. An analytical method, based on a Couette analogy (Choplin and Marchal, 1999; Aït-kadi *et al.*, 2002; Lacoste *et al.*, 2005), allows the quantitative extraction of absolute rheological data from torque rotation data for this non-conventional geometry.

The liquid to solid ratio used was chosen to ensure good mixing and easy placement in the measurement tool when the parallel plates were used. The value of water to concrete (w/c) is thus different for the two geometries and for the two materials. It was limited to a small range in each case because lower w/c result in poor mixing and/or problems while placing the sample in the measurement tool and higher w/c result in sedimentation or samples too liquid to be placed within the parallel plates.

Mixing and transfer both induced strains which were relaxed before the data were recorded by operating a high amplitude strain (5%) at high frequency (50 rad/s) for 45 seconds. Experiments were then performed using a 0.002% shear strain and a 5 rad/s frequency.

During cement paste hydration, the shear storage modulus (G′) quickly becomes much more important than the shear loss modulus (G″) and, therefore only the G′ modulus was analysed subsequently. Hydration degree was monitored by isothermal calorimetry on part of the samples. The hydration was followed by a high sensitivity (0.1 μW) isothermal Tian–Calvet type or a TAM air isothermal calorimeter (Thermometrics) (Damidot *et al.*, 1990).

5.2.2 From mixing to setting

About one minute after the end of mixing, the material changes from a liquid character $G'' > G'$ to a solid character $G' > G''$ (Fig. 5.8). This transition is due to

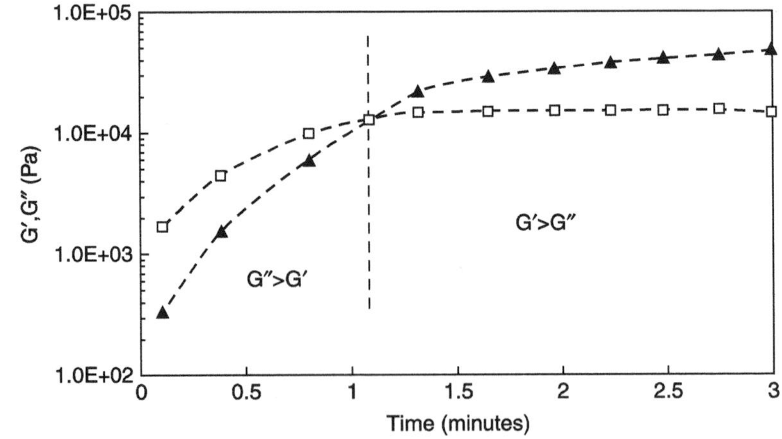

5.8 Evolution of elastic moduli G′ and G″ of the viscous cement paste.

the rapid coagulation of grains. The increase in moduli of one or two orders of magnitude, visible in logarithmic scale, shows the important structure development in the opening minutes. However, this related structure offers very little resistance as a low mechanical stress such as the one applied during mixing destroys it. This evolution is reversible as this structure rebuilds spontaneously when stress ceases.

This stiffening can be associated with the strengthening of the structure previously described by the CSH precipitation at contact points between particles forming the aggregate. It has indeed been shown that the germination of CSH takes place on the surface of anhydrous grains (Garrault and Nonat, 1999). Moreover, the probability of germination is greater as supersaturation is high. Because of the proximity of two surfaces of grains in contact, it is in these pseudo-contact zones where supersaturation is highest and where the first seeds are formed preferentially. When the structure formed can not be destroyed by the mechanical stress applied during mixing, this evolution becomes irreversible. Figure 5.9 shows the temporal evolution of the elastic modulus superimposed on the progress of the hydration of cement paste. The modulus G' increases relatively more than the hydration progresses; it increases with the amount of CSH that form between anhydrous grains. What is called 'setting' is thus not a transition. The so-called initial setting and final setting refer only to a given mechanical strength established by standards (Lootens *et al.*, 2009). The strength keeps on evolving continuously before and after these landmarks.

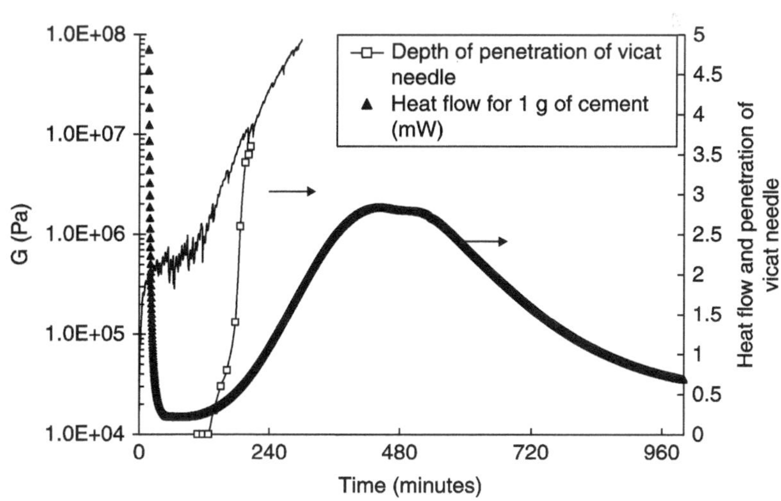

5.9 Evolutions of storage modulus, heat flow and vicat needle penetration vs. time during hydration of cement paste (Nachbaur, 1997).

5.2.3 Forces at the origin of cohesion

The cohesion of cement paste is therefore due to the formation of a dense network of CSH nanoparticles that are highly 'linked' together via short range physical forces, induced by ionic correlations occurring in the system (Pellenq *et al.*, 1997). These forces depend on the composition of pore solution and the CSH surface.

The surface of CSH is characterized by a high density of electric charge. Indeed, there are oxygen atoms on the ends of the silicate tetrahedra. If these oxygen atoms are not engaged in connecting two tetrahedra or coordinating a calcium ion, they are connected to protons and form silanol groups > Si-OH. The surface silanol groups are ionized by:

$$\mathrm{Si-OH+OH^{-}\rightarrow>SiO^{-}+H_2O} \qquad [5.10]$$

At high pH, as is the case in cement paste, most sites are ionized resulting in a very high surface charge density. The high power density surface of CSH is the source of hardened cement paste cohesion (Pellenq *et al.*, 1997; Jönsson *et al.*, 2004; Labbez *et al.*, 2007).

5.2.4 Critical strain of cement paste

Several different tests are used in dynamic mode rheology (cf. Chapter 3). The first is a strain sweep in which the sample is submitted to an increasing oscillating strain. For small strains, the moduli are independent of the strain. When the strain exceeds a critical value (critical strain), the moduli decrease as the strain increases. This is the end of the linear viscoelastic domain of the material. Dynamic mode rheometry results can be easily interpreted only within this linear domain. From a structural point of view, the linear domain is a sign that there is some kind of structure in the material, corresponding to well-defined interparticle forces and distances (Mewis and Spaull, 1976). When the applied strain exceeds the critical strain, the observed structure is destroyed. This test is therefore necessary to determine the conditions under which the dynamic mode measurements are non-destructive.

There are two critical strains for cement paste. Each can be associated with a specific physical and chemical phenomenon involved in the behavior of the paste in the fresh state. The largest critical strain is of the order of a few percent and can be linked to the network of colloidal interactions between cement particles. The time needed to form this network is of the order of a couple seconds. The smallest critical strain is of the order of several hundredths of a percent and can be associated, as described above, to CSH which forms preferentially at the contact points between aggregated cement grains. CSH develops as soon as a flocculated network of particles has formed in the mixture. The development of this network of cement particles interacting via hydrates interactions (described in Section 5.2.1) can be decomposed in two successive steps: a rigid percolation followed by a rigidification.

These have reversible macroscopic consequences as long as the available mixing power is sufficient to break the rigid links between cement particles.

5.2.5 Rheology evolution versus degree of hydration: mechanical efficiency of CSH

Figure 5.10 shows a plot of the evolution of the elastic modulus G′ as a function of the progress of hydration (i.e. the amount of hydrates formed). Two successive regimes can be distinguished in this plot: for low similar amounts of hydrates, the elastic modulus G′ increases relatively more than the amount of formed hydrates. The first regime corresponds to the strengthening of the structure coagulated by CSH precipitating at the pseudo-contact points; each formed hydrate particles contributes to the elastic modulus and to the stiffening of the structure. The second regime corresponds to a clogging of the pores. Precipitating hydrates do not necessarily create new contacts between anhydrous grains but contribute to filling the inter-particles porosity.

The mechanical efficiency of calcium silicate hydrates ($R = \delta_{G'}/\delta_Q$) can be defined as the increase in elastic modulus $\delta_{G'}$ corresponding to the formation of a certain amount of CSH, estimated by the released hydration heat δ_Q. It is computed in the first regime of the evolution of G' versus time. It can therefore be associated with the precipitation CSH, which develops inter-particle contacts and bridges the not yet hydrated cement grains.

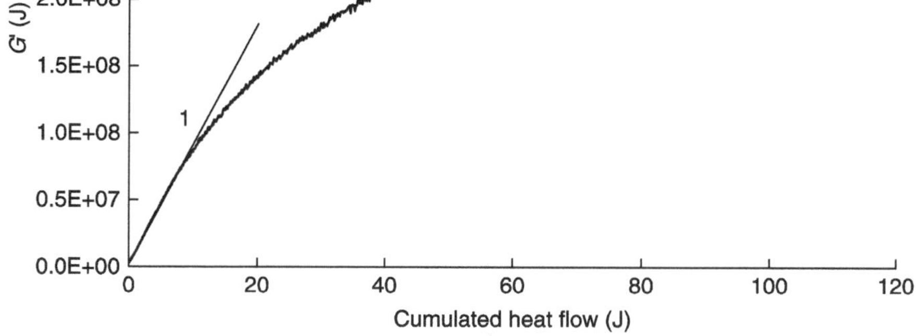

5.10 Storage modulus increase vs. degree of hydration of tricalcium silicate paste. 1 – Coagulation of structure by CSH; 2 – Clogging of the pores by CSH.

5.3 Parameters influencing mechanical efficiency of calcium hydrosilicate (CSH)

It is possible to use the concept of mechanical efficiency to assess the influence of various parameters on the cohesion of the coagulated structure. This section examines the influence of temperature and how phases other than calcium silicates, which are also present in cement, act on paste cohesion.

5.3.1 Influence of temperature

The results obtained during the hydration of Ca_3SiO_5 paste at different temperatures (Fig. 5.11) show that the higher the temperature of hydration, the lower the mechanical efficiency of CSH. This is true up to 25°C. Beyond this temperature, the ratio between the increase in modulus and the advancement of hydration remains almost constant and may even increase.

Several facts may explain this effect. Firstly, during hydration, CSH, which precipitates and develops near the pseudo-contact points between anhydrous grains will increase the number of contacts between grains. It was shown that, if the hydration is conducted at 5°C, the number of blocks of growth (i.e. initial nucleation sites) was about three times larger than in the case of hydration at 45°C. Under these conditions, for the same degree of hydration, the probability of

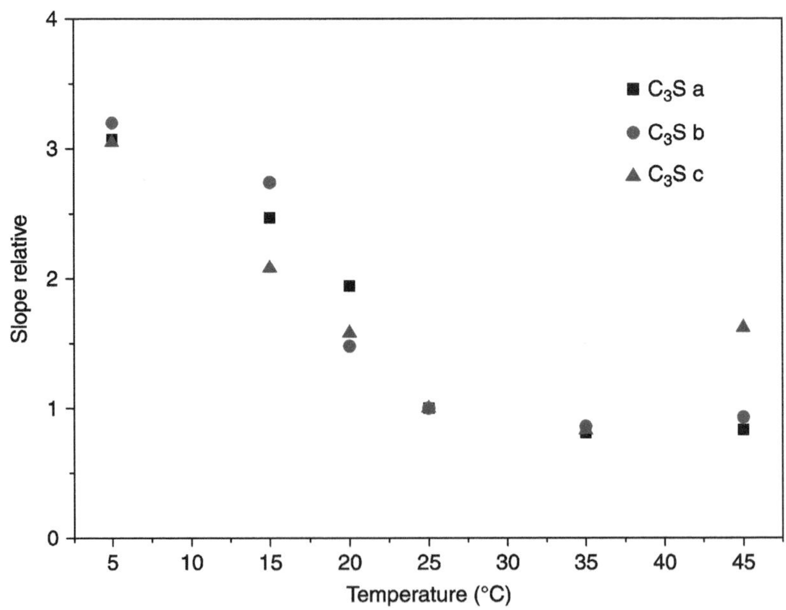

5.11 Evolution of the mechnical efficiency vs. the degree of hydration of tricalcium silicate paste for different temperatures.

creating new percolating paths increases and the storage modulus is greater. Once the grains are fully covered by the layer of CSH formed during the accelerated period, this layer becomes the percolating phase. The contact surface between the grains, and therefore the moduli, increases with the thickness of this layer. Numerical simulations of the kinetics of hydration at an early time showed that the CSH layer is thicker at 5°C than at 25 or 45°C, which is consistent with the observed increase (Zajac, 2007).

The increase in elastic modulus measured by rheometry is linked to the modification of the coagulation of cement grains or anhydrous C_3S and the formation of the initial layer of CSH, which becomes the percolating phase. A variation of the initial number of nuclei or the mode of growth of hydrates can change this trend and thus affect the cohesion of the paste of C_3S and cement.

5.3.2 Influence of presence of tricalcium aluminate

During the first few minutes, the elastic modulus of alite (impure tricalcium silicate)–C_3A mix increases very quickly and is always higher than moduli measured when only alite is suspended in water (Garrault *et al.*, 2007). The elastic modulus increase is relatively higher than the C_3A content in the mixture. It is known that this initial and fast increase of moduli is due to the hydration of the aluminate phase which causes a stiffening of the paste, commonly called 'flash set'.

If we look at the evolution of the storage modulus versus cumulated heat flow in Fig. 5.12 (representative of the degree of hydration of alite), the intercept of

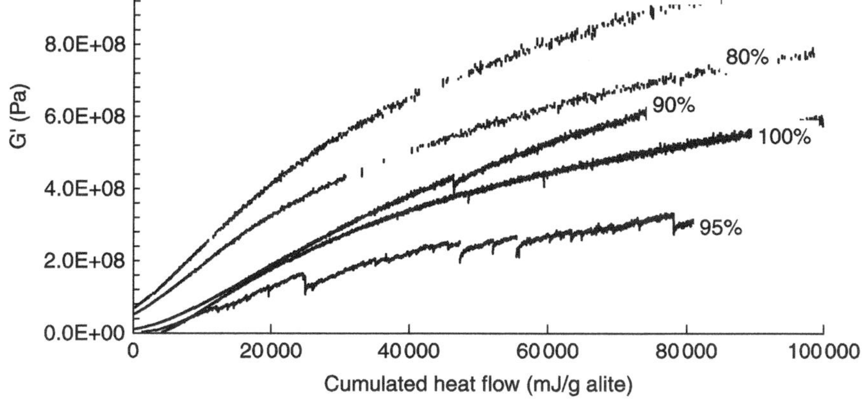

5.12 Evolution of storage modulus vs. cumulated heat flow for hydration, with W/C = 0.5, of different alite–C3A mixes: 100%, 95%, 90%, 80%, 75% alite.

each mixture represents the modulus value obtained at the end of the period of the hydration of tricalcium aluminate (about 120 minutes). This quantity increases relatively more than the amount of C_3A in the mix. However, the increase in the modulus due to hydration of C_3A in the mixture is relatively low compared with the increase in modulus due to the hydration of alite.

When the percentage of C_3A in the mixture increases, the mechanical efficiency of CSH evolves non-monotonically (Fig. 5.13). It is lower than in the case of pure alite additions of 5 and 10% and higher additions of 20 and 25% of C_3A. The origin of the variation in mechanical efficiency is to be found in the physico-chemical process as the w/c ratio is constant and the density of the blends is almost constant.

The decrease in the mechanical efficiency of CSH is consistent with what we expect: the hydration of C_3A provides aluminate ions in solution that have a negative effect on the mechanical efficiency of CSH. However, the positive effect can be attributed to the presence of hydration products of tricalcium aluminate. The hydration products of tricalcium aluminate are used to support the growth of CSH, the hydrates connection is greater and moduli increased relatively more than C_3A content in the mix.

5.3.3 Mechanical efficiency of CSH in the alite–C_3A–gypsum system

To obtain the evolution of the storage modulus with respect to degree of hydration for such a complex system (deduced from calorimetric measurements), it is

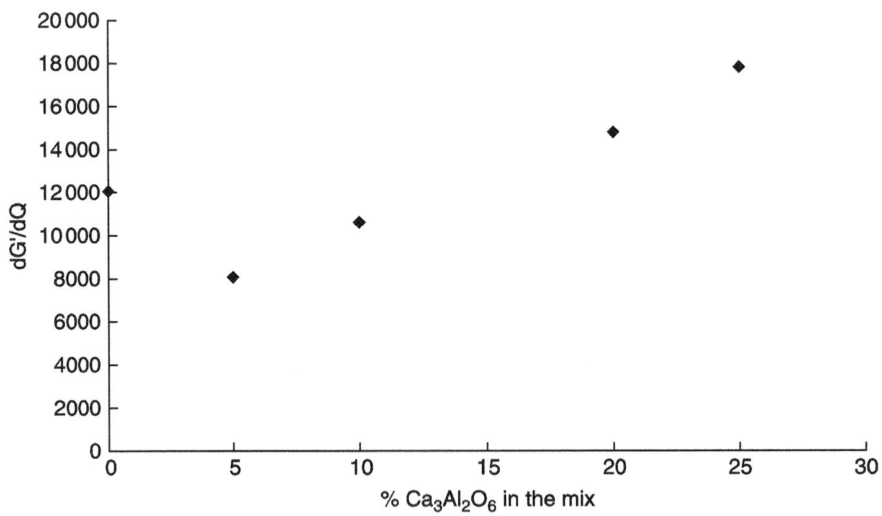

5.13 Evolution of mechanical efficiency of CSH vs. percentage of C_3A in the different studied mixes.

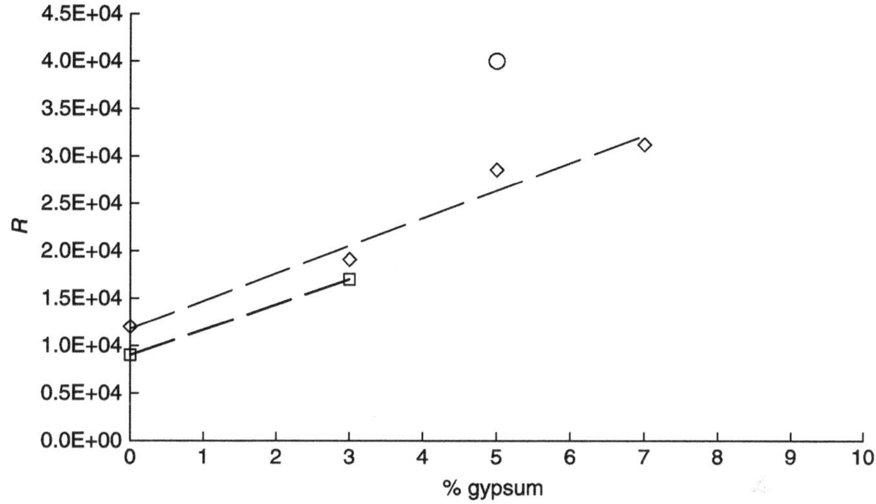

5.14 Evolution of mechanical efficiency of CSH vs. gypsum percentage in the system 76% alite–17%C_3A–gypsum–portlandite (◊), 75% alite–20%C_3A–gypsum (o) C_3S–gypsum (□).

necessary to consider the heat corresponding to ettringite formation, which continues during alite hydration. This quantity is subtracted from total heat flow. Figure 5.14 presents the deduced evolution of mechanical efficiency of CSH for various quantities of gypsum; results for the 'C_3S–gypsum' system are also reported. For the mixes containing the same quantity of C_3A and alite, and for C_3S, the mechanical efficiency of CSH increases with gypsum quantity. For the alite–C_3A–gypsum system, this increase seems to be more important when the quantity of C_3A increases. Although it has been shown that sulfate ions can modify CSH growth (Garrault *et al.*, 2003), here sulfate ions are consumed by ettringite formation; the increase in mechanical efficiency (*R*) could be attributed to the solid gypsum presence which is not consumed when *R* is determined. It has been shown than the rate of ettringite formation is constant and does not depend on the initial gypsum quantity (Medala *et al.*, 2006).

5.4 References

Aït-kadi, A., Marchal, P.H., Chrissemant, A.S., Bousmina, M. and Choplin, L. (2002), 'Quantitative analysis of mixer-type rheometers using Couette analogy', *Can. J. Chem. Eng.*, 80, 1166–1174.

Barret, P. and Bertrandie, D. (1986), 'Fundamental hydration kinetic features of the major cement constituents: Ca_3SiO_5 and bCa_2SiO_4', *J. Chim. Phys.*, 83(11/12), 765–775.

Barret, P., Ménétrier, D. and Bertrandie, D. (1983), 'Mechanism of C_3S dissolution and problem of the congruency in the very initial period and later on', *Cem. Concr. Res.*, 13, 728–738.

Choplin, L. and Marchal, P.H. (1999), 'Rheo-reactor for in situ rheological follow-up of chemical or physical processes', *Ann. Trans. Nordic Rheol. Soc.*, 7, 5–12.

Damidot, D., Nonat, A. and Barret, P. (1990), 'Kinetics of tricalcium silicate hydration in diluted suspensions by microcalorimetric measurements', *J. Am. Ceram. Soc.*, 73(11), 3319–3322.

Famy, C., Scrivener, K. and Crumbie, A.K. (2002), 'What causes differences of C-S-H grey level in backscattered electron images?', *Cem. Concr. Res.*, 32(9), 1465–1471.

Gallucci, E., Mathur, P. and Scrivener, K. (2007), 'Microstructural studies of early age hydration of OPC using TEM', 12th International Congress on the Chemistry of Cement, 8–13 July 2007, Montréal, Canada.

Garrault, S. and Nonat, A. (1999), 'Experimental investigation of calcium silicate hydrate (C-S-H) nucleation', *J. Cryst. Growth*, 200(3–4), 565–574.

Garrault, S. and Nonat, A. (2001), 'Hydrated layer formation on tricalcium and dicalcium silicate surfaces: experimental study and numerical simulations', *Langmuir*, 17, 8131–8138.

Garrault, S., Nachbaur, L. and Sauvaget, C. (2003), 'Influence of alkali sulfates on the rheological properties of tricalcium silicate pastes', *Ann. Chim.*, 28 (Suppl. 1), S43–S50.

Garrault, S., Minard, H. and Nonat, A. (2007), 'Hydration of silicate phase and mechanical evolution in 'alite-tricalcium aluminate-gypsum' complex system', 12th International Congress on the Chemistry of Cement, 8–13 July 2007, Montréal, Canada.

Gauffinet, S., Finot, E., Lesniewska, E. and Nonat, A. (1998), 'Observation directe de la croissance d'hydrosilicate de calcium sur des surfaces d'alite et de silice par microscopie à force atomique', *C. R. Acad. Sci., Ser. IIa: Sci. Terre Planetes*, 4(327), 231–236.

Jennings, H.M. (2008), 'Refinements to colloid model of C-S-H in cement: CM-II', *Cem. Concr. Res.*, 38, 275–289.

Jiang, S.P., Mutin, J.C. and Nonat, A. (1995), 'Studies on mechanism and physico-chemical parameters at the origin of the cement setting. I. The fundamental processes involved during the cement setting', *Cem. Concr. Res.*, 25(4), 779–789.

Jiang, S.P., Mutin, J.C. and Nonat, A. (1996), 'Studies on mechanism and physico-chemical parameters at the origin of the cement setting. II. Physico-chemical parameters determining the coagulation process', *Cem. Concr. Res.*, 26(3), 491–500.

Jönsson, B., Wennerström, H., Nonat, A. and Cabanne, B. (2004), 'Onset of cohesion in cement paste', *Langmuir*, 20, 6702–6708.

Labbez C., Nonat, A., Pochard, I. and Jönsson, B. (2007), 'Experimental and theoretical evidence of overcharging of calcium silicate hydrate', *J. Colloid Interface Sci.*, 309(2), 303–307.

Lacoste, C., Choplin, L., Cassagnau, P. and Michel, A. (2005), 'Rheology innovation in the study of mixing conditions of polymer blends during chemical reaction', *Appl. Rheol.*, 15, 314–325.

Lootens, D., Jousset, P., Martinie, L., Roussel, N. and Flatt, R.J. (2009), 'Yield stress during setting of cement pastes from penetration tests', *Cem. Concr. Res.*, 39, 401–408.

Louis-Tisserand, B. (1995), *Rhéologie des suspensions de sulfate de calcium. Analyse de l'influence de la forme et des caractéristiques morphostructurales des particules*, PhD thesis, University of Bourgogne, France.

Medala, M., Minard, H., Nocun-Wczelik, W., Garrault, S. and Nonat, A. (2006), 'Hydration of C_3A-$CaSO_4$-H_2O system revisited – influence of highly concentrated salt solutions', Polish Ceramic Bulletin, 92, 275–282.

Mewis, J., Spaull, A.J.B. (1976), 'Rheology of concentrated dispersion', *Adv. Colloid Interface Sci.*, 6, 173–200.

Minard, H., Garrault, S. and Nonat, A. (2007), 'Mechanisms and parameters controlling the tricalcium aluminate reactivity in presence of gypsum', *Cem. Concr. Res.*, 37(10), 1418–1426.

Nachbaur, L., Mutin, J.-C., Nonat, A. and Choplin, L. (2001), 'Dynamic mode rheology of cement and tricalcium silicate pastes from mixing to setting', *Cem. Concr. Res.*, 31(2), 183–192.

Nachbaur, L. (1997), *Etude de l'influence d'électrolytes sur l'hydratation et la prise du silicate tricalcique, composant principal du ciment Portland. Caractérisation des interactions à l'origine de la prise*, PhD thesis, University of Bourgogne, France.

Nonat, A. (2004), 'The structure and stoichiometry of C-S-H', *Cem. Concr. Res.*, 34(9), 1521–1528.

Nonat, A. and Mutin, J.-C. (1992), 'From hydration to setting', in *Hydration and Setting of Cements*, proceedings of the international RILEM workshop, 3–5 July 1991, Dijon, ed. A. Nonat and J.C. Mutin, E & FN Spon, London, pp. 171–191.

Pellenq, R., Crespin, M., Lequeux, N., Menager, C., Costalin, L., Delville, A., *et al.* (1997), 'A Monte-Carlo study of the stability of charged interfaces, application to cement and clay minerals', in *2nd International RILEM Symposium on Hydration and Setting*, ed. A. Nonat, RILEM Publications, Bagneux, France, pp. 63–86.

Richardson, I.G. (1999), 'The nature of C-S-H in hardened cements', *Cem. Concr. Res.*, 29(8), 1131–1147.

Schultz, M.A. and Stuble, L. (1993), 'Use of oscillatory shear to study the flow behavior of fresh cement paste', *Cem. Concr. Res.*, 23, 273–282.

Scrivener, K., Patel, H.H., Pratt, P.L. and Parrott, L.J. (1985), 'Analysis of phases in cement paste using backscattered electron images, methanol adsorption and thermogravimetric analysis', *Mater. Res. Soc. Symp. Proc.*, 85, 67–76.

Taylor, H.F.W. (1997), *Cement Chemistry*, 2nd edn., Thomas Telford Publishing, London.

Tennis, P.D. and Jennings, H.M. (2000), 'A model for two types of calcium silicate hydrate in the microstructure of Portland cement pastes', *Cem. Concr. Res.*, 30(6), 855–863.

Zajac, M. (2007), *Etude des relations entre vitesse d'hydratation, texturation des hydrates et résistance mécanique finale des pâtes et micro-mortiers de ciment Portland*, PhD thesis, University of Bourgogne, France.

Part II
Mix design and the rheological behaviour of concrete

6
Particle packing and the rheology of concrete

X. CHATEAU, Université Paris-Est, France

Abstract: Mixtures of particles of different size are widely used in order to design concrete mixtures containing the maximum amount of solid particles possible that can be readily transported, placed and finished. This chapter provides an overview of the main results of the effects of particle properties on packing characteristics and presents three different models that aim to predict the packing density of a particle mixture. The dependence of the overall rheological properties of highly concentrated suspensions of polydisperse particles on the particle size distribution and particle blend composition is then investigated.

Key words: maximum packing density, particle size distribution, high concentration suspension, non-Newtonian rheology.

6.1 Introduction

It is well known in concrete technology that the grading of aggregate is one of the main factors influencing the hardened concrete strength (Neville, 1995; Kumar Mehta and Paulo Monteiro, 2006). Aggregate is added to cement paste for various reasons: aggregate is cheaper than cement and it confers to concrete higher strength and better durability than cement paste. In a first approach, it is widely admitted that adding as much as possible aggregate to the cement paste is desirable in order to produce better concrete.

However, the rheological properties of a suspension (viscosity, yield stress, etc) are increasing functions of the volume fraction occupied by the particles (Chong *et al.*, 1971; Coussot, 2005; Stickel and Powell, 2005). Adding particles to a paste also reduces its workability; consequently it is necessary to reduce this effect in order to design concrete mixtures containing the maximum amount of solid particles possible that can be transported, placed and finished easily. This chapter will first explain that mixtures of particles of different size give less porous packing than mixtures of particles of the same size. It will also be shown that the particle size distribution that gives the highest packing density also gives the most flowable suspensions when the particle volume fraction is prescribed. This is why mixtures of particles of different size have been used in concrete technology for a long time in order to produce granular packing of maximal compacity.

The packing density of dry particle mixtures is also of interest in many other industrial fields (ceramic, pharmacy, civil engineering, sintering, food processing, etc) and the properties of dry particle mixtures have attracted the interest of mathematicians, physicists and engineers for a long time. Although much work has been done or is in progress on this topic, no model exists that is able to predict the packing density of a set of particles as a function of the shape of the particles, their size distribution, the container geometry and the mixing process. Given the complexity of the problem, it is outside the scope of this chapter to detail all the results available in the literature on this topic and interested readers are referred to Aste and Waire (2000), Cumberland and Crawford (1987) and German (1989) for a survey.

This chapter is restricted to the main results useful for those interested in the influence of particle size distribution on the overall rheological properties of a suspension. The best way of addressing this question is to start from the simplest situation of mono-size spherical particles (even if it is far from being obvious that stacking mono-size spheres is a simple problem) and then to look at problems of increasing complexity. This is done in the first part of the chapter. In the second part, three different models that aim to predict the packing density of a particle mixture as a function of its composition are presented. Finally, the dependence of the overall rheological properties of highly concentrated suspensions of polydisperse particles on the particle size distribution and particle blend composition is investigated.

6.2 Compacity and porosity

To start with, it is necessary to recall how particle packing compacity can be characterised. When particles are poured into a container, they do not completely fill the inner space and some voids remain between the particles. The packing density φ (sometime also called the packing fraction or compacity) is the ratio of the volume of the particles V_s to the total volume of the packing V_p, while the porosity π is the ratio of the volume of voids V_v to the total volume of the packing. These two quantities are linked by the relation $\varphi + \pi = 1$. In soil mechanics, the void index e (defined as the ratio of the volume of the voids V_v to the solid volume V_s) is widely used. φ, π and e are linked by the simple equation $e = \pi/\varphi$. For practical purposes, mass apparent density or mass concentration γ (defined as the ratio of the mass of solids over the total volume of the packing) is easier to measure than packing density; γ can be experimentally determined by weighting a container of known volume full of particles. The mass apparent density depends on the particle density γ_s and the compacity φ through the equation $\gamma = \gamma_s \times \varphi$. Care must be taken with the various definitions of density used in the literature. For example, some authors use the number density $\nu = \varphi/\upsilon$ with υ the volume of one particle to measure the density of a stack of particles. In the sequel only the packing density φ is used to characterise the particle concentration.

6.3 Packing of mono-size spheres

Let us start with a simple example in two dimensions. When a set of coins of the same size are put on a table and pushed together, the coins form a disordered packing. Then, if the coins are shacked gently, the surface occupied by the packing decreases regularly until the coins arrange themselves (at least locally) to a triangle arrangement with density $\pi / 2\sqrt{3} = 0.91$. It can be shown that this triangular close packing is the densest possible arrangement of an equal size disc in two dimensions providing no wall effects disturb the coins' arrangement (Aste and Weaire, 2000). Even if such regular packing of mono-size spheres exists in three dimensions, the 3D regular packing can not be built as easily as in two dimensions.

6.3.1 Regular packing of mono-size spheres

Regular packings of mono-size spheres have been studied for a long time, particularly in solid state physics. The simplest regular packing is certainly the cubic packing in which the centres of the particles each occupy the vertex of a cube, the length of which is twice the sphere radius. The packing density of the cubic packing is $\pi/6 = 0.5236$. Of course such a packing is far from stable and, even if one is able to stack rigid spheres in a cubical lattice, any small perturbation is likely to destroy the structure.

Packing with the maximum packing density can be realised by pyramidal stacking of spheres. As there are two options for the laying down of one layer of particles onto another, there are many different arrangements of the maximum packing density. The most well-known high density packings are face centred cubic packing and hexagonal packing. Both have a packing density $\pi / 3\sqrt{2} = 0.7404$.

Despite the study of the regular packing of mono-size spheres having held the attention of mathematicians for a long time, some problems remain unsolved. For example, the rigorous proof proposed by Hales of 0.7404 being the maximum density of packed mono-size particles (Aste and Weaire, 2000) is currently under verification. The fact that there are packings of mono-size spheres with different compacity reveals that particle packing density depends the way the particles are geometrically distributed in space.

6.3.2 Random packing of mono-size spheres

It is everyday experience that it is impossible to obtain a regular packing of mono-size spheres in three dimensions by merely shaking a pile of beads or kneading a bag of marbles. To build a three-dimensional regular packing of spheres, it is necessary to take time and care to stack the particles one by one – as done by greengrocers. It is also unlikely that particles dispersed in a fluid arrange themselves in a regular packing.

Particles are distributed randomly in most of the situations of practical interest. The random packing of balls was first studied by Bernal (1959). He found experimentally that the packing density of random packing is roughly 0.64, or slightly less if the stack is not kneaded enough. In addition, it appears that the random packing falls well short of the maximum density that can be achieved in an ordered packing. Nevertheless, it is recognised as the best random packing even if a rigorous definition of this notion remains to be found. When beads are poured in a container, wall effects modify the random close packing fraction. As an example, Scott (1960) experimentally measured random packing close to $0.64 - 0.37 \times N^{-1/3}$ (with N the number of beads) by kneading beads poured in a ball. In addition, the larger (with respect to the particle size) the container, the greater the packing density.

6.3.3 Factors affecting the packing density of mono-size spheres in practice

The packing density value 0.64 corresponds to hard spheres (*i.e.* non-Brownian and non-colloidal particles) that are gently kneaded enough. When mono-size spheres are poured into a container or on a surface, the packing density depends mainly on two factors: the method used to deposit the particles and the small gap inter-particle forces (colloidal versus non-colloidal particles). As colloidal force intensity depends mainly on particle size, particle density also depends on particle size.

Bournonville (2003) performed a simple experiment evidencing this point. Mono-size glass beads were poured in a cup of volume 100 ml and stirred gently by hand. The cup was then shaken softly until the free surface was flat and horizontal. When the same procedure was used to fill the cup, it was observed that the packing density depended on the particle size; packing density values of 0.63, 0.62 and 0.58 ± 0.005 respectively were measured for particles of 320, 148 and 50 μm in size respectively, i.e. packing density decreases when particle size decreases. The lower value of the packing density measured for smaller particles can be ascribed to the repulsive electric forces exerted between charged particles which become significant and can no longer be neglected with respect to particle weight.

The method used to pour the particles into the container also affects the packing density of mono-size spheres of a diameter large enough that colloidal effects do not play a role. If the packing volume is measured just after particles were gently laid down within the container, packing density values close to 0.60 are obtained. It suffices to shake the container erratically for a few seconds to obtain packing densities close to 0.64. The value 0.60 corresponds to random loose packing. In this situation, most particles occupy the first equilibrium position they find after falling into the container. It is likely that configurations of these particles correspond to a local minimum of the potential energy but not a global one. Shaking the container allows the particles to move around their initial equilibrium position so that they can reach more stable equilibrium configuration, which in turn, induces a decrease in the packing density (particles moves downward in order to decrease their gravity potential energy).

It is outside the scope of this chapter to summarise all the results of the literature on these topics. Suffice to say that it has been experimentally shown that the packing densities of beads depend, among others, on the deposition intensity (number of particles falling by unit time per unit area), the velocity of the particles (which depend on the fall height), the particles' mass and size, friction law, the container size and the container shape It is also worth noting that three main packings of interest exist with mono-size sphere:

- ordered (particles stacked one by one at the right position to minimise the porosity);
- random loose (particles deposited gently and no perturbation applied onto the stack);
- random close (particles pile is shacked in order to let the particle to minimise their potential energy).

It is widely accepted that random close packing density of mono-size spheres is equal to 0.64 and that random loose packing density of mono-size spheres is close to 0.60. It is likely that, when the space between mono-size beads is not filled by air, the random packing density should depend on the rheological properties of the fluid. This question is addressed in Section 6.8.

6.4 Packing of identical particles

Packing density depends on the shape of the particles. First, it is obvious that non-spherical particles have more parameters (shape, orientation, size) than spherical ones. Therefore, the analysis and modelling of non-isotropic particles are more complicated than for spheres. Some simple examples from the literature illustrate this point (Cumberland and Crawford, 1987; German, 1989; Aste and Weaire, 2000). First, it is well-known that parallelepiped can be stacked so that they completely fill the space (like bricks in a wall). It is also possible to obtain a packing density equal to 0.91 by laying down layers of disks with triangle arrangement. Therefore, regular particles can give packing densities higher than spheres. Nevertheless, it is generally observed that non-spherical particles give a lower random packing density than spherical particles. The influence of a particle's shape on random close packing density has been studied by several authors. Some results from the literature for particles of different shapes are gathered in Table 6.1.

Even if much work aiming at relating particle shape to packing density has been performed, such a correlation (if it exists) remains to be found. Size ratio (also called aspect ratio) is often considered one of the main parameters influencing the particle packing density. Nevertheless for a particle with a size ratio less than ten, random close packing does not clearly depend on the particle size ratio (Fig. 6.1). Particles with a regular geometric shape can give densities larger than 0.64 (for example, packing densities as high as 0.74 have been obtained by vibrating prolate spheroids for a long time). These high density values can be reached because particles with a regular shape and flat faces arrange themselves locally in a regular arrangement more easily than particles with an irregular

Table 6.1 Random close packing density for particle of different shapes

Particle shape	Random close packing density
Spheres	0.60–0.64
Cubes	0.76
Parallelepiped (4.0 < L/D < 8.0)	0.51–0.67
Disks (L/D = 1)	0.63
Spheroids (0.6 < L/D < 1.3)	0.58–0.61
Rounded aggregates	0.59–0.63
Crushed aggregates	0.50–0.57
Fibres (L/D = 10)	0.48
Fibres (L/D = 167)	0.03

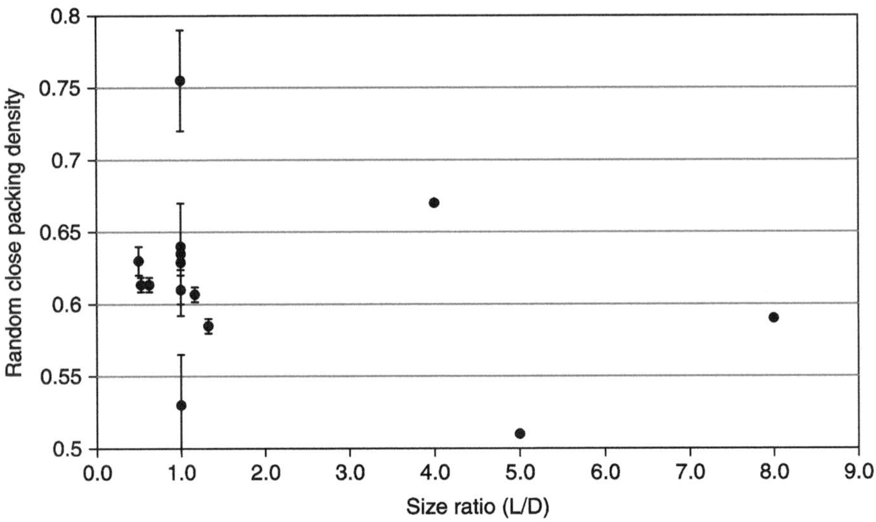

6.1 Random close packing density vs. particle size ratio for particles with size ratio close to one. D = diameter; L = length.

geometric shape or rounded boundary. Nevertheless, densities much lower than 0.60 have been obtained for spheroid particles.

It also seems that the particle's angularity has a strong influence on packing density: de Larrard (1999) reported that rounded aggregates give a packing density close to 0.60 while crushed aggregates packing gives values between 0.50 and 0.57. Such results suggest that packing density can be related to a 'roundness' factor and some attempts to do this have been described in the literature (German, 1987). The roundness of a particle is usually defined as the ratio of the surface area of a sphere with the same volume as the considered particle to the actual surface of the particle. Some authors report having observed a good correlation between random close density and roundness ratio. Nevertheless, such a relation can not apply to any particle shape because particles with a roundness factor lower than one can be randomly stacked with a density lower than 0.64 (e.g. crushed aggregates) or higher than 0.64 (e.g. cubes, parallelepiped).

It must be recognised that simple particle shape characterisations such as size ratio or roundness fail to account for all the effects playing a role in particle packing density. It seems it is not yet possible to accurately predict the packing density of mono-size particles from knowledge of their shape. Consequently it must be concluded that an experimental approach remains the most accurate method of determining the packing density of particles.

The situation is more satisfactory for fibres and rods. First, it is observed that random packing densities of elongated particles are low when particle aspect ratios are high (Nardin *et al.*, 1985; Philipse, 1996). Besides, the random packing ratio depends strongly on size ratio (Fig. 6.2). More precisely, it has been observed that for sufficiently thin fibres ($L/D > 15$), the quantity $\varphi L/D$ is constant. Then the random packing density of fibres is given as:

$$\varphi = c_f\,(D/L) \qquad [6.1]$$

where D denotes the diameter of the fibre, L the length of the fibre and c_f a numerical factor.

Philipse (1996) gave a theoretical justification of this result using a random contact model in which the average number of contacts per fibres is constant (equal to c_f) whatever the fibre size aspect ratio. By fitting experimental data to Eq. 6.1 he found that $c_f = 5.4$. More recently, Martinie *et al.* (2010) measured the random close packing and random loose packing of fibres with a size ratio

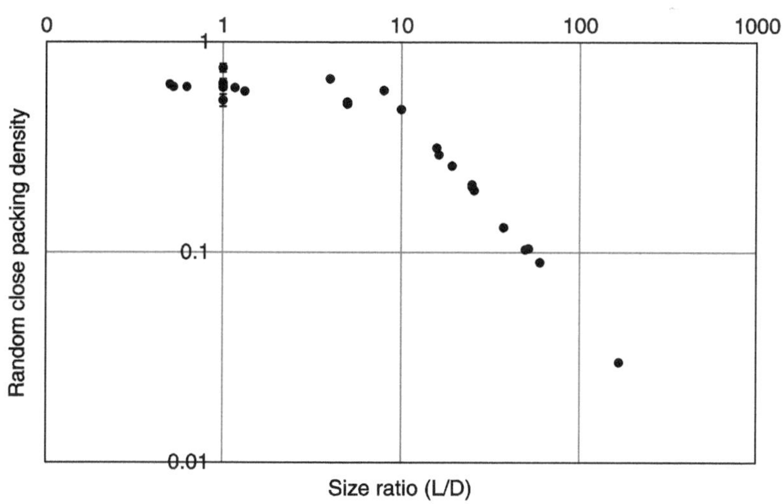

6.2 Random close packing density vs. particle size ratio for particles with size ratio larger than one.

belonging to [17, 100]. They showed that both close and loose packing densities can be described by Eq. 6.1; from a fit to their experimental data they determined that $c_f = 4$ for close packing density and $c_f = 3.2$ for loose packing density.

6.5 Packing of non-identical particles

We now turn our attention to the situation where non-identical particles sizes are mixed. The aim is to provide tools allowing to packing density and the composition of the particle mixture to be linked. A detailed approach to this problem would distinguish the situation of particles having the same shape (e.g. spherical) but a different size from the situation of particles with a different shape. As it is not possible to take into account all the properties of the particles in a tractable approach to this problem, each class of particles belonging to a mixture is characterized only by its size and its packing density.

6.5.1 Size distribution

To define the composition of a particle mixture, it is necessary to divide the particles into size groups and to determine their relative proportion. In practice, the grain size distribution of particles is often determined by passing them through sieves with successively smaller mesh size. The particles are sorted into classes containing particles with diameters ranging from the hole diameter of one sieve to the hole diameter of the following sieve. The grain size distribution can be described by plotting the proportion (by volume or, more easily by weight) of particle passing through the sieve as a function of the particle size (grading curve) or by plotting the distribution of particle as a function of the particle diameter (particle size distribution).

However, sieve analysis does not give any information about particle shape. The result of a sieve analysis is the separation of an aggregate sample into fractions of particles of the same size, or more precisely, into classes of particles of size larger than one sieve opening and smaller than the next sieve opening. For simplicity, each class of particles is generally characterised by only one size (the opening size of the retaining sieve or the mean opening size), the exact definition of which is of no importance. Let $d_1 > d_2 > \ldots > d_i > \ldots > d_n$ denote the particle sizes and $\xi_i = V_{si}/(V_{s1} + \ldots V_{sn})$ the volume fraction occupied by particles i where V_{si} is the volume occupied by the particles i in the particle mixture under consideration.

6.5.2 Bidisperse mixtures of spheres

First, consider the case of a bidisperse bead mixture. Bournonville (2003) used the same procedure as described above to measure the packing density of binary mixes of particles of diameter 50 and 320 μm poured into a 100 ml container. The

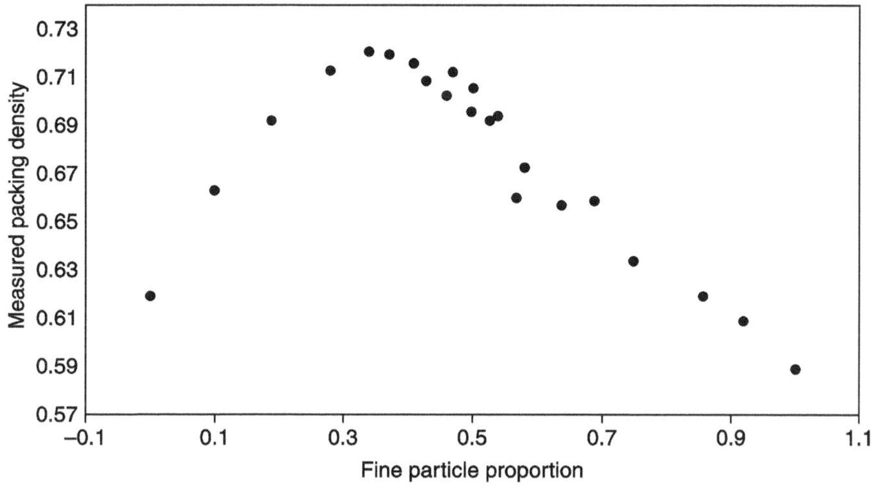

6.3 Packing density vs. fine particle proportion for a mix of 50 and 320 μm beads.

measured packing density is plotted in Fig. 6.3 as a function of the fine particle proportion $\xi_f = V_{sf}/(V_{sf} + V_{sc})$ where V_{sf} and V_{sc} denote the fine and coarse particle volume respectively. It is observed that the packing density is an increasing then decreasing function of the fine particle proportion. The measured maximum packing density $\varphi = 0.72$ is obtained for $\xi_f = 0.34$. As the random close packing density of fine and large particles is 0.59 and 0.62 respectively, this simple experience reveals that mixing beads with different size allows an increase in compacity up to 22%. Similarly, packing with density equal to 0.77 has been obtained by mixing 26% of particles with a diameter of 50 μm, 14% of particles with a diameter of 148 μm and 60% of particles with a diameter of 320 μm (Bournonville, 2003). These results provide clear evidence that the packing density of a mixture of particles depends on its composition.

A simple sketch in two dimensions allows us to explain why mixtures of particles with different sizes can give a larger compacity than mono-size packings. Let us consider the situation where large discs are packed regularly with packing density $\varphi_L = 0.91$. It is possible to build a packing with larger compacity by adding small disks of a diameter chosen so that they just fill the pores without distorting the original packing as depicted in Fig. 6.4. The optimal (i.e. larger) compacity is obtained when as many small disks as pores have been added to the original packing.

Let $\xi^{opt} = 0.046$ denote the optimal small disk fraction. It is readily seen that the optimal packing density is $\varphi^{opt} = 0.95$. A small disk fraction lower than ξ^{opt} gives a packing density lower than φ^{opt} because there are not enough small disks to fill all the pores of the large disk packing. A small disk proportion larger than ξ^{opt}

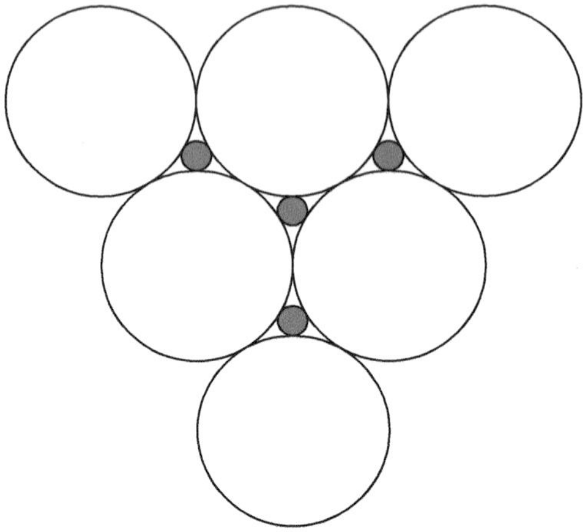

6.4 Bidisperse regular packing of disks. The small disks just fill the pores between the large disks without distorting the original packing.

produces a packing density lower than φ^{opt} because an excess of fine particles triggers off distortion of the original regular packing. Such a procedure can of course be repeated as many times as desired to obtain a multimodal mixture of disks with maximum packing density. Such a bidimensional packing is an example of an Appolonian packing.

6.4.3 Appolonian packing

In the search to define the densest possible granular packing, Appolonian packing retains the attention of theoreticians and practitioners for a long time. The principle is rather simple: each void in a packing of spheres is filled with a sphere having the greatest possible radius so that it touches all the neighbouring particles (kissing particle). In this procedure, the size of the inserted particles become smaller and smaller while the number of added particles become greater and greater. When this process is repeated indefinitely, packing density tends toward one. Appolonian packing is a classical example of fractal structure composed of similar parts that scale over an infinite range. Even if Appolonian packing does not exist in the real world, some workers proposed that the ideal particles mixture for concrete should have a grading curve close to that of Appolonian packing. The grading curve of an Appolonian packing in two dimensions is depicted in Fig. 6.5.

The packing density of an Appolonian packing with particles sizing from d_{min} to d_{max} is given as:

$$\varphi^{app} = 1 - \left(d_{min}/d_{max}\right)^{d-d_f} \quad [6.2]$$

where d_f is the fractal dimension and d is the space dimension. Numerical simulations give $d_f = 1.305$ in two dimensions and $d_f = 2.474$ in three dimensions.

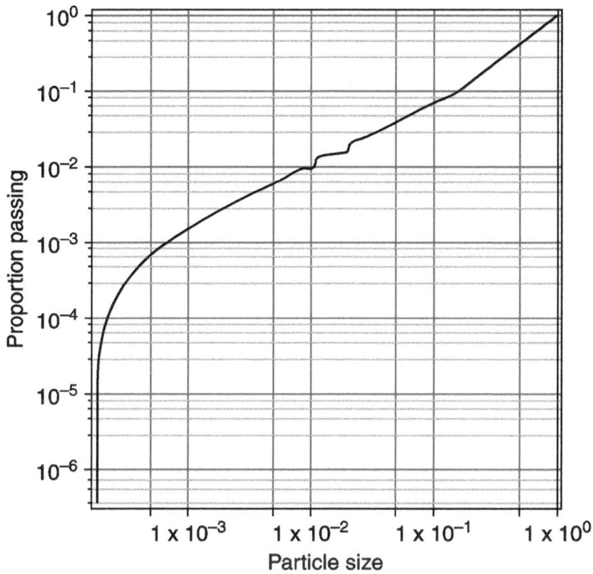

6.5 Grading curve of an Appolonian packing of disk.

6.6 Particle packing models

In most of the situations of practical interest, it is not possible to choose the particles one by one (or class by class) as done to define Appolonian packings. The particle properties (size and packing density) are given or must be chosen from a limited set of values because of practical constraints (price, allowance, limited number of classes that can be included in the mixture, etc). Of course, it is always possible to experimentally optimise the composition of the particle mixture by weighting samples of several mixtures. Nevertheless, such an experimental approach does not allow the definition of a general concrete proportioning design method. Particle packing models have therefore been developed in order to optimise design methods for granular materials.

6.6.1 Furnas model

When mono-size particles are packed randomly, it is not possible to compute the correct radius value for each 'kissing sphere' touching each particle located at the boundary of the void. It is also not possible to define a random Appolonian packing or to determine rigorously the packing density as a function of the particle size distribution and then to define the densest random packing grading curves. Nevertheless, there is at least one situation in which it is possible to perform exact computations (Furnas, 1931).

Consider a binary mixture containing particles with size $d_2 << d_1$. When such a condition is fulfilled, there is no geometric interaction between the two classes of particles: fine beads can fill the voids between coarse particles without causing the coarse particles to move away and it is possible to put a coarse particle within a fine particle packing by just removing the fine particles located in the domain to be occupied by the coarse particle. The packing density of such a mixture can be easily computed. In order to do so, it is necessary to consider two situations: either coarse particles are embedded in a matrix of fine particles (Fig. 6.6) or fine particles are located within the voids of a coarse particle packing (Fig. 6.7). Fine particles are dominant in the first situation, whereas coarse particles are dominant in the second situation.

Let φ_1 and φ_2 denote the packing densities of mono-size packing of particles with size d_1 and size d_2 respectively, ξ_2 being the fine particle proportion. When coarse particles are embedded in a fine particle matrix, the packing density of the mixture is:

$$\varphi_2^{mix} = \varphi_2 / (\xi_2 + (1 - \xi_2)\varphi_2) \tag{6.3}$$

which is a decreasing function of ξ_2, the value of which is equal to φ_2 when $\xi_2 = 1$ and is equal to one when $\xi_2 = 0$ (Fig. 6.8). When fine particles fill the voids between coarse particles, the packing density of the mixture is:

$$\varphi_1^{mix} = \varphi_1 / (1 - \xi_2) \tag{6.4}$$

6.6 One coarse particle embedded in a matrix of fine particles (Furnas model).

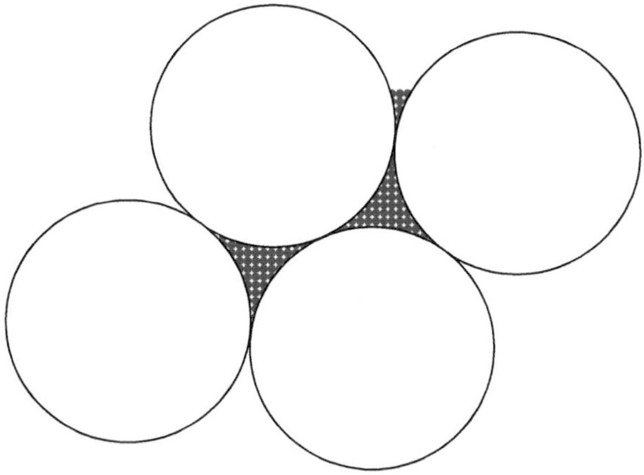

6.7 Fine particles located in the voids between coarse particles (Furnas model).

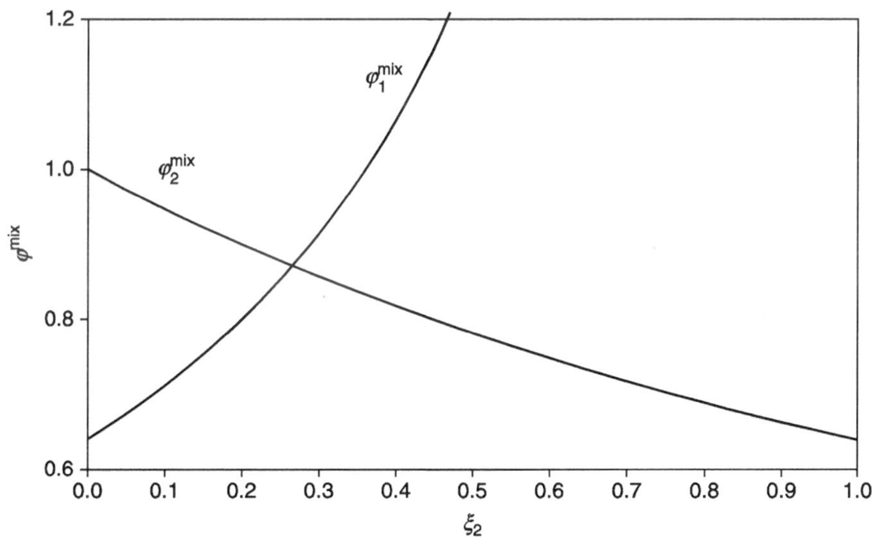

6.8 Packing density of a bidisperse mixture of spheres as a function of the fine particle fraction predicted by the Furnas model.

which is an increasing function of ξ_2, equal to φ_1 when $\xi_2 = 0$ and equal to infinity when $\xi_2 = 1$.

As the coarse particle volume can not be larger than $\varphi_1 \times V_c$ in the first situation and the fine particle volume can not be larger than $\varphi_2 \times (1 - \varphi_1) \times V_c$ in the second situation, with V_c the volume of the container, it is readily seen that the packing density of a mixture of particles φ^{mix} is the minimum of φ_1^{mix} and φ_2^{mix}. Then, the packing density of a bidisperse mixing of particles with coarse particles much larger than fine particles is given as:

$$\varphi^{mix} = \min(\varphi_1^{mix}, \varphi_2^{mix}) \qquad [6.5]$$

The variation of packing density φ^{mix} with fine particle proportion ξ_2 is plotted in Fig. 6.8. The densest packing density $\varphi_{\max}^{mix} = \varphi_1 + \varphi_2 - \varphi_1 \times \varphi_2$ is obtained when fine particles with packing density φ_2 completely fill the voids in a coarse particle packing with density φ_1. For spherical particles with $\varphi_1 = \varphi_2 = 0.64$, the densest packing density is $\varphi_{\max}^{mix} = 0.87$. This optimal mixture is a blend of 63.5% of large particles and 26.5% of fine particles (both in volume fraction).

A higher value of the density is attainable by mixing more than two classes of particles. When n classes of particles with size $d_1 >> d_2 >> \ldots >> d_n$ and monodisperse packing density $\varphi_i = \varphi_{\text{mono}}$ for all $i \in [1, n]$ are mixed, the larger attainable packing density is:

$$\varphi_{\max}^{\text{mix}}(n) = 1 - (1 - \varphi_{\text{mono}})^n \qquad [6.6]$$

which tends toward one when n tends to infinity. It is worth noting that this model applies whatever the particle shape. If the homogeneous close packing of each class composing the mixture is not equal, the maximal packing density is given as:

$$\varphi_{\max}^{\text{mix}}(n) = 1 - \prod_{i=1}^{n}(1 - \varphi_i) \qquad [6.7]$$

where φ_i denotes the homogeneous packing density of particles belonging to the ith class.

We now consider the case of a mixture of n classes of particles whose composition is given. ξ_i denotes the volume fraction occupied by the particles i in the packing. Let us first assume that class i is dominant. Then the packing can be seen as a matrix of particles i with particles belonging to classes $j \in [1, i-1]$ embedded within the matrix and particles belonging to classes $j \in [i+1, n]$ located in the pores of the packing of the dominant class particles. A straight generalization of Eq. 6.3 and Eq. 6.4 gives:

$$\varphi_i^{mix} = \frac{\varphi_i}{1 - (1 - \varphi_i)\sum_{j=1}^{i-1}\xi_j - \sum_{j=i+1}^{n}\xi_j} \qquad [6.8]$$

Computing the quantities φ_i^{mix} for $i \in [1, n]$, the maximum packing density of the mixture is:

$$\varphi^{mix}(\xi_1, \ldots, \xi_n, \varphi_1, \ldots, \varphi_n) = \min_{i \in [1,n]}(\varphi_i^{mix}) \qquad [6.9]$$

For particles with a large size difference, the Furnas model enables us to compute the packing density of a polydisperse mixture as a function of the number of classes n, the composition of the mixture defined by ξ_i, $i \in [1, n]$ and the homogeneous packing density of each class φ_i, $i \in [1, n]$. The Furnas packing density estimate is exact for a mixture of particles with size ratios tending towards infinity. For 'real life' particles, the larger the particle size ratio, the closer the packing density of the mixture will approach the Furnas estimate (Eq. 6.9). When the particle size ratio is not large enough (say lower than ten), it is no longer possible to neglect the geometrical interactions between particles belonging to different classes.

Various models aiming at improving the Furnas estimate have been proposed in the literature. Roughly speaking, three different approaches have been explored:

- continuous modelling of particle mixtures;
- discrete and continuous modelling combined using self-consistency assumptions;
- computer simulations of particle motion.

Moreover, various mathematical forms have been proposed to fit experimental data. Two models of the literature (a continuous theory and a self consistent approach) are summarised below.

6.6.2 De Larrard model

The model of de Larrard (1999) is a continuous model that aims to take the geometrical interactions between particles of different size into account. Let us start with the bidisperse case. Two different configurations are distinguished as in the Furnas model. The fine particle class is said to be dominant when coarse particles are embedded in a fine particle matrix and the coarse particle class is dominant when fine particles fill the empty spaces between coarse particles. As this model aims to take into account geometrical interactions between particles of a different size, two physical effects are also described. When one fine particle is inserted into a packing of coarse particles (which are thus the dominant class) and if the fine particle is not small enough to occupy the empty space between the coarse particles, there is a loosening of the coarse particle packing which induces a decrease of the overall coarse particle density (Fig. 6.9). This effect is referred to as the 'loosening effect' below. Similarly, when one large particle is inserted into a local packing of fine particles (dominant fine particle configuration), the fine particle density is decreased near the coarse particle. This wall effect induces a decrease in the overall fine particle density as depicted in Fig.6.10. Both effects are considered a linear function of the packing density of the dominant class.

To compute the maximum packing density of a bidisperse mixture of particles of respective diameters d_1 and d_2 with $d_1 > d_2$, one can first calculate the coarse dominant particle density φ_1^{mix} defined by:

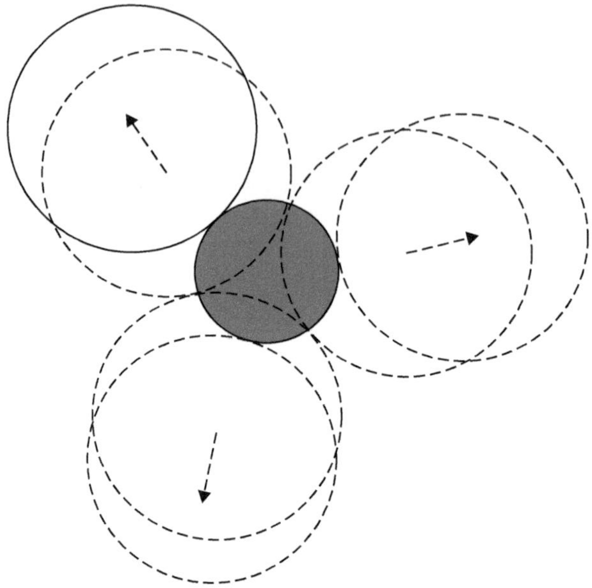

6.9 Loosening effect taken into account in the de Larrard model.

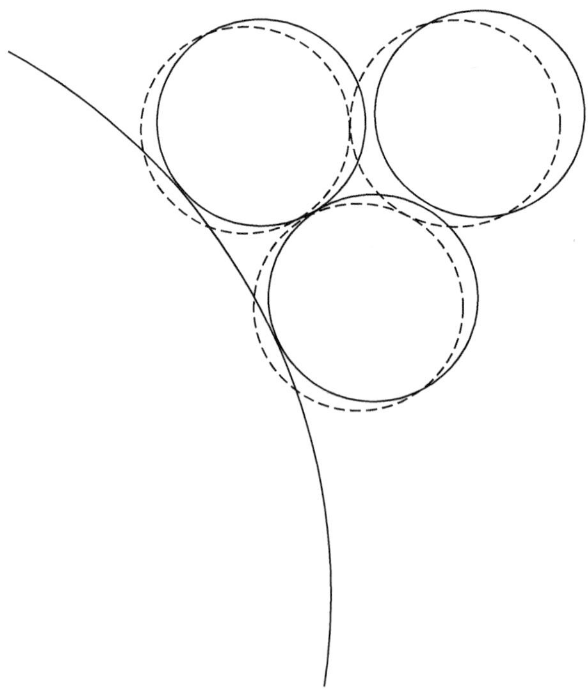

6.10 Wall effect taken into account in the de Larrard model.

$$\varphi_1^{mix} = \frac{\varphi_1}{1-\xi_2\left(1-a_{12}\dfrac{\varphi_1}{\varphi_2}\right)} \qquad [6.10]$$

where a_{12} is the loosening effect coefficient. The infinite size ratio Eq. 6.4 is recovered from Eq. 6.10 when $a_{12} = 0$. On the other hand, putting $a_{12} = 1$ and $\varphi_1 = \varphi_2$ into Eq. 6.10 yields $\varphi_1^{mix} = \varphi_1 = \varphi_2$ whenever $\xi_2 \in [0, 1]$.

When the fine particle class is dominant, the packing density φ_2^{mix} is defined by:

$$\varphi_2^{mix} = \frac{\varphi_2}{1-\xi_1\left(1-\varphi_2+b_{21}\left(\varphi_2-\dfrac{\varphi_2}{\varphi_1}\right)\right)} \qquad [6.11]$$

where b_{21} is the wall effect coefficient. Putting $b_{21} = 0$ and $\xi_1 = 1 - \xi_2$ into Eq. 6.11 yields the Furnas Eq. 6.3. The packing density $\varphi_2^{mix} = \varphi_1 = \varphi_2$ of a mixture of identical particles is again recovered from Eq. 6.11 when $b_{21} = 1$ and $\varphi_1 = \varphi_2$.

As for the case of particles with a large size difference, the packing density of a bidisperse packing of given composition is $\varphi^{mix} = \min(\varphi_1^{mix}, \varphi_2^{mix})$.

Thanks to the linearity of Eq. 6.10 and Eq. 6.11 describing loosening and wall effects, the Furnas model for a polydisperse mixture with n classes of particles can be easily generalised to take into account interactions between particles. Eq. 6.8 is replaced by:

$$\varphi_i^{mix} = \frac{\varphi_i}{1-\sum_{j=1}^{i-1}\xi_j(1-\varphi_i+b_{ij}(\varphi_i-\varphi_i/\varphi_j))-\sum_{j=i+1}^{n}\xi_j(1-a_{ij}\ \varphi_i/\varphi_j)} \qquad [6.12]$$

with $d_1 > d_2 > \ldots > d_n$ while Eq. 6.9 is still valid.

In the original work of de Larrard, the 'virtual' packing density of a mixture is calculated using Eqs 6.9 to 6.12. De Larrard defined 'virtual' packing density as the maximum packing density attainable for a given mixture. The virtual packing of mono-size spheres is 0.74 and that of parallelepipeds of same shape is 1. Consistently, the quantities φ_i and φ_i^{mix} in Eqs 6.9 to 6.12 are 'virtual' packing densities too. In the de Larrard approach, the 'real' packing density is calculated from the 'virtual' packing density using a compressible packing model accounting for the packing process. As explained in Section 6.4, the 'virtual' packing density is known only for few particle shapes. Once the compressible packing model is chosen, one has to determine not only coefficients a_{ij} and b_{ij}, but also the 'virtual' packing densities φ_i and a compaction index with which to calibrate the model. Details of the compressible model and the determination of parameters a_{ij}, b_{ij} and φ_i can be found in de Larrard (1999).

It has been shown that accurate estimates for the real packing density of a particle mixture can be obtained from a simplified version of this model (Bournonville, 2003). In the simplified approach, both φ_i and φ_i^{mix} are real packing densities and the compressible packing model is unnecessary. Homogeneous packing densities φ_i are measured and parameters a_{ij} and b_{ij} are determined so that the simplified estimate fits experimental data accurately. For a bead mixture, good agreement between the model and experimental data is obtained with:

$$a_{ij} = \left(1 - \left(1 - \frac{d_j}{d_i}\right)^{1.13}\right)^{0.57} \quad \text{and} \quad b_{k\ell} = \left(1 - \left(1 - \frac{d_k}{d_\ell}\right)^{1.79}\right)^{0.82} \qquad [6.13]$$

where $i < j$, $(d_i > d_j)$ and $k > \ell$ $(d_k < d_\ell)$.

Model predictions for a bidisperse mixture are shown in Fig. 6.11, where the packing density φ^{mix} is plotted as a function of the fine particle proportion ξ_2 for several values of the size ratio $\lambda = d_1/d_2$. Note that, whatever the value of the size ratio, the maximum packing density is a singular point of the packing density versus fine particle proportion curve. Even if such a singular point was not experimentally observed for the bidisperse packing of dry particles, it seems to be the hallmark of a theoretical model distinguishing two regimes. In the original de Larrard (1999) model, this undesirable effect is corrected by introducing the scalar index accounting for the packing process.

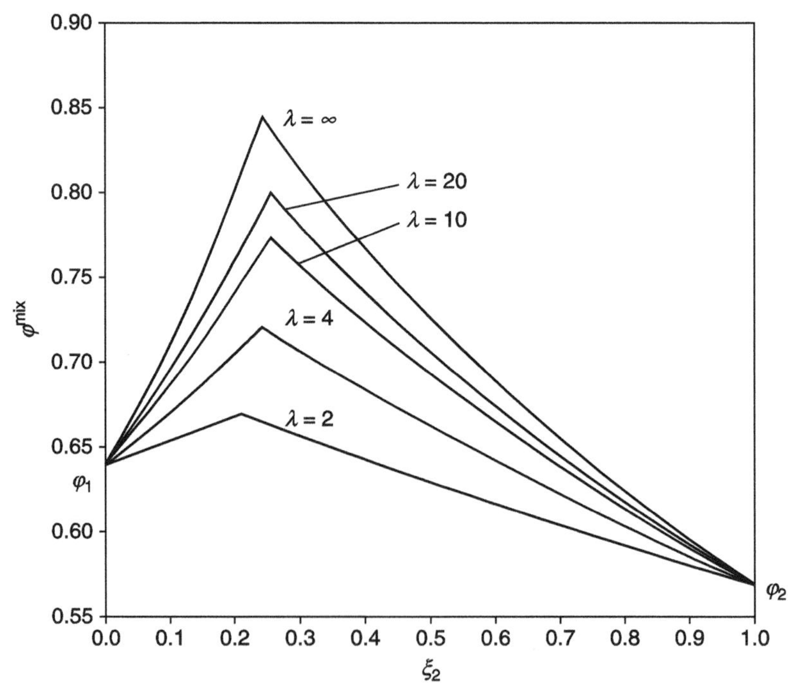

6.11 Packing density of a bidisperse packing of spheres predicted by the modified de Larrard model as a function of the fine particle proportion for several size ratio, $\lambda = d_1/d_2$.

6.6.3 Ouchiyama and Tanaka model

Ouchiyama and Tanaka (1981, 1986) proposed a model to estimate the packing density of a sphere packing. Their approach consists of replacing the real packing of spheres by a homogenised one as follows. Each sphere of the original packing is considered successively as the 'reference' sphere. Its neighbours in contact are all replaced by the same number of 'mean' spheres of the same diameter $\bar{d}$ equal to the mean diameter of the packing. Then it is assumed that the sphere of influence of the reference particle is a spherical shell having an outer diameter $\bar{d} + d$ and an inner diameter $\tilde{d}$ (Fig. 6.12), with $\tilde{d} = d - \bar{d}$ if d is greater than $\bar{d}$ and $\tilde{d} = 0$ if d is lower than $\bar{d}$. As all the particles in the packing have a sphere of influence, each particle has to share this space with the other particles. If $V_m(d)$ is the volume of the sphere of influence of a particle with size d, the volume of the space allocated to this particle is $V_c(d) = \pi \tilde{d}^3/6 + V_c(d)/\bar{n}$, where $\bar{n}$ is an average of the number of particle sharing the sphere of influence of the particle under consideration. The total volume of the space occupied by the packing is given as:

$$V_T = \int_{d_{\min}}^{d_{\max}} V_c(x) N f(x)\,\mathrm{d}x \qquad [6.14]$$

with N the total number of particles within the packing, $f(d)$ the frequency size distribution function, and $d_{\min}$ and $d_{\max}$ the smaller and larger particle

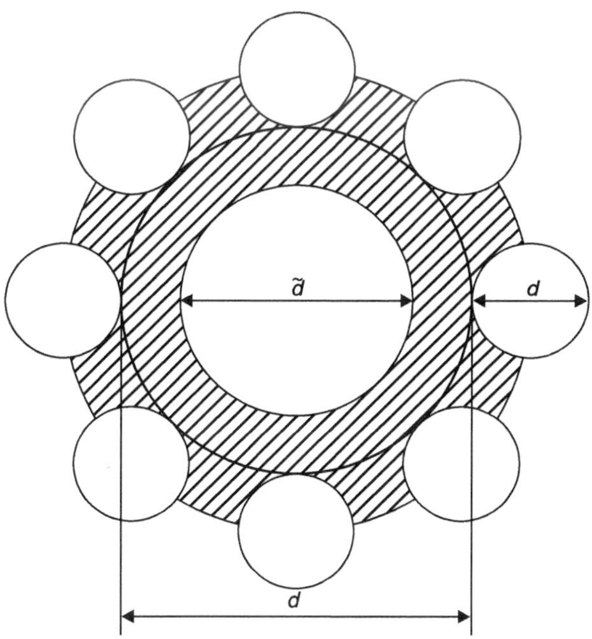

6.12 Self-consistent model of a packing of spheres proposed by Ouchiyama and Tanaka (1981, 1986).

size respectively. Then, it is readily seen that the packing density can be written as:

$$\varphi^{mix} = \frac{1}{V_T}\int_{d_{min}}^{d_{max}} Nx^3 f(x)\mathrm{d}x \qquad [6.15]$$

To obtain an estimate for $\bar{n}$ as a function of the particle size distribution function, the coordination number of the reference sphere is expressed as:

$$c(d) = 4\varphi_a (d+\bar{d})^2/\bar{d}^2 \qquad [6.16]$$

Equation 6.16 is exact only if $\bar{d}$ is much smaller than $d + \bar{d}$ (then $1 - \varphi_a$ is the surface porosity at the spherical surface of diameter $d + \bar{d}$). However, Ouchiyama and Tanaka assumed that Eq. 6.16 applies whatever the particle size and moreover that the surface porosity does not depend on the particle size. Application of the model to the packing of mono-size spheres enables us to evaluate the quantity φ_a.

The packing density of a disordered packing of spheres can be estimated from the packing of a mono-size sphere φ_s and the size distribution function $f(d)$ from the following equations:

$$\varphi^{mix} = \frac{\bar{n}\int_{d_{min}}^{d_{max}} d^3 f(d)\mathrm{d}d}{\int_{d_{min}}^{d_{max}} ((d+\bar{d})^3 - (\bar{n}-1)\tilde{d}^3) f(d)\mathrm{d}d} \qquad [6.17]$$

with:

$$\bar{n} = 1 + \frac{4}{13}(7 - 8\varphi_s)\bar{d}\frac{\int_{d_{min}}^{d_{max}} (d+\bar{d})^2\left(1 - \frac{3}{8}\frac{\bar{d}}{d+\bar{d}}\right) f(d)\mathrm{d}d}{\int_{d_{min}}^{d_{max}} (d^3 - \tilde{d}^3) f(d)\mathrm{d}d} \qquad [6.18]$$

and:

$$\bar{d} = \int_{d_{min}}^{d_{max}} d \times f(d)\mathrm{d}d \qquad [6.19]$$

The evolution of φ^{mix} as a function of the fine particle proportion ξ_2 is plotted in Fig. 6.13 for several values of the size ratio $\lambda = d_1/d_2$. The curves plotted in Fig. 6.13 are calculated based on the discrete form of Eqs 6.17 to 6.19, considering that $\varphi_s = 0.64$ for both mono-size packings. The general feature is an increase in the packing fraction φ^{mix} as the size ratio λ is increased, with a maximum of φ^{mix} even more pronounced as λ is larger. The value of ξ_2 for which this maximum is reached depends slightly on λ varying non-monotonically from roughly $\xi_2 = 0.37$ for $\lambda = 2$ up to $\xi_2 = 0.24$ for $\lambda = 20$. This trend was also predicted by the modified de Larrard model (Fig. 6.11).

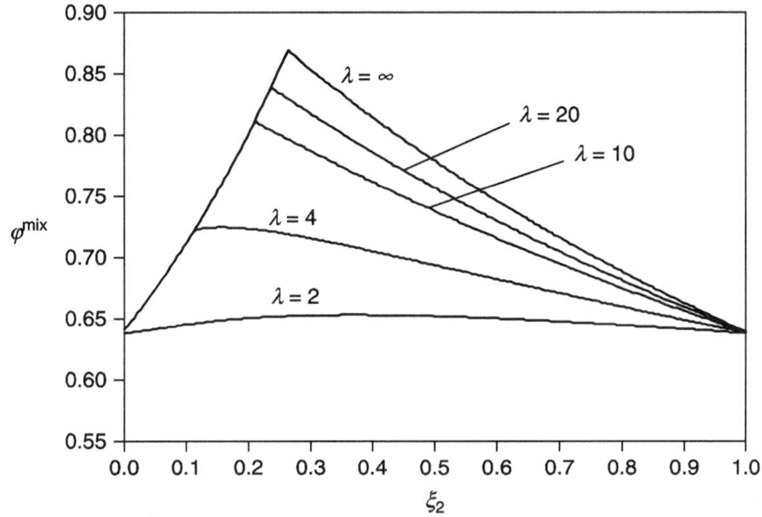

6.13 Packing density of a bidisperse packing of spheres predicted by the Ouchiyama and Tanaka model as a function of the fine particle proportion for several size ratio, $\lambda = d_1/d_2$.

It is clear from Fig. 6.11 that the effective maximum packing fraction of the bidisperse mixture of any size ratio and composition is always higher than the value of a mono-size packing. For smaller values of λ, Ouchiyama and Tanaka packing density is a smooth function of the fine particle proportion whereas a sharp point is always predicted by the de Larrard model.

Other models have been proposed in the literature. Liu and Ha (2002) developed an extended Furnas model which closely resembles the de Larrard model, the only difference being that the extended Furnas model accounts only for a loosening effect. They also proposed an extended Westman model, which consists of fitting the bidisperse packing density–fine particle fraction relationship to a conic equation. To summarise, it must be said that for bidisperse packings, most of the packing models predict the same behaviour as depicted in Fig. 6.11 and 6.13.

6.6.4 Numerical simulations

Numerical simulations can of course provide useful results, in particular to study the influence of some parameters (particle shape, angularity) on the overall properties of the packing. Substantial progress has been made in this field, partly thanks to the increase in the power of computers. Nevertheless, it is believed that discrete numerical simulation tools are not yet able to take into account the complexity one has to face in solving the packing problems encountered in concrete technology.

6.7 Fibre and grain mixture

Fibres are sometimes added to concrete to increase its mechanical properties in hardened state. If fibres and aggregates are very different in size, the packing density of the mixture can be predicted using the Furnas model. However, geometric interactions between particles and fibres with a length of the same order as the particle size strongly affect the packing density. As the de Larrard and Ouchiyama and Tanaka models are valid only if the shape of the particles is not too different, they do not apply in this situation.

In this situation, de Larrard (1999) suggests that the packing density of a mixture of one class of particles and one class of fibres can be written as:

$$\varphi_{ave} = \varphi_a(1-\xi_f(1+k_f)) \qquad [6.20]$$

when particles are dominant, where φ_a is the aggregate packing density (with no fibres), ξ_f the fibre volume proportion and k_f an adjustable parameter. This equation was obtained by assuming that the volume of the perturbed zone around each fibre in which the aggregates packing is modified depends on the fibre length and the particle size. A good agreement between experimental data and model prediction was obtained by de Larrard (1999) for two kinds of fibres: a cylindrical one and a prismatic one. To the best of our knowledge, there is no theoretical model accounting for the interaction between the fibres and multimodal particles mixing. Of course, the situation is even worse if stiff fibres are dealt with.

6.8 Effect of particle size distribution on rheology

Concrete can be seen as a suspension of polydisperse particles (aggregates) suspended in a non-Newtonian fluid (the cement paste). Knowing and predicting the rheological properties of such suspensions is thus a major issue for both concrete mix design and the modelling of deformation and flow of fresh concrete. Let us start with the case of a dilute Newtonian suspension of non-interacting particles. If the particles are identical in shape and in orientation, the overall viscosity of the suspension depends on two main parameters: the total volume fraction of the particles ϕ and the viscosity of the continuous phase. When spheres are suspended in a Newtonian fluid, overall viscosity is expressed by the Einstein (1906) equation $\mu^{\text{dilute}} = \mu \times (1 + 2.5\phi)$ (see Chapter 2). As the volumetric concentration of the particle increases to more than 0.1, interparticle hydrodynamic interactions begin to influence the viscosity. The overall viscosity of the suspension also depends on the morphology of the particles as shown by Batchelor and Green (1972). As the solid volume fraction increases, the suspension viscosity increases more and more strongly until approaching infinity as the solid volume fraction approaches some maximum value ϕ_m. This strong increase in viscosity is ascribed to the decrease in interparticle distance, which in turn induces a strong increase in interparticle hydrodynamic forces (the lubrication force tends to infinity when the particle distance approaches zero); once the particle volume fraction reaches ϕ_m, the suspensions can no longer be sheared because of the percolation of the particle contact network. Thus, it appears that the maximum particle volume fraction ϕ_m equates to the maximum packing density φ_m for a given suspension composition.

The viscosity of concentrated suspensions can be related to ϕ and ϕ_m by the phenomenological Krieger–Dougherty (1959) equation $\mu^{KD} = \mu \times (1 - \phi/\phi_m)^{-[\eta]\phi_m}$ or by the Quemada (1977) equation $\mu^Q = \mu \times (1 - \phi/\phi_m)^{-2}$. Note that for hard spheres, the Krieger–Dougherty parameter $[\eta] = 2.5$ so that the Krieger–Dougherty equation yields the Einstein equation when ϕ approaches zero. It has been observed that experimental data for isotropic suspensions of spheres (low shear rate viscosity, or elastic modulus) agree pretty well with the Krieger–Dougherty equation whereas the Quemada equation is more appropriate for anisotropic suspension of spheres (high shear viscosity). Accordingly, the measured maximum solid volume fraction also depends on the morphology of the particles; it seems to be equal to $\phi_m^{iso} = 0.57$ for isotropic dispersions (Mahaut *et al.*, 2008) and to $\phi_m^{ani} = 0.605$ for anisotropic dispersions (Ovarlez *et al.*, 2006). The fact that the solid volume fraction at which viscosity diverges is not the same for low shear rate and high shear rate is not at all surprising.

As the packing density of a stack of dry particles depends on the way they are placed, the maximum solid fraction of spheres dispersed in a fluid depends on the applied shear rate.

For non-Brownian fibres dispersed in a Newtonian fluid, the concentration-dependence of the low shear viscosity has been reported to follow the Quemada equation with $\phi_m = \varphi_m^{fib}$ given by Eq. 6.1 (Wierenga and Philipse, 1998).

When particles are dispersed in a fluid obeying the Herschel–Bulkley law, the suspension behaves as a Herschel–Bulkley fluid with the same exponent n' as that of the suspending fluid (Chateau *et al.*, 2008). Furthermore, for mono-size spheres, the suspension yield stress/solid volume fraction and the suspension consistency/solid volume fraction relationships are very well fitted to the simple laws $\tau_c^{hom} = \tau_c(1-\phi)(1-\phi/\phi_m)^{-1.25\phi_m}$ and $\eta^{hom} = \eta(1-\phi)^{(1-n')/2}(1-\phi/\phi_m)^{-1.25(n'+1)\phi_m}$, where τ_c, τ_c^{hom}, η and η^{hom} denote respectively the suspending fluid yield stress, the suspension yield stress, the suspending fluid consistency and the suspension consistency (Chateau *et al.*, 2008; Mahaut *et al.*, 2008).

The behaviour of polydisperse (particularly bidisperse) Newtonian suspensions has attracted considerable interest from rheologists and several studies (theoretical or experimental) have been devoted to this subject in the last few decades (Chang and Powell, 1994; Gondret and Petit, 1997; Servais *et al.*, 2002). Different theoretical relationships that seek to predict the viscosity of polydisperse suspensions as a function of the solid volume fraction ϕ and of the particle size distribution have been proposed in the literature (Farris, 1968; Chong *et al.*, 1971; Storms *et al.*, 1990; Gondret and Petit, 1997; Bournonville *et al.*, 2004).

Farris (1968) investigated the relative viscosity of bimodal suspensions in the framework of a rigorous theoretical approach, valid only if the coarse particle size is much larger than the fine particle one. When this condition is fulfilled, one can consider that the coarse particles interact with the Newtonian fine particle suspension and, therefore, the relative viscosity of the bimodal suspension is equal to the product of the relative viscosities of each monodisperse suspension. The Farris approach can be generalised to polydisperse suspensions or to non-Newtonian suspending fluids provided that estimates of the monodisperse suspension overall properties exist. If n classes of particles with size $d_1 >> d_2 >> \ldots >> d_n$, particle volume fraction ϕ_i, $i \in [1, n]$ and maximum particle volume fraction ϕ_{mi}, $i \in [1, n]$ are dispersed in a Newtonian fluid, the overall suspension viscosity Farris estimate is:

$$\mu^{\text{Farris}} = \mu \times \prod_{i=1}^{n} g\left(\frac{\phi_i}{\phi_{mi}} \frac{1}{1 - \sum_{j=1}^{i-1} \phi_i} \right) \qquad [6.21]$$

with $g(\cdot)$ any viscosity model. The quantity between the brackets in Eq. 6.21 is the ratio of the particle volume fraction occupied the particles of class i dispersed in a suspension made up of the fluid and the $n - i + 1$ classes of particles with size smaller than d_i to the ith class maximum volume fraction ϕ_{mi}. As the Farris model and Furnas model rely on the same assumptions, the Farris approach to the overall rheological properties suffers from the same limitations as the Furnas model of the density of particle packing. Like the Furnas approach, the Farris model does not account for interaction between particles.

In most of the models in the literature devoted to polydisperse Newtonian suspensions, the influence of the particle size distribution is taken into account through the concept of maximum packing fraction (Chong *et al.*, 1971, Storms *et al.*, 1990, Gondret and Petit, 1997, Stickel and Powell, 2005). Hence, ϕ_m can be estimated using models developed to estimate the packing density of dry particles. For example, Gondret and Petit (1999) have measured the low shear viscosity of suspensions of bidisperse spheres in the low shear regime. They found that their experimental data are well fitted to the law $\mu^{\text{bid}} = \mu \times (1 - \phi/\phi_m)^{-1.5}$ with $\phi_m = \varphi_{\text{OT}}^{\text{mix}}$ calculated from the model of Ouchiyama and Tanaka.

Bournonville *et al.* (2004) proposed a modified version of the Farris model that aimed to take into account geometrical interactions between spherical particles of a different size. This model generalises the reasoning of Farris using the de Larrard model instead of the Furnas model to describe the geometrical interaction between particles. The theoretical predictions of this model accurately fit experimental data for polydisperse suspensions of beads in a Newtonian fluid.

The elastic modulus and yield stress of isotropic bidisperse suspension of non-colloidal particles in a yield stress fluid have been studied by Vu *et al.* (2010). As is classically observed for particles suspended in a Newtonian fluid, the elastic modulus and yield stress of bidisperse suspensions are lower than the same quantities measured for a monodisperse suspension of the same solid volume fraction. The elastic modulus and yield stress of the bidisperse suspension can be estimated as $G^{\text{hom}} = G \times (1 - \phi/\phi_m)^{-1.43}$ and $\tau_c^{\text{hom}} = \tau_c\sqrt{1-\phi}(1-\phi/\phi_m)^{-0.72}$ where $\phi_m = \varphi_{\text{dL}}^{\text{mix}}$ is calculated using the modified de Larrard model Eqs 6.9 to 6.13.

All these results show that it is sufficient to determine the dependence of the rheological properties on ϕ to predict the behaviour of bidisperse suspensions using a dry packing model. This should remain true for more complex polydisperse cases.

6.9 Conclusions

Polydisperse suspensions of particles suspended in a fluid are encountered in various domains. In concrete technology this topic is of paramount importance because fresh concrete is a suspension of particles with different size in a non-Newtonian fluid (the cement paste). Even if models allowing for accurate predictions of overall properties of such materials from its mix design do not yet exist, substantial progresses have been achieved in the last 20 years. It has been shown in particular that the effect of particle size heterogeneity on the overall rheological properties of a suspension can be described by means of a packing model developed to estimate random packing of assemblies of dry particles. If there is an estimate of the overall property of the suspension as a function of the solid volume fraction, it is then sufficient to determine the maximum solid volume fraction to predict the overall properties of polydisperse suspensions.

6.10 References

Aste T and Weaire D (2000), *The pursuit of perfect packing*, Bristol, Institute of Physics.

Batchelor G K and Green J T (1972), Determination of bulk stress in a suspension of spherical-particles to order c2, *J. Fluid. Mech.*, 56, 401–427.

Bernal J D (1959), Geometrical approach to the structure of liquids, *Nature*, 183, 141–147.

Bournonville B (2003), *Influence de la granulométrie des particules sur la rhéologie des suspensions concentrées*, Rapport interne, LMSGC, Champs sur Marne.

Bournonville B, Coussot P and Chateau X (2004), Modification du modèle de Farris pour la prise en compte des interactions géométriques d'un mélange polydisperse de particules, *Rhéologie*, 7, 1–8.

Chang, C and Powell R L (1994), Effect of particle size distribution on the rheology of concentrated bimodal suspensions, *J. Rheol.*, 38, 85–98.

Chateau X, Ovarlez G and Luu Trung K (2008), Homogenization approach to the behavior of suspensions of noncolloidal particles in yield stress fluids, *J. Rheol.*, 52, 489–506.

Chong J S, Christiansen E B and Baer A D (1971), Rheology of concentrated suspensions, *J. Appl. Polym. Sci.*, 15, 2007–2021.

Coussot P (2005), *Rheometry of pastes, suspensions and granular materials*, New York, John Wiley & Sons.

Cumberland D J and Crawford R J (1987), *The packing of particles*, Amsterdam, Elsevier.

Einstein A (1906), Eine neue bestimmung der moleküldimensionen, *Ann. Phys.*, 19, 289–306.

Farris K J.(1968), Prediction of the viscosity of multimodal suspension from unimodal viscosity data, *Trans. Soc. Rheol.*, 12, 281–301.

Furnas C C (1931), Grading aggregates: I; mathematical relations for beds of broken solids of maximum density, *Ind. Eng. Chem.*, 23, 1052–1058.

German R M (1989), *Particle packing characteristics*, Princeton, Metal Powder Industry Federation.

Gondret P and Petit L (1997), Dynamic viscosity of macroscopic suspensions of bimodal sized solid spheres, *J. Rheol.*, 41, 1261–1274.

Krieger I M and Dougherty T J (1959), A mechanism for non-Newtonian flow in suspensions of rigid spheres, *Trans. Soc. Rheol.*, 3, 137–152.

Kumar Mehta P and Paulo Monteiro J M (2006), *Concrete: microstructure, properties, and materials*, 3rd edition, New York, McGraw-Hill.

de Larrard F (1999), *Concrete mixture proportioning: a scientific approach*, London, E & F Spon.

Liu S and Ha Z (2002), Prediction of random packing limit for multimodal particle mixtures, *Powder Tech.* 126, 283–296.

Mahaut F, Chateau X, Coussot P and Ovarlez G (2008) Yield stress and elastic modulus of suspensions of noncolloidal particles in yield stress fluids, *J. Rheol.*, 52, 287–313.

Mahaut F, Mokéddem S, Chateau X, Roussel N and Ovarlez G (2008), Effect of coarse particle volume fraction on the yield stress and thixotropy of cementious materials, *Cem. Concr. Res.*, 38, 1276–1285.

Martinie L, Rossi P. and Roussel N. (2010), Rheology of fiber reinforced cementious materials: classification and prediction, *Cem. Concr. Res.*, 40, 226–234.

Nardin, M, Papirer E and Schultz J (1985), Contribution à l'étude des empilements au hasard de fibres et/ou de particules sphériques, *Powder Tech.*, 44, 131–140.

Neville A M (1995), *Properties of concrete*, 4th edition, Essex, Pearson.

Ouchiyama N. and Tanaka T (1981), Porosity of a mass of solids having a range of sizes, *Ind. Eng. Chem. Fundam.* 20, 66–71.

Ouchiyama N. and Tanaka T (1986), Porosity estimation from particle size distribution, *Ind. Eng. Chem. Fundam.* 25, 125–129.

Ovarlez G, Bertrand F and Rodts S (2006), Local determination of the constitutive law of dense suspension of noncolloidal particles through magnetic resonance imaging, *J. Rheol.*, 50, 259–292.

Philipse A P (1996), The random contact equation and its implication for (colloidal) rods in packing, suspensions and anisotropic powders, *Langmuir*, 12, 1127–1133.

Quemada D (1977), Rheology of concentrated disperse systems and minimum energy-dissipation principle. 1. Viscosity-concentration relationship, *Rheol. Acta*, 16, 82–94.

Scott G D (1960), Packing of equal spheres, *Nature*, 188, 908–909.

Servais C, Jones R and Roberts I (2002), The influence of particle size distribution on the processing of food, *J. Food Eng.*, 51, 201–208.

Stickel J J and Powell R L (2005), *Fluid mechanics and rheology of dense suspensions*, Ann. Rev. Fluid. Mech., 37, 129–149.

Storms J J, Ramarao B V and Weiland R H (1990), Low shear rate viscosity of bidispersed suspensions, *Powder Technology* 63, 247–259.

Vu T S, Ovarlez G and Chateau X (2010), Macroscopic behavior of bidisperse suspensions of noncolloidal particles in yield stress fluids, *Journal of Rheology* 54, 815–833.
Westman A E R (1936), The packing of particles: empirical equations for intermediate diameter ratios, *J. Am. Ceram. Soc.*, 19, 127–129.
Wierenga M W and Philipse A P (1998), Low-shear viscosity of isotropic dispersions of (Brownian) rods and fibres; a review of theory and experiments, *Colloids Surfaces A: Physicochem. Eng. Aspects* 137, 355–372.

7 Superplasticizers and the rheology of concrete

R. FLATT, Eidgenössische Technische Hochschule Zürich (ETHZ), Switzerland and I. SCHOBER, Sika Technology AG, Switzerland

Abstract: Superplasticizers are a fundamental ingredient for lowering the yield stress of concrete. This chapter provides an overview of the chemical classes of superplasticizers and their performance, with a selective review of those aspects that we consider important in understanding superplasticizers. We also demonstrate the consistency in findings from various recent studies in terms of the structure function behaviour of polycarboxylates. We show the importance of adsorption of these polymers on the cement surface and the polymer structural parameters as well as the cement properties that may influence this. We also include comments about typical issues in experimental procedures and possible errors in their too rapid interpretation.

Key words: superplasticizers, comb copolymers, polycarboxylic ethers, lignosulfonates, polynaphthalene sulfonates, polymelamine sulfonates, adsorption, dispersion, organo-aluminate, rheology, slump life.

7.1 Introduction

Superplasticizers are polymeric dispersants used in cementitious materials either to reduce yield stress at constant solids content or to increase the solids content at constant yield stress (Spiratos *et al.*, 2003). They are essential tools for formulating concrete with a reduced environmental impact. They can for example be used to increase service life by making more durable concrete (less water means lower porosity and in particular lower permeability). They can also be used to reduce the clinker content per unit volume of concrete while maintaining a constant (or low enough) water to clinker ratio to ensure durability. This makes it possible to use blended cements with higher levels of clinker replacement or simply to use more sand and aggregates (Hermida *et al.*, 2009).

Superplasticizers were introduced already in the 1930s. Since then the market has witnessed the appearance of different generations of products with improving performance. At the start of this chapter we provide an overview of the chemistry of these products. The last superplasticizer generation, polycarboxylates, brought the most radical changes. Introduced in the mid-1980s, they have gone through drastic improvements and diversification over the last decade. For this reason, we devote most of the chapter to describing the key features of their working mechanisms and factors that affect them.

We deliberately do not provide a detailed discussion of the basics of interparticle forces in cementitious systems. We rather attempt to 'put some meat on the bone' of this general theory, emphasizing the factors that may cause cementitious systems to diverge from their expected performance. We also have included recent results that show under which conditions the overriding principles of the general theory can be demonstrated.

For readers unfamiliar with the notions of interparticle forces and their role in the rheology of cementitious systems, we first provide the following brief reminder. Superplasticizers adsorb onto particles, reducing or eliminating the attractive interparticle forces that cause yield stress. These attractive forces are generally attributed to van der Waals origin (Flatt, 2004a), but may also include electrostatic forces (Flatt and Bowen, 2003; Plassard *et al.*, 2005). Adsorbed superplasticizers, in particular polycarboxylates, provide steric hindrance that prevents close contact between particles, reducing the magnitude of the attractive forces (Lewis *et al.*, 2000; Flatt, 2001). The resulting interparticle force is calculated by adding the different forces, in agreement with the Derjaguin–Vervey–Landau–Overbeek (DLVO) hypothesis. In the case of cementitious systems, calculations lead to the conclusion that dispersion is steric for polycarboxylates and probably electro-steric for the other polymers (Flatt *et al.*, 2000). This contrasts with long-standing belief that earlier generation polymers act solely by electrostatic repulsion. It supports criticism formulated early on by Banfill about this theory, but that was long overlooked (Banfill, 1979). The fundamental point here is that cementitious systems have a high ionic strength that drastically reduces the range of electrostatic repulsion (Flatt and Bowen, 2003). Specific aspects about electrostatic repulsion are discussed in Section 7.3.2 on zeta potential measurements.

While elegant, these theories are quite difficult to verify in practice for the simple reason that cement particles are large enough to segregate in well-dispersed systems. This means that, when the particle surfaces are saturated with superplasticizers, cement particles tend to sediment. Unfortunately, this is the one situation in which interparticle forces can be most reliably calculated. Therefore, if one works at a high superplasticizer dosage, the forces can be calculated well, but the experiments are confronted with a massive risk of artefacts due to sedimentation. Conversely, if the dosage is lower, the experiments become more reliable, but the interparticle forces cannot be calculated accurately. Nevertheless, in specific cases it has been possible to demonstrate a direct relationship between interparticle forces and rheological behaviour, showing that the theories from colloidal science do also apply to cementitious materials.

In situations of greater practical relevance than the corresponding experiments, the situation is rather more complex. Indeed, one is then often faced with situations where polymer dosages are insufficient to saturate the particle surfaces. In such situations, calculations become more complex. More importantly, the system behaviour becomes quite sensitive to numerous other phenomena that perturb the dispersion power of superplasticizers by modifying the extent of adsorption. Examples are a modification of the aluminate phase reactivity, competitive adsorption and temperature-driven desorption – all of which are discussed in this chapter.

Taking a step closer to practice, it is important to recognize that most commercially available products are formulations, being composed of several types of molecules. On the one hand this offers a way for us to adapt to different cements or application demands. On the other, it allows us to match very different requirements from varying applications. Think for example of a car. Its main function, to provide transportation, comes from its engine and wheels. However, its safety comes from other elements and its comfort from yet others. In a hot country, temperature regulation would be an important aspect and no one designing a car for such a country would try to achieve that requirement by optimizing the car wheels rather than by making sure it has an adequate air-conditioning system or proper windows. In the same way, a person formulating a chemical admixture for hot countries would generally include a set retarder to make sure the open time of concrete remains long enough. The need for admixtures to match various requirements is often either overlooked or underappreciated in academic research. It therefore seemed useful to introduce a small section devoted to formulations in this chapter.

It is our hope that this chapter provides a synthetic overview on the most significant steps that have been accomplished in research on chemical admixtures over the past decade. We also hope that it helps to highlight the most important open questions to which answers are needed to allow better use and further development of superplasticizers. It is our belief that superplasticizers and chemical admixtures in general have a great future in front of them. An important reason for this is the greater demands that they will be asked to meet to enable broader use of blended cements with higher levels of clinker replacements. Although this brings us to considering the impact of admixtures on hydration kinetics, it is not a subject we have addressed as this book is devoted to rheology. Readers interested in those subjects and in the outstanding challenges in this area are referred to the recent review by Cheung *et al.* (2011).

7.2 Chemical characteristics of superplasticizers

7.2.1 Chemical families and production

Lignosulfonates

Lignosulfonates (often abbreviated as LS) were one of the first dispersants added as an admixture to concrete. Lignosulfonates have been used since the 1930s as plasticizers or water reducers. Mixed with concrete, their use has afforded the beneficial effects of retarding the setting time and reducing the amount of mixing water required. Furthermore, greater workability of plastic concrete during its placement in formworks and increased air entrainment have resulted from their use. As by-product of the pulping industry, lignosulfonates are of low price and despite their limited performance they find a large field of applications for improving the quality of concrete.

Lignin is part of wood and is, after cellulose, the second most abundant organic molecule on Earth. Wood contains about 20–30% lignin. During sulfite pulping of wood, the lignin is sulfated, the bonds between lignin and the polysaccharides as well as bonds within the lignin are broken, and the molecular weight of the lignin is reduced. This process renders the molecule water-soluble. The liquid phase separated from the raw pulp contains lignosulfonates, salts, sugars and sugar derivatives (oligomers and monomers from the cellulose and hemicelluloses of wood) and is called 'spent liquor'. Most of it is burnt for energy production and the recovery of chemicals. Only a small part is used as a dispersant. Lignosulfonates are highly branched macromolecules and are formed from phenylpropane units connected in a non-regular manner by ether or C–C bonds, the latter between the aromatic rings. The molecular weight of lignosulfonates ranges from 10^3–10^6 g/mol. A schematic drawing of part of the structure of a lignosulfonate is given in Fig. 7.1, which shows possible bonds and substitution pattern in the molecule. Depending on the source of the lignosulfonate (softwood or hardwood) and the pulping and isolation process, a great variety of different types with regard to structural features such as the number of functional groups, degree of sulfonation, molecular weight and counter ions are available.

Lignosulfonates retard the setting of cement. While this retardation can occasionally be advantageous, under certain weather conditions (e.g. in cool weather) the setting time may be extended to an impractical length of time. Approximately a quarter of the total solids in spent liquor is sugars, which can cause a strong retarding effect. Therefore, the removal of some or all of the sugars

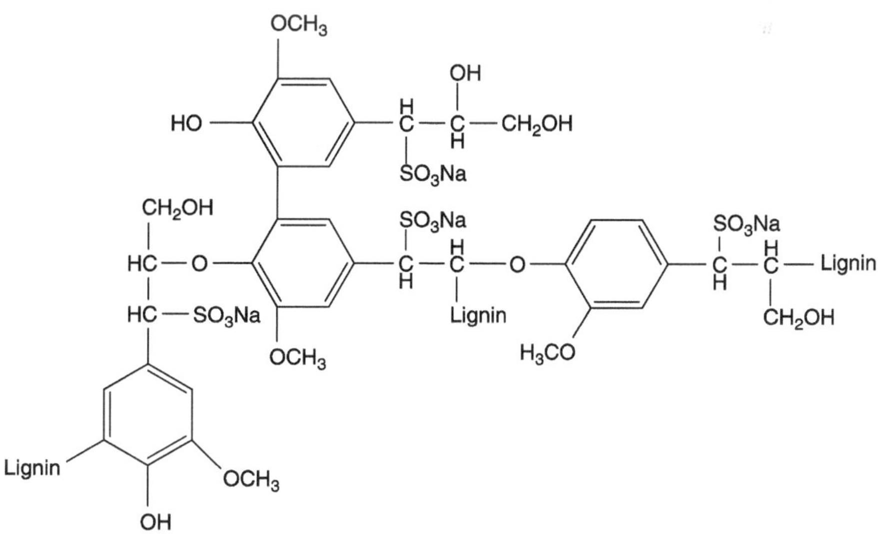

7.1 Schematic drawing of the main building blocks of the lignosulfonate molecule.

is usually desirable. Different methods such as precipitation, alkaline heat treatment or ultrafiltration are used to reduce the sugar content in the spent sulfite liquor. This purification reduces the retarding effect. However, it does not disappear as sugar-free lignosulfonate shows retarding effects. In short, commercially available lignosulfonates show retardation, which is due to the lignosulfonates themselves, possibly but not always incremented by retardation from residual sugars.

Lignosulfonates are available as calcium, sodium and magnesium salts. In formulations with other polymers, sodium salts are preferred. Calcium would cause precipitation of gypsum with sulfate ions that may be present in other formulation constituents, e.g. polynaphthalene sulfonate (PNS) or polymelamine sulfonate (PMS). Lignosulfonates are sold either as a brown, aromatic-smelling powder or as brown 30–40% solutions.

Another important factor affecting the performance of lignosulfonates is their molecular weight. Tests with lignosulfonates with different molecular weights have shown that higher molecular weight fractions show better plasticizing effects (Reknes and Gustafsson, 2000). Although lignosulfonate is the most used material for the formulation of water-reducing products, it is hardly or never used in the design of high-performance concrete. However it is reported that lignosulfonates of high molecular weight, from which sugars and lower molecular weight compounds are removed in a special treatment, can be used for water reduction up to 20% and with moderate retardation (Reknes and Petersen, 2003; Banfill *et al.*, 2007).

Polynaphthalene sulfonates

Polynaphthalene sulfonates (PNS) are naphthalene sulfonic acid formaldehyde condensates. The dispersing ability of PNS is higher than that of lignosulfonates and therefore higher water reduction is possible. PNS can be called 'high range water reducers' or superplasticizers. First introduced into the admixture market in the 1960s they are widely used nowadays.

The polycondensates are produced in two steps as shown in Fig. 7.2. First naphthalene is sulfonated with sulfuric acid to give β-naphthalene sulfonic acid. In the following condensation reaction with formaldehyde in aqueous medium the linear, anionic polymer is formed. The molecular weight of these polymers is dependent on the reaction time and can be thus regulated (Pieh, 1987).

PNS are mostly linear anionic polymers that are commercially available as sodium or calcium salts. They are sold as aqueous solutions with 20–45% solids or as brown powder. The sodium salts may still contain remarkable amounts of residual sodium sulfate from the production, whereas the amount of sulfate in the calcium salts is usually lower due to the poor solubility of gypsum which precipitates and can be removed form the solution.

PNS dispersants have relatively low molecular weights. Their dispersing ability starts with about five repeating units with a maximum dispersing power at higher

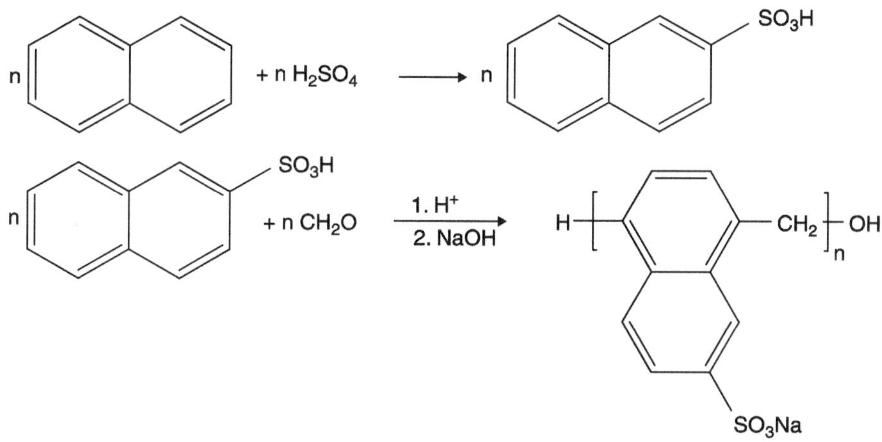

7.2 Reaction scheme for production of PNS (Pieh, 1987).

molecular weights between 1000 and 20 000 (Pieh, 1987; Aitcin *et al.*, 2001). The optimal molecular weight of PNS dispersants is, however, dependent on the type of the cement. High molecular weight PNS are more effective in fluidizing pastes made with cements with a high alkali content. However, the molecular weight has little effect in fluidizing pastes made from a low alkali cement. Differences in the molecular weight of PNS do not appear to significantly affect cement hydration (Kim *et al.*, 1999).

Not all cements are compatible with all PNS. This is a serious problem for concrete production. PNS dispersants usually greatly increase the fluidity of fresh concrete. However, the dispersing effect can be of short duration and can result in a rapid slump loss depending on the particular combination of cement and PNS used. Because of these incompatibility problems, the mode of action and behaviour of PNS in different cements or concretes have been extensively studied.

Soluble alkalis were shown to increase the fluidity and the duration of the fluidity of a cement paste superplasticized with PNS (Jiang *et al.*, 1999). This effect is explained by the fact that alkalis increase the concentration of sulfate ions, which in turn reduce the adsorption of PNS on aluminate phases causing more PNS to remain in the solution (Kim *et al.*, 2000; Aitcin *et al.*, 2001). This is congruent with the findings that the incompatibility is related to the type and amount of gypsum in the cements (Nakajima and Yamada, 2004). More detailed explanations on the influence of soluble alkalis and sulfates on the performance of superplasticizers are given later.

Polymamine sulfonates

Comparable to the PNS dispersants are the sulfonated melamine formaldehyde condensates (PMS) which have been used since the mid-1970s as superplasticizers

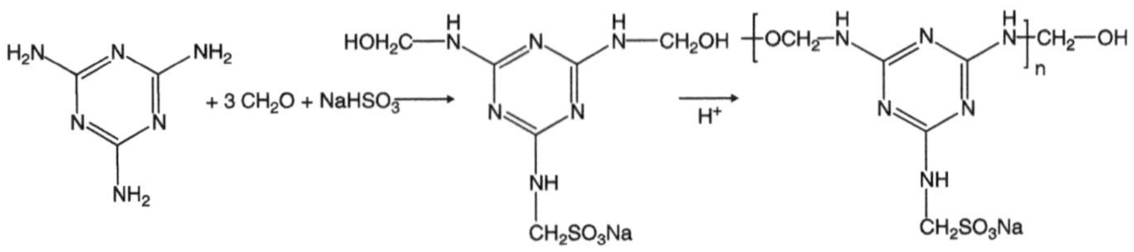

7.3 Production of PMS polycondensates.

for concrete. They are produced by a condensation reaction of melamine (2, 4, 6-triamino-s-triazine) with sulfite and formaldehyde (Bürge *et al.*, 1981; Pieh, 1987) as shown in Fig. 7.3. In the first step, melamine, formaldehyde and sulfite are reacted under alkaline conditions to give methylolated melamine with sulfonic acid groups. In the second step the polycondensation is performed under acidic conditions. Depending on the reaction conditions and the amount of reactants, a more or less linear polymer chain with sulfonic acid groups is obtained.

Mineral acids are usually used to adapt the pH to acid conditions for the condensation. In most cases sulfuric acid is added (Aignesberger and Michaud, 1966). The resulting sulfate ions can cause gypsum precipitation when calcium ions are added. This is the case if the condensates are later blended with plasticizers in calcium salt form. To avoid this sulfate precipitation, sulfuric acid can be replaced by sulfamic acid or sulfanilic acid. This can further improve the performance (Bürge *et al.*, 1981). The molecular weight of the polymer and the degree of branching are additional factors that may influence their performance.

The condensation reaction of melamine with formaldehyde is an equilibrium determined reaction. Therefore a small amount of free formaldehyde is always found in the aqueous solution of these plasticizers. In some countries the content of free formaldehyde is limited by regulation or even banned. Therefore, although it is possible to reduce the amount of free formaldehyde below the demanded level, PMS polymers lost market share and are replaced nowadays in many areas by the new polycarboxylic ether (PCE) polymers.

Sulfonated melamine formaldehyde condensates (PMS) are often the preferred superplasticizers in the precast industry because they do not, or only marginally, retard the setting of cement. Compared to PNS the water reduction or plasticizing ability is about the same, but the slump life with PNS is slightly better. However, the retardation of PMS is lower.

Vinyl copolymers

Vinyl copolymers used as superplasticizers are polymers produced via radical (co)polymerization but which are not PCE polymers (see later). These linear polymers, which contain anionic groups, were developed in the 1970s and had their maximum usage in the 1980s and 1990s. Later on they were mostly replaced by the PCE-type superplasticizers.

In contrast to the PNS and PMS polymers that are polycondensates, the vinyl polymers are produced by radical copolymerization. The molecular weight of polycondensates depends on the reaction time – provided the correct stoichiometry and purity of the raw materials are maintained. In contrast, the molecular weight of polymers produced by radical polymerization is determined by the amount of initiator and chain transfer agent, and by the reactivity of the monomers. Compared to the polycondensates the molecular weight of the vinyl copolymers is generally higher.

The dosage of the vinyl copolymers, in solids by mass of cement, is up to half the dosage of PNS or PMS superplasticizers for the same plasticity or water reduction respectively. In addition, the slump life of concrete can be strongly improved but in most cases with slightly increased retardation.

Different polymers were produced bearing sulfonate or carboxylate groups, as well as combinations thereof. Some examples are copolymers of acrylic acid with hydroxypropyl acrylate (Bradeley and Szymanski, 1985), or copolymers of sulfonate modified maleic acid with vinyl pyrollidone, vinyl acetate or others (Bürge *et al.*, 1990). Figure 7.4 shows the structure of two vinyl copolymers that contain sulfonate and carboxylic acid groups and were investigated in more detail by Flatt *et al.* (1997) and Mäder *et al.* (1999).

Polyethylene oxide (PEG) phosphonates

Superplasticizers with phosphonate groups were developed at the beginning of the 1990s. In contrast to the polycarboxylic ethers (PCE) which are anionic

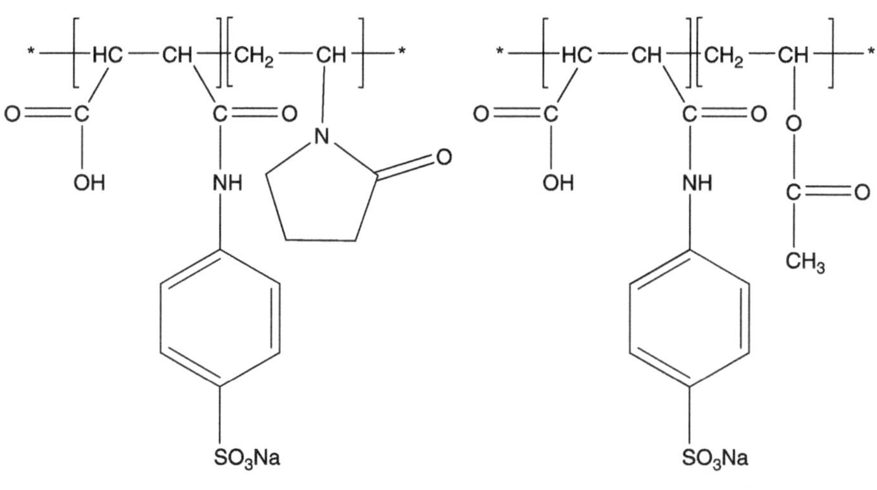

7.4 Molecular structure of two vinyl copolymers (Flatt, 2001).

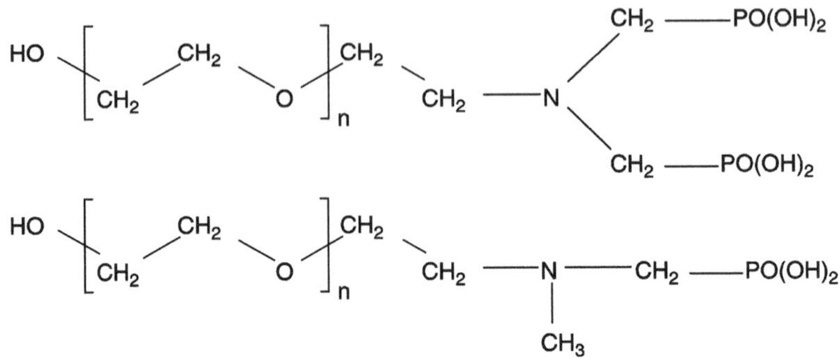

7.5 Structure of PEG-phosphonates.

polymers with polyethylene oxide side chains, these polymers are built of only one polyethylene oxide chain with phosphonate groups on one end. Thus only one or two phosphonate groups are bound to one single polyethylene oxide of medium to high molecular weight. The structures of such superplasticizers are described by Mosquet *et al.* (1997) and Guicquero *et al.* (1999), and are shown in Fig. 7.5.

The complexing ability of phosphonate groups with calcium is higher than that of carboxylic acid or sulfonate groups. Thus, since calcium ions play a major role in adsorption of superplasticizers on cement, the mono-phosphonate adsorption is stronger than that of carboxylate or sulfonate. So, one or two phosphonate groups seem to be enough to force the molecule to adsorb on cement or on calcium carbonate (Mosquet *et al.*, 1997).

PEG-phosphonates show increased slump life in cementitious compositions (Guicquero *et al.*, 1999). However, they strongly retard the hydration of C_3S and thus also of cement (Comparet *et al.*, 2000). The PEG-chain must be long enough to take advantage of the superplasticizing effect with an acceptable level of set retardation. Due to the retarding effect, this type of superplasticizer is particularly used for hot weather concreting or where long slump life is required (Mosquet *et al.*, 2003). A typical area of applications requiring strong retardation is that of oil well cementing.

Polycarboxylic ethers (PCEs)

In the mid-1980s a new generation of superplasticizers was developed in Japan (Tsubakimoto *et al.*, 1984). They are not linear ionic polymers like the previous synthetic superplasticizers or bulky natural polymers like the lignosulfonates, but have a comb-like structure. They consist of a so-called backbone polymer to which side chains are connected in a more or less regular manner. If such a polymer is drawn with all side chains pending to the same direction, as shown in Fig. 7.6, it looks like a comb. This is why these types of polymers are called

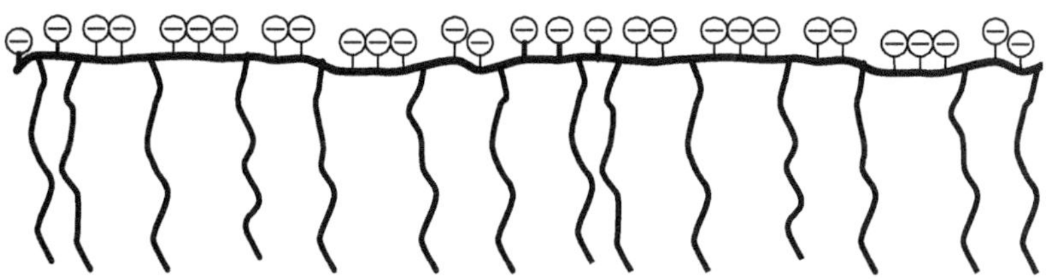

7.6 Schematic drawing of a comb polymer with an anionic backbone and attached flexible side chains.

'comb copolymers'. The prefix 'co' indicates that the monomers in the backbone and the side chain are not the same.

Since the first development of these new superplasticizers, their market grew rapidly and a huge number of different structures with regard to backbone and side chains appeared on the market. However, all of them are based on the same principle: a polymeric backbone built mainly from monomers bearing carboxylic acid groups and attached non-ionic side chains made of polyethers. These main components are the reasons for the name of this class of superplasticizers used in the construction business: polycarboxylic esters or polycarboxylic ethers, or, not quite correct but often used, polycarboxylates.

This new class of superplasticizers has replaced the former generations of superplasticizers (PNS, PMS and the vinyl copolymers) in many applications. It was PCEs that really enabled the concept of self-consolidating concrete (SCC) to find its place in the market. Only PCEs made it possible to reduce the water/cement (W/C) ratio of concretes to values as low as 0.20 while keeping a good workability with prolonged slump life (Danzinger *et al.*, 2003). This low W/C is roughly the amount of water that is necessary for the chemical hydration of cement.

Although only few basic structural units are used to form PCE polymers (i.e. a carboxylic group containing monomers in the backbone and polyethylene oxide units for the side chains), a huge variety of polymers adapted to the different customer needs was developed. Based on the various possibilities of changes in the PCE structure, special polymers were developed for different applications from low to extreme water reduction, from slump keeping polymers to polymers for the precast industry. PCEs are, however, not limited to use in concrete plants but are powerful dispersants for other applications such as gypsum-based products and are also successfully used as grinding aids (e.g. for cement). PCEs are available as aqueous solutions or as powders (e.g. for dry mortars). An overview of the special types of PCEs is given in the next section.

7.2.2 Overview of PCEs

All PCE polymers are composed of an anionic backbone with mostly carboxylic groups and side chains of polyethers covalently attached to the backbone. The carboxylic groups drive the polymer to adsorb on the cement surface whereas the non-ionic polyether chains, which must not adsorb on the cement, tangle into the solution. The backbone brings a negative charge to the cement surface. This charge, however, is much lower than that of PNS, PMS or LS polymers and depends on the number of free carboxylic groups in the backbone. PCEs therefore do not lead to major charging of cement surfaces, as was shown by zeta potential measurement. This contrasts with PNS or LS polymers which induce a strong negative zeta potential on particles (Perche *et al.*, 2003). The dispersing effect of PCEs is clearly dominated by steric repulsive hindrance which is caused by the non-adsorbing side chains of the comb copolymer. Figure 7.7 shows schematically such adsorbed polymers on the surface of two particles. The polyethylene oxide chains do not want to overlap and thus induce repulsive forces if the surfaces come too close to each other. The length and the amount of chains on the surface are responsible for the effectiveness of the steric stabilization of a PCE in cement suspensions (Nawa *et al.*, 2000; Flatt *et al.*, 2009a).

The main monomers building the anionic component of the backbone are acrylic acid, methacrylic acid and maleic acid. In the majority of the PCE polymers, polyethylene oxide is used as side chains. Another synonymous name for polyethylene oxide is polyethylene glycol – both expressions are found in the literature, abbreviated as PEO or PEG respectively. These side chains are connected to the backbone via ester, amide or ether bonds. The amide and ester groups come from esterification or amidation of acrylic, methacrylic or maleic units by mono-functional polyethylene oxides, while the ether bonds result from the copolymerization of mono allyl or vinyl or isoprenyl polyethylene oxide

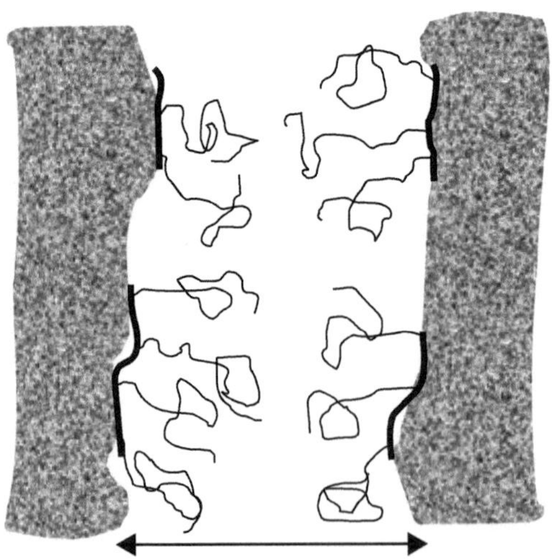

7.7 Idealised drawing of the steric stabilization of powder suspensions by PCE polymers.

macro-monomers. Polyethylene oxide with its unlimited water solubility represents up to 90% of the mass of the polycarboxylic ether. Patent literature reports that part of the side chains may be replaced by the more hydrophobic polypropylene oxide or polybutylene oxide units.

The main parameters that influence the properties of a PCE polymer are:

- chemistry of the backbone (acryl, methacryl, vinyl, allyl, malein, etc);
- length of the backbone (degree of polymerization);
- amount of anionic groups (mostly carboxylic acids or salts) with respect to the side chains and with respect to the backbone;
- number of side chains with respect to ionic groups and with respect to the backbone;
- length of the side chains;
- type and stability of the bond between side chain and backbone;
- overall charge density of the polymer.

Quite different polymer structures and resulting different performance as superplasticizers can be generated by these variations. Some of the many possible structures are shown in Fig. 7.8. There are basically two production routes for PCE polymers (Fig. 7.9). One is the radical copolymerization of monomers bearing carboxylic acid groups with monomers bearing the PEG chain. Another production method is the esterification or amidation of the pre-formed backbone polymer bearing carboxylic acid groups with mono-functional polyethylene oxide.

Structure of the polymer backbone and connections of side chains

In acrylate-based PCEs, the backbone is formed by acrylic acid – acrylic ester or acrylic amide units. Although acrylic acid and methacrylic acid are chemically very close, their polymerized esters show quite different behaviour in the pore solution of cementitious systems. The pH of the aqueous phase in a cementitious system is about 12.5–13. At this high pH the ester bonds of acrylic ester units in the polymer are prone to cleavage with time, whereas that of methacrylic ester units are more stable. Thus, in PCE polymers in which the PEG chains are bound

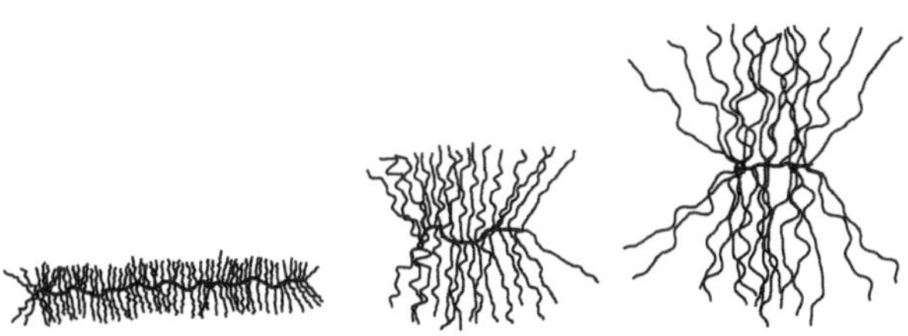

7.8 Structural varieties of PCE polymers (Plank *et al.*, 2010a).

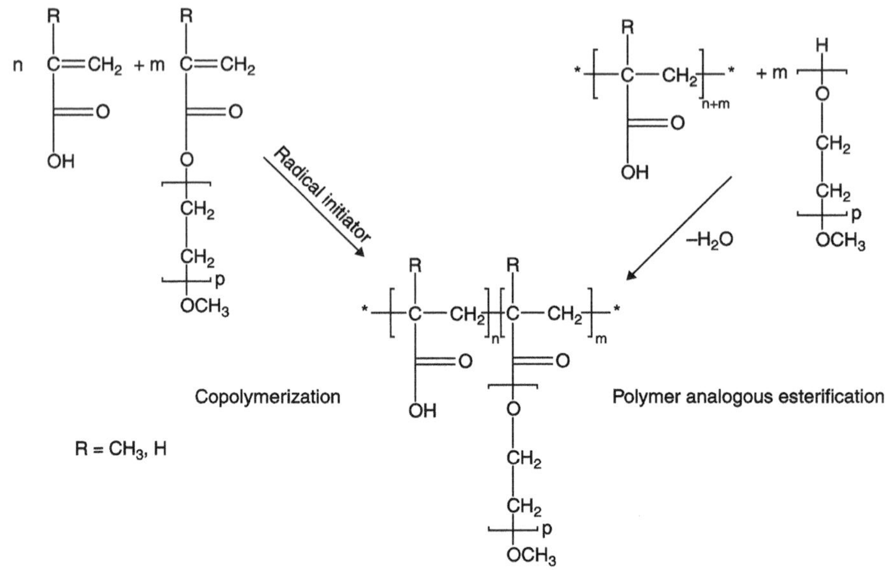

7.9 Production of (meth)acrylate based PCE-polymers: copolymerization (left) and polymer analogous esterification (right). R=CH_3 corresponds to methacrylic and R=H to acrylic acid/ester.

by ester bonds to acrylic units, these ester bonds will be cleaved with time, releasing the PEG chains into the solution and increasing the number of carboxylic groups in the backbone. This ester hydrolysis is the back-reaction of the polymer analogous reaction shown in Fig. 7.9.

This increase in carboxylic groups strongly influences the performance. Indeed, PCE polymers with a higher number of anionic charges (e.g. carboxylic acid groups) show a higher initial plasticizing effect than the same type of polymers with fewer carboxylic acid groups. This is because polymers with higher ionic strength adsorb more on a cement surface than polymers with a lower ionic strength. Thus PCEs that contain a high amount of units derived from acrylic esters are not very reactive with respect to adsorption within the first minutes of contact with the cement. However, upon hydrolysis of the ester bonds with time, adsorption of the now more ionic PCEs that are still free in the solution can take place. A higher adsorption leads to higher repulsion of the cement particles, at least with this type of polymers, and thus a better flow of the paste. This counteracts the loss of plasticity of a cement paste with time. Therefore polymers that increase in ionic strength with time are ideal slump keepers. However, if the polyethylene oxide chains in the PCE are connected via amide or imide linkages as shown in Fig. 7.10, these bonds will be stable against hydrolysis. A deeper insight into the effect of adsorption and addition time of the superplasticizers on the rheology of a cement paste or concrete is given later in this chapter.

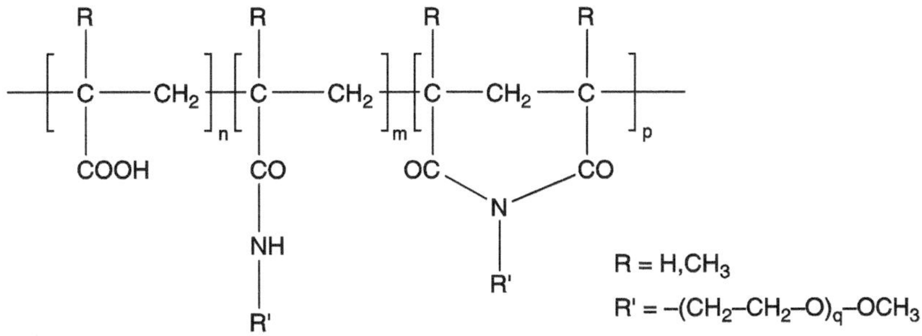

7.10 Chemical structure of a PCE with the side chain connected by amide and imide groups (Arfaei *et al.*, 1995).

Other co-monomers may be incorporated into the polymer to modify the performance of PCEs. For example methallyl sulfonic acid has been described as being used as a co-monomer for the polymerization of acrylic-based PCEs (Yamada *et al.*, 2000a). The structure of such a polymer is shown in Fig. 7.11.

Besides the (meth)acrylic PEG-esters, other macro-monomers can be used to introduce the side chains into the comb polymer. For example vinyl ethers, allyl ethers or other alkenyl ethers of PEG with a different chain length can be copolymerized. Since ether bonds are stable in the alkaline environment of a cement paste, these side chains will not be cleaved off in the concrete. Figure 7.12 shows the structure of a copolymer of acrylic acid, polyethylene oxide monovinyl ether and maleic acid. Other examples of variations of the PCE structure as described by Ohta *et al.* (1997) are shown in Fig. 7.13, demonstrating further the variety of different polymers that can be produced.

The type of the monomer units used to form the backbone of a PCE will influence its performance as dispersant. As mentioned above, acrylic or maleic esters will be cleaved over time in the alkaline pore solution of cementitious system. This will strongly influence the ionic character of the PCE and thus its performance over time (the details are discussed later). The use of allyl, vinyl, acrylic and maleic units in the backbone will result in more flexibility of the

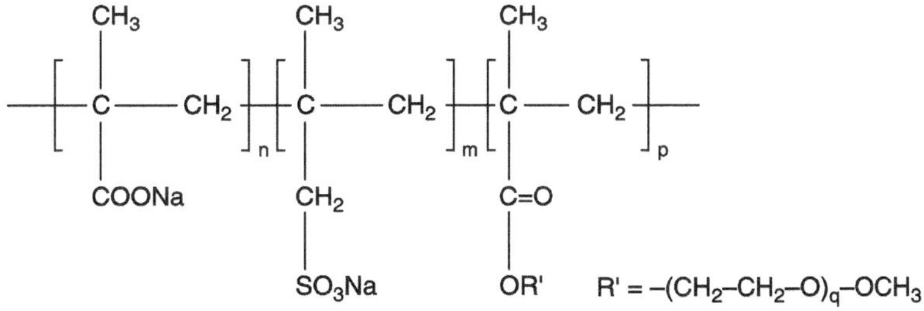

7.11 Structure of a PCE polymer containing methallyl sulfonic acid.

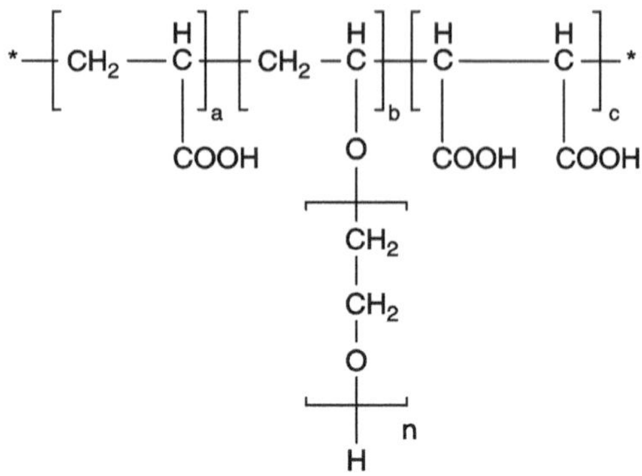

7.12 Structure of a copolymer of acrylic acid, polyethylene oxide mono-vinyl ether and maleic acid.

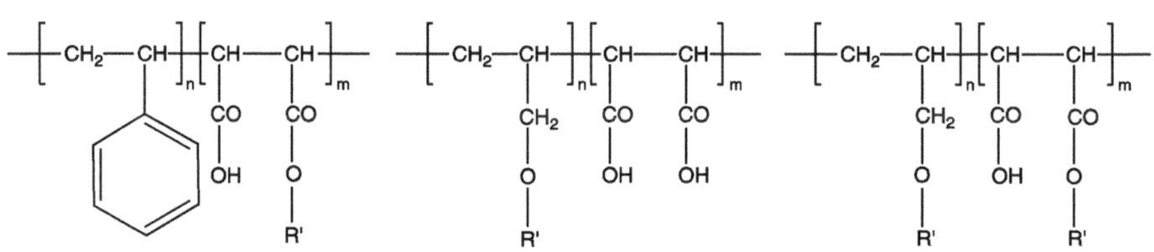

7.13 Copolymer of styrene and maleate (left), copolymer of allyl ether and maleic acid and copolymer of allyl ether and maleate (right).

backbone than the use of methacrylic or methallylic units where a methyl group may sterically influence the motions of the polymer chain. Maleic structural units with their vicinal carboxylic groups have other chelating or complexing ability to ions and may have other affinity to the cement surface than acrylic or methacrylic polymers, which have a methylene bridge between the carboxylic groups. However, the backbone alone will not explain the different performance of PCE polymers and the polymer as a whole must be taken into account.

Structure of the side chain

The variations in the PCE structure described above are more or less related to the backbone. Variations in the side chain play an equally important role. It is the side chain and the number of side chains that determines the steric repulsive forces of a PCE polymer adsorbed on the surface.

The majority of PCE polymers contain side chains built solely of ethylene oxide units. The length of the side chains of the PCEs depends on the desired performance and may therefore vary. The typical range of molecular weight is between 500 and 10 000 g/mole, which corresponds to 11–230 ethylene oxide units in the chain. The majority of commercial polymers contain side chains of molecular weights of about 750–5000 g/mole, which corresponds to 17–110 units. Polyethers of different lengths are combined in some polymers to optimize the balance of ionic strength and grafting degree.

The side chains may also contain propylene oxide units, which can be incorporated in the polymer chain randomly or in blocks. Furthermore, a combination of polyethylene oxide chains and polypropylene oxide chains is possible. Since polypropylene oxide is less hydrophilic than polyethylene oxide, the incorporation of such units can influence the performance of the polymer with respect to its action as plasticizer but also to its air entrainment effect. The incorporation of hydrophobic parts like polypropylene oxide units or longer alkyl chains is said to reduce the air entrainment of some types of PCE polymers (Arfaei and Lambert, 1995; Tsuyoshi *et al.*, 2000).

The end group of the side chains in (meth)acrylate ester based PCEs is usually an alkyl group (e.g. a methyl group). The production of a mono methyl polyalkylene oxide is shown in Fig. 7.14. This only mono-functional polyalkylene oxide is necessary for the production of PCE superplasticizers of the types shown in Fig. 7.9. Only one end of the PEG chain should react during the esterification of the (meth)acrylic acid to produce the macro-monomer or in the polymer analogous esterification of the poly(meth)acrylic acid. A bi-functional polyethylene oxide would lead in both cases to cross-linking of the polymer, resulting in a network of interconnecting chains that gels in water but is insoluble. Copolymers of vinyl or allyl PEGs, however, contain hydroxyl functions at the end of the PEG side chains. Plank *et al.* (2008) investigated the effect of either a methoxy or a hydroxy group at the end of the PEG chain on the performance of PCE superplasticizers. They produced a macro-monomer by direct ethoxylation of methacrylic acid and copolymerized this PEG methacrylate with methacrylic acid. No significant difference in the performance as dispersant was observed compared with a methacrylate-based PCE containing methoxy groups at the end of the side chain. Side chains based on ethoxylated polyamides, which are additionally incorporated in small amounts together with the polyethylene oxide side chains, have shown a positive influence on the performance of the PCE polymers (Amaya *et al.*, 1999).

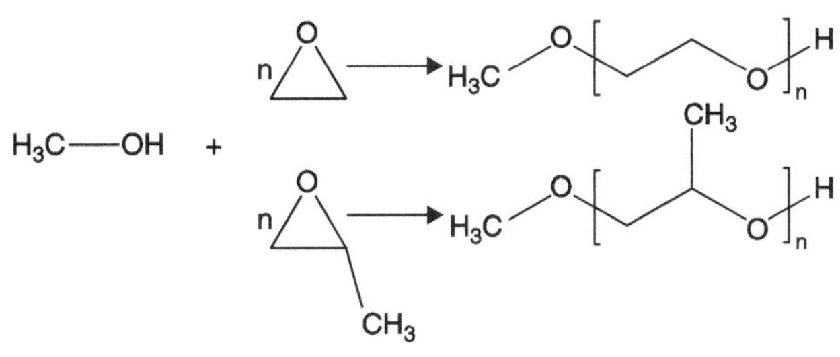

7.14 Synthesis of mono-functional polyethylene oxide or polypropylene oxide.

Researchers continuously try to vary the PCE structures and find new monomers that can be introduced such as phosphonic acids (Hamada, 2005), fatty alcohols (Karl *et al.*, 2009) or silyl groups (Witt and Plank, 2009). However, the PCEs based on the previously described basic chemicals dominate the market and show best performance.

The high degree of flexibility for chemical modifications of the polycarboxylate-based superplasticizers makes a comprehensive overview almost impossible. Therefore, we will focus in this chapter on PCEs with mainly carboxylic acid groups in the backbone and side chains of polyethylene oxide.

7.2.3 Formulations

When a concrete producer buys a superplasticizer, it seldom consists of just a pure solution of one of the previously described superplasticizer polymers. Commercial superplasticizers are usually formulated products adapted for the special needs of the respective market, application or area. A ready-mix concrete producer needs a superplasticizer guaranteeing a good plasticity of the concrete until it is casted, which may span from one up to several hours. This long time may be necessary in the case of traffic jams or especially in tunnel construction where the concrete may be transported through several kilometres of tunnel to the job site (including stops and delays) until it may be cast. A precast concrete producer needs a concrete with good workability but only for up to about 30 minutes. After that the concrete should harden as fast as possible. In countries with a hot climate or in summer, the concrete should not stiffen too fast due to the high temperature whereas in Nordic countries the setting and hardening should not be delayed too long in spite of the cold temperatures. A cold climate further demands constructions that are resistant to freeze/thaw cycles as well as to de-icing salt. Part of the different demands can be satisfied by selecting the right polymer, especially in the PCE range as huge structural variations are possible. But formulation is often also the key here to achieve the best solution for the customer. This formulation know-how is vital and is based on the broad experience of the admixture companies.

Superplasticizers can lead to increased air content when the mortar or concrete is mixed. This air may improve the workability to some extent but air contents of more than 3% for a non air-entrained concrete reduce durability and must be avoided. Therefore most superplasticizer formulations contain a defoamer – especially the ones based on lignosulfonates and PCEs.

Superplasticizer solutions are solutions of organic substances and thus, although not all are readily biologically degradable, they are nevertheless prone to microbial attack. This may result in bad odour and depositions, but in some cases, especially in formulations containing retarders like sugar derivatives, the degradation of the retarder by microorganisms may dramatically influence performance. Therefore, many liquid concrete admixtures also contain some biocide.

Admixture solutions may further contain retarders and accelerators. This depends on the application of the specific admixture. Admixtures for hot countries or summer time and for the ready-mix business often contain retarders, especially if they are based on PMS or PNS, which do not or only slightly retard the setting of cement themselves. Admixtures based on lignosulfonates or PCEs contain retarders only in cases where higher retardation is necessary since lignosulfonates themselves retard the setting and PCE polymers can be designed to produce an elongated concrete slump life. Commonly used retarders are hydroxycarboxylic acids like sodium gluconate or carbohydrates such as hydrolyzed cornstarch or molasses (Collepardi, 1984), which themselves show slight rheology improvement.

Vice versa, admixtures for cold countries, winter time or precast concrete often contain accelerators to speed up early hydration. Examples of accelerators are calcium chloride, calcium nitrate, calcium nitrite, calcium formate, thiocyanates, triethanolamine, hydroxy amino compounds, and blends thereof. The use of calcium chloride is limited since chlorides induce corrosion of steel bars (Ramachandran, 1984).

A further improvement of concrete can be achieved by using so-called viscosity modifying admixtures (VMAs). These are most often high molecular weight polymers, in many cases based on modified polysaccharides such as hydroxyethyl cellulose, welan or xanthan gums, or synthetic copolymers. They help to obtain concrete with higher tolerance to slight changes in the water content, and reduce bleeding and segregation if either superplasticizer is overdosed or the water content is too high. Special applications where these viscosity modifiers are used in higher amount are under water concreting or oil well cementing (Plank *et al.*, 2010b). A detailed discussion of viscosity modifiers is found in Chapter 8.

Another common procedure is the blending of dispersing polymers to optimize the performance of the admixture. Thus PNS or PMS can be used alone or blended with lignosulfonates. This presents two advantages: one is the low cost of lignosulfonates, the second is the slight retardation of the sugars present in lignosulfonates and that helps the concrete to keeps its workability longer.

Different PCEs can be blended to adapt the performance of the admixture to the user's demand. A very useful blend is the combination of a PCE with strong water reduction with a so-called slump-keeping PCE. This results in a low water content and increased workability time (Yamashita *et al.*, 1999; Velten *et al.*, 2001; Hamada *et al.*, 2003; Mäder and Schober, 2003; Lorenz *et al.*, 2010). Such slump-keeping PCE polymers are, for example, copolymers that contain acrylic esters which hydrolyse with time in the alkaline pore solution and release carboxylic groups. The so-formed new polymer has a higher number of adsorbing groups and thus its adsorption is increased with time. This higher adsorption results in an increase of plasticity of the cement paste which counteracts the stiffening of the paste. Other slump-keeping PCEs are polymers with a high grafting degree. These polymers adsorb slower than polymers with a high content of carboxylic or other ionic groups and thus there is still enough superplasticizer in the solution after the first hydration reactions have taken place. These 'slow adsorbing polymers' are not intercalated or adsorbed on most reactive cement surfaces and can induce the plasticizing effect when the fast adsorbing polymers loose their effectiveness.

As one can imagine, the blending and formulation to create optimal concrete admixtures requires extensive knowledge of concrete, cement and cement–superplasticizer interactions. Long experience and a large number of test formulations are also necessary to find the correct admixture formulation. Blends of PNS with PCE polymers were found to exhibit negative synergy with regard to slump; PMS–PCE blends behaved somewhat better but the cost/benefit ratio was not favourable (Zhang *et al.*, 2006). Competitive adsorption of the components may also influence the performance (Plank and Winter, 2008; Plank *et al.*, 2010b). In addition, incompatibility of the components in the admixture itself may cause problems. For example polymer incompatibilities may increase the viscosity of the admixture itself resulting in problems with dosing, sedimentation or phase separation. Another problem can occur when calcium salts are blended with polymers that contain sulfates. For example, if calcium lignosulfonates are blended with sodium PNS, calcium sulfate may precipitate since the PNS may still contain sulfate form the polycondensation.

Skilled researchers in admixture companies, together with experienced personnel from concrete and cement producers, can determine the optimal concrete formulation for almost all job site demands. Experience and know-how (theoretical and practical) are crucial to being effective in the process of formulation.

7.3 Physical characteristics of superplasticizers

7.3.1 Adsorption

Measuring adsorption

Adsorption is one of the most fundamental processes involved in the working mechanisms of superplasticizers as it allows these products to move from the aqueous phase of the system to the surface of particles, where they can then reduce or eliminate attractive interparticle forces.

Adsorption is most often measured by so-called solution depletion. In this procedure the admixture is added first to a paste or suspension of the particulate solid under investigation. After a given time, the liquid phase is separated either by sedimentation, centrifugation or filtration. The collected liquid phase is then analysed to determine the amount of admixture remaining. Various techniques may be utilized, depending on the product. For visible light or ultraviolet (UV) absorbing admixtures (e.g. PNS, PMS or lignosulfonates), spectrometry is a convenient choice. For polycarboxylates, one often resorts to measuring the total organic carbon in the aqueous phase. This step in the analysis can also be coupled with chromatography if one wants to distinguish the relative adsorption of different fractions of the admixtures (Flatt *et al.*, 1998).

Once the amount of admixture remaining in the aqueous phase has been determined, one calculates by difference with the initial amount, the fraction that has been 'consumed' by the solid phase. This includes the materials that are truly adsorbed, but also any material that is not more soluble and would have remained with the solid because of its association in micelles or precipitates. The distinction is important because only the 'adsorbed' part is expected to be active in terms of dispersion. Unfortunately, it is very difficult to distinguish between these different parts, and most of the time, one is just left to assume that all the 'consumed' material is 'adsorbed'.

Measurements of solution depletion have been performed since the various first mechanistic studies on superplasticizers (Daimon and Roy, 1978). While they have been useful in progressing in the understanding of the working mechanisms, they are clearly only one of various elements needed. In particular, they are insufficient to determine the extent to which interparticle forces are changed or why some polymers may adsorb more extensively than others.

Fundamentals of polymer adsorption

Adsorption of a polymer cannot be directly related to the adsorption of the monomers it is made up of. Indeed, it is well-known that many monomers or oligomers do not adsorb while large chains of the same materials can show a pronounced adsorption. This has been confirmed in various studies on superplasticizers, the best documented of which concern PNSs (Piotte, 1993; Jolicoeur *et al.*, 1994).

The simplest explanation is that the adsorption energy of a monomer is of the order of the thermal energy (kT) and therefore its adsorption is not favoured. In contrast, a polymer will have an adsorption energy that is some multiple of kT. The exact extent of the multiplication factor is the result of a non-trivial energy minimization of the overall adsorption process involving change in polymer conformation and solvent desorption. To this end, it is worth noting that solvent desorption plays an important role. Indeed, each water molecule 'released' will lead to an entropy gain with a characteristic energy of the order of kT. Very schematically still, the polymer 'immobilization' on the surface will on its side lead to an increase in entropy of the same order. Globally this explains why larger polymers will tend to adsorb and smaller ones with the same chemical basis will not. The details are, however, much more complex.

Adsorption isotherms

Adsorption isotherms refer to a series of adsorption measurements performed at a given temperature and whose results are plotted as a relationship between adsorbed and non-adsorbed amounts. The shape of such isotherms holds a lot of information regarding the nature of the adsorption process. In some cases, it is more convenient to plot the adsorbed amount with respect to the added admixture rather than the one remaining in solution. Both representations are found, although strictly speaking only the first really represents an adsorption isotherm.

Adsorption isotherms generally show a plateau at high dosages, which corresponds to complete surface coverage. Multiple layer adsorption can (in some cases) be seen as different steps. The absence of a plateau and the continuous increase of 'adsorbed' admixture with dosage are considered to correspond to a process of precipitation, either in the bulk or on the surface. This has been seen for a specific polycarboxylate polymer in suspensions of magnesium oxide (Perche *et al.*, 2003). In this specific case (Fig. 7.15), the process was very much dependent on the specific chemical structure of the admixture (Fig. 7.16), as other polymers of the same family and with rather similar compositions did not show this behaviour. It was also dependent on the substrate, as this was not seen in cement suspensions prepared with the same polymers.

An intriguing situation is the one where a plot of adsorbed amount versus dosage is initially linear but the slope is different from unity over a fair range of

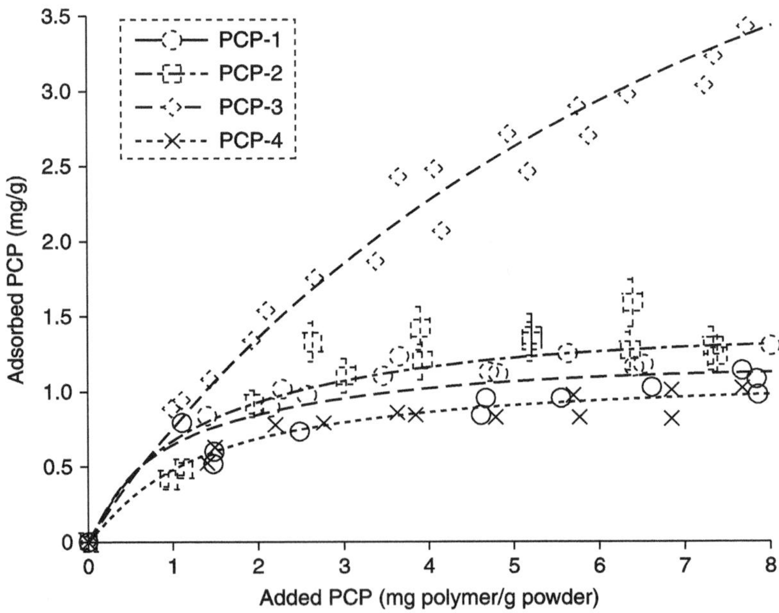

7.15 Adsorption of polycarboxylate polymers. PCP-1, 2 and 4 show the typical adsorption plateau at high dosage, while PCP-3 shows a continued increase. (Perche *et al.*, 2003).

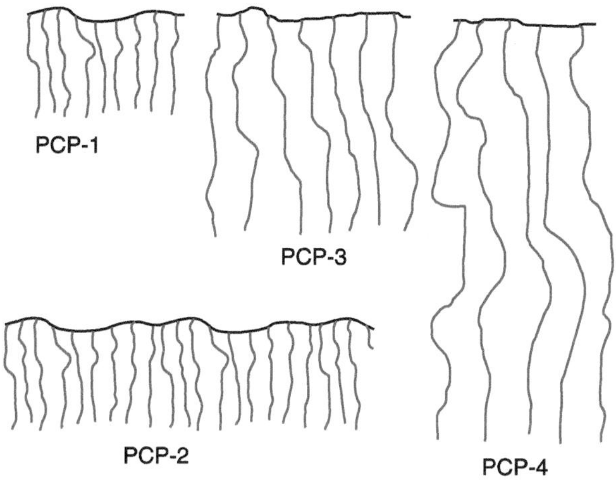

7.16 Schematic drawing of the polymers used by Perche *et al.* (2003) and Schober and Mäder (2003).

dosages. Such behaviour implies that, regardless of the added amount of admixture, the same fraction is always adsorbed. This could be the case in a polycarboxylate superplasticizer that would contain some amount of non-grafted and non-adsorbing side chains. It could also be the case for a polymer containing residuals of monomers or oligomers too small to adsorb. Perche has, however, observed this behaviour with polycarboxylate polymers that had been ultrafiltrated (Perche *et al.*, 2003). In this case, the linear relation between dosage and adsorption can neither come from non grafted PEG nor from residual monomers nor from non adsorbed oligomers. Therefore, Perche's experiments probably point to the fact that there is a distribution in the molecular structure of these polymers; within this distribution some of the polymers would adsorb more strongly than others. Thus one can imagine that upto a certain dosage, one fraction of the polymer is completely adsorbed while the other remains in solution. This could come about, for example, from differences in the grafting or polymerization statistics and would explain the linear increase in adsorption versus dosage with a slope different to unity.

When attempting to rationalize the shape of an adsorption isotherm, the Langmuir representation is often used for polymers. While the adequacy of its hypothesis has been extensively questioned for polymers, it nevertheless often provides a decent fit to many data. To this end, we must emphasize that adsorption measurements generally have large error bars. In addition, any function that starts from zero and goes to a plateau at high dosages would probably also give a decent representation. The quality of the fit is therefore no good reason to further assume that the fitting parameters correspond to their initial meaning as defined by Langmuir.

However because this representation is so much used, it is worthwhile to recall its funding hypothesis. For this, we note the adsorbing species SA, the free surface sites A and the occupied surface sites SA. A chemical equilibrium between them can be written as:

$$A + S \rightleftarrows SA \qquad [7.1]$$

The chemical equilibrium is then written as:

$$K = \frac{[SA]}{[A][S]} \qquad [7.2]$$

The total number of surface sites is:

$$S_\infty = S + SA \qquad [7.3]$$

Substitution into Eq. 7.2 to express leads to:

$$K = \frac{[SA]}{[A]\,([S_\infty] - [SA])} \qquad [7.4]$$

So that the number of occupied adsorption sites is:

$$[SA] = \frac{K[A][S_\infty]}{1 + K[A]} \qquad [7.5]$$

At high values of A, the adsorbed amount is equal to the maximum amount of available sites, which corresponds to reaching a plateau. The slope of the isotherm is given by:

$$\frac{\partial[SA]}{\partial[A]} = \frac{K[S_\infty]}{(1 + K[A])^2} \qquad [7.6]$$

which at low dosages gives $K[S_\infty]$. The equilibrium constant K can therefore be obtained from the initial slope $K[S_\infty]$ and the value of the adsorption plateau $[S_\infty]$.

Probably the most criticizable fact in using this approach for polymers is the rigid definition of adsorption sites. While this was acceptable for the situation with ions or small molecules, for which the model was initially proposed, it is very questionable for polymers. Polymers are indeed known to have various interaction points with a surface. Moreover, it is also generally accepted that the polymer does not remain static on the surface, so that each polymer does not permanently adhere to the same points. This in turn makes the use of adsorption sites at the polymer scale very questionable (this is also true even for small molecules). Evidence for the mobility of the polymer comes from notion of the reversibility of adsorption discussed in the next section.

Reversibility of polymer adsorption

Because of large adsorption energies, polymer desorption is most often an unfavourable process. Another way of understanding this is to say that all adsorbed parts of the polymer would have to desorb simultaneously in order for the polymer to leave the surface. Moreover, as the surface coverage decreases, each polymer has the possibility to increase its number of interaction points with the surface. All of this means that it is difficult, if not impossible, to completely remove all adsorbed polymers by repeatedly rinsing the system with fresh water.

However, the polymers can exchange with polymers in the bulk solution. This was shown by Pfefferkorn *et al.* (1985) (cited in de Gennes, 1987), who adsorbed labelled polymers onto grains and then washed the system with a solution containing identical unlabelled polymers. The kinetic law for the decay of radioactivity is proportional to the coverage by labelled polymers Γ^* and to the bulk concentration of unlabelled polymers, c_b.

$$\frac{d\Gamma^*}{dt} = -k_D \Gamma^* c_b \qquad [7.7]$$

As discussed by de Gennes (1987), this law indicates an exchange process and confirms the existence of a restoring force leading to a constant surface coverage Γ. Below this surface coverage, the conformation of the polymers is (can be) much more stretched out over the surface. Above it, the excess polymers tends to be spilled out into the solution.

The desorption behaviour of superplasticizers has been extensively studied by Platel (2009), who prepared cement suspensions in presence of polycarboxylates of various structures. These suspensions were then centrifuged and the liquid phase analyzed. His first observation was that the adsorption isotherms tended to have two distinct domains, which he described as being of high and low affinity. The first zone is characterized by an almost vertical rise, while the second shows a linear increase of the adsorbed amount with the amount of admixture in solution (Fig. 7.17). More importantly, Platel also attempted to wash out the adsorbed admixtures. For this, after the centrifugation steps, he replaced a portion of the liquid phase by an equivalent portion of fresh water. He then mixed the suspension again before determining again the admixture content in the new liquid phase. The data obtained were then seen to lie on the same line in the zone of low affinity. However, he was not able to reduce the adsorbed amount below the 'plateau' of the high affinity zone.

These data confirm that most polymers adsorbed are not irreversibly adsorbed but that complete desorption is difficult. However, such studies would need to be repeated with pure phases. Indeed, cementitious systems contain various phases of different reactivity and adsorptive behaviour. It is for example known that aluminate phases have a high adsorption capacity. This differentiation between zones of high and low affinity may therefore just translate the multimineral nature of cement phases. It should also be noted that some of the polymers used in

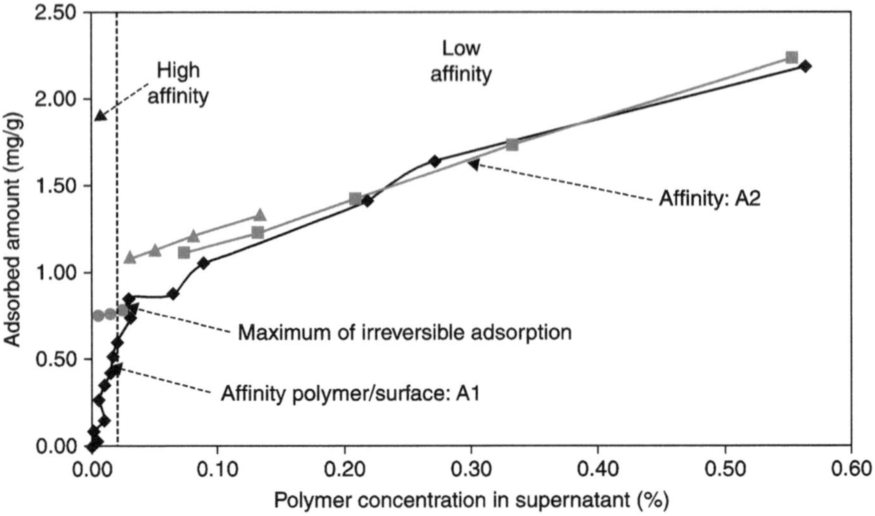

7.17 Adsorption isotherm as obtained by Platel (2009).

Platel's study have a very large density of ionic groups (Platel, 2005). This is generally accepted to be unfavourable for dispersion, probably as a result of precipitation. Therefore, while Platel's study defines many interesting ideas and addresses the important issue of desorption, some of its conclusions suffer from the too broad range of the polymer structures selected and the fact that this is not well enough integrated in the discussion of the results.

The other important study in relation to the reversibility of polymer adsorption was by Yamada and Hanehara (2001), who were examining the role of sulfates on the performance of a polycarboxylate comb copolymer. They carried out adsorption tests and flow measurements after successive additions of $CaCl_2$ and Na_2SO_4. Results showed that the dispersion capacity of the polycarboxylate superplasticizer decreased when the sulfate concentration was increased by adding Na_2SO_4. Conversely, they saw that when this concentration was reduced by adding $CaCl_2$ (to precipitate gypsum), the flow performance was recovered. They also saw that these changes were well correlated to changes in the amount of adsorbed polycarboxylate. High sulfate concentrations led to limited adsorption, while low concentration led to large extents of adsorption.

The detailed reasons for this behaviour could be diverse. One interpretation involves a process of competitive adsorption between sulfates and polymers for surface sites (Flatt *et al.*, 2009b). The other point of view first states that calcium ions are needed for polycarboxylate adsorption. It then argues that the competition between polymers and sulfates is for calcium ions rather than for surface sites. To some extent this ends up relatively similar, as a high amount of calcium is concentrated around the surface of the cement particles.

Role of C/E in polycarboxylate adsorption

For polycarboxylate polymers the density of ionic groups in the polymer backbone has an important role on adsorption. This is well summarized in Fig. 7.18 in which the two series of data overlay surprisingly well. They suggest that there is a critical value of carboxylate to ester (C/E) groups on the order of two or slightly lower, below which adsorption drops strongly. While the study by Regnaud *et al.* (2006) reported only polymers with one side chain length, the one by Winnefeld *et al.* (2007) used polymers with various chain lengths. Their data suggest that this critical value is the same for all their polymers, although they did not push the C/E values down low enough to truly confirm this.

Role of temperature

Temperature is another factor that can affect adsorption. The first detailed studies we are aware of on this subject were performed on suspensions of MgO and $Mg(OH)_2$ taken as model suspensions because of their low reactivity and slightly positive zeta potential at high pH (Flatt *et al.*, 1997). Since these studies were performed, it has been shown that MgO is not as inert as initially expected (Perche, 2004), so that the result must be treated with care.

Nevertheless, their importance lies much more in the difference in adsorption among the polymers tested than in absolute values. In particular, it was found that two strongly ionic vinyl copolymers did not show any change in their adsorption

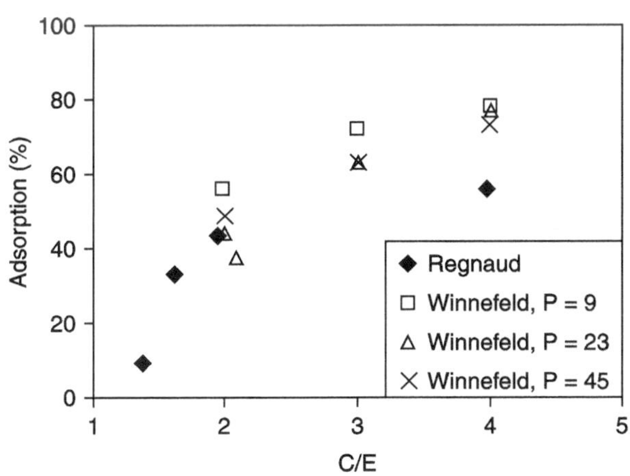

7.18 Adsorption data from Regnaud *et al.* (2006) and Winnefeld *et al.* (2007). Results are given as the fraction of adsorbed polymer with respect to the amount added (0.16% and 0.2% respectively) in cement pastes. The *x*-axis (C/E = N – 1) is the ratio of carboxylate to ester groups in the polymer backbones. The value of P corresponds to the length of the side chains.

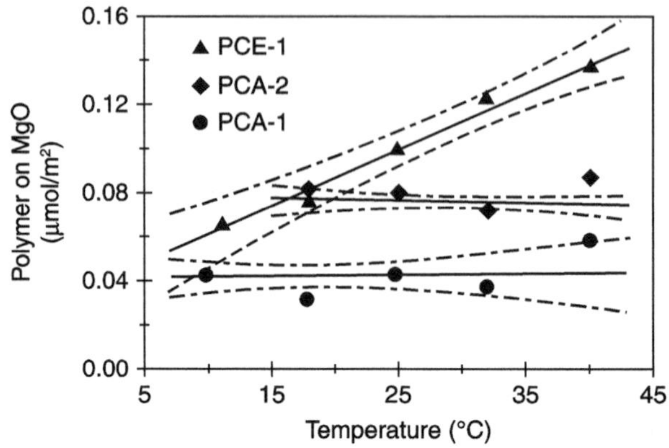

7.19 Adsorption plateau on MgO as a function of temperature.

plateau as a function of temperature (Fig. 7.19). In contrast a polyacrylate polymer, with very short side chains, had a pronounced increase in its adsorption plateau as temperature increases. The interpretation given to this was that the polyacrylate, having a lower ionic charge density, had a lower adsorption energy. Consequently it would be more sensitive to competitive adsorption, which in this case was supposed to involve hydroxyl ions. For entropic reasons it was argued that the ions would be able to compete at low temperature for surface sites, but not at higher temperature. Moreover, it was stated that the vinyl copolymers would not be subject to competitive adsorption because of their higher adsorption energy. More recently, Plank *et al.* (2010a) obtained results on three PCEs that also show an increase in the adsorption plateau as temperature is increased. In their case, the suspensions were of calcite and the polycarboxylates were of the allyl ether type. They also interpreted these results by discussing the relative contributions of enthalpy and entropy to adsorption energy.

In cementitious systems, temperature accelerates hydration reactions. Thereby, the total specific surface increases, which raises the need for superplasticizer adsorption to achieve a given fluidity. In situations where not enough polymer is available to saturate the surface, this will imply a reduction in fluidity. It could also be accompanied by an apparent increase in the total adsorbed amount due to a displacement of the equilibrium between the solution and the surfaces.

7.3.2 Zeta potential

Definition of zeta potential

Particles suspended in aqueous media can acquire a certain surface charge density (s_0) owing to the dissociation of surface groups and/or the specific adsorption of charged species initially present in the bulk solution (mineral or organic ions, including ionic polymers). Counter ions are attracted to this surface and form a first layer of long residence time, which can be considered to move along with the particle. The centre of the plane of these fixed counter ions is the Stern layer. Somewhat further away from this plane is the shear plane, which defines the zone of liquid that still moves along with the particle under normal conditions. It is the plane at which the zeta potential is measured. This is the electrokinetic potential that is the value most readily accessible from experiments.

Measurement techniques and their limitations

The oldest techniques for measuring zeta potential are of the micro electrophoresis type. They use very dilute solutions and follow the motion of individual particles in an electrical field. For cementitious systems it is particularly important to regulate the ionic concentration of the large amount of liquid for it to be representative of the liquid phase in a paste. Each measurement is therefore a time snapshot for the selected iconic concentrations. More problematic, however, is the fact that particles larger than a few microns in diameter tend to sediment and cannot be measured reliably.

Over the past 10–20 years techniques based on acoustic excitation and electrical detection (and *vice versa*) have become very popular. In particular, they are able to measure zeta potentials in concentrated suspensions. They are also capable of handling the background signal from the electrolytes that has been shown to account for a fair part of the signal (Flatt and Ferraris, 2002). In addition, they offer options to include the effect of particle sizes. However, it must be emphasized that these corrections become questionable for a large fraction of the particles in ordinary cement. The loss of signal in these techniques is indeed an intrinsic characteristic of the system. In a specific case, it was in fact calculated that for ordinary cement, 60% of the signal comes from the 10% smallest particles and that 85% of the signal is due to the finest 20%. Therefore, even if corrections can be applied, considerable care is needed in the quantitative interpretation of these results. This statement is enhanced by the fact that the finer fractions of cement particles are generally enriched in calcium sulfate.

An additional word of care must be given for blended cements. In this case, the suspensions contain particles of different densities. This parameter enters in the equations used to calculate the zeta potential, bringing an additional uncertainty concerning the measured zeta potential values. Moreover, the case of flyash can be expected to be particularly problematic. Indeed, flyash contains a fraction of hollow particles, of which the apparent density can be smaller than unity. This means that, in an acousto-electric system, the motion of these particles could be opposite to that of the other particles. The result of this would be additional signal attenuation.

The above comments are intended to make users aware of the physical limitations intrinsic to zeta potential measurements. Bearing this in mind, such measurements can bring useful information, particularly in comparative studies.

More specifically, measurements involving the titration by an admixture can indicate if charged species are adsorbing as well as the dosage range where surface saturation is reached. However, the somewhat 'black-box' nature of the zeta potential calculation offered in these instruments implies a high risk of over-interpretation in the case of cementitious systems.

Results in cementitious systems

For cement suspensions without polymer additions, the zeta potential values reported are always small. This is probably the combination of a potential that is indeed small and probably also of the lack of signal from the larger fraction of particles. Owing to the various uncertainties mentioned above and to variability in cement composition, it is also not surprising that values both slightly positive and slightly negative have been reported.

What is, however, unanimous is that the addition of a strong polyelectrolyte such as an PNS leads to strong negative zeta potentials. Most probably, this is not only the result of the charges brought to the surface from polymer adsorption, but the fact that some of these are located at the exterior of the adsorbed layer (Lewis *et al.*, 2000). In contrast addition of polycarboxylates leads to much smaller zeta potentials. Here the number of charges brought is lower per polymer, but these are also located close to the surface, while the shear plane is extended towards the bulk solution due to the non adsorbing side chains (Lewis *et al.*, 2000). To this end it is worth noting that the adsorbed layer thickness is on the order of 2.5 nm at least, while the Debye length κ^{-1} in cementitious systems is of the order of 0.7 nm (Flatt and Bowen, 2003). The potential decrease with distance D from the surface is about $\exp(-\kappa D)$. This implies that the potential that could be measured outside a 2.5 nm layer would be only about 3% of the potential inside at the inner side of this layer, provided that is where the charge would be localized.

The above simple numerical example emphasizes the difficulty to interpretate unambiguosly zeta potential measurements in presence of polycarboxylates. For example, Plank *et al.* (2006a) measured an increase rather than a decrease of the zeta potential for polycarboxylates with side chains having more than seven ethylene oxide (EO) units. They proposed this to be a result of a perpendicular adsorption of the polymer. whilst interesting, this interpretation needs to be considered with care owing to the delicate nature of these measurements and numerous factors involved.

7.3.3 Conformation of PCEs

The conformational behaviour of the comb copolymers has been examined in solution using light scattering and viscosity measurements (Borget *et al.*, 2005; Kauppi *et al.*, 2005) and by small angle neutron scattering (Giraudeau *et al.*, 2009). These studies showed that these polymers could be well described by the model developed by Gay and Raphaël (2001) for comb homopolymers in solution. This model proposes a phase diagram with different types of conformations (Fig. 7.20). The location in this phase diagram is deduced using the following notation for the polymer structural parameters:

P = number of monomers in each side chain
n = number of repeating units, or number of side chains
N = number of monomers in the backbone per repeating unit.

The different types of conformation are as follows:

DC = decorated chain
FBW = flexible backbone worm
SBW = stretched backbone worm
SBS = stretched backbone star
FBS = flexible backbone star.

It should be emphasized that owing to the nature of the scaling law approach, the position of the boundaries is not exact. Nevertheless, it is important to note that most studies on superplasticizers, in particular those involving admixture companies, operate with polymers in the FBW domain as shown in Fig. 7.21.

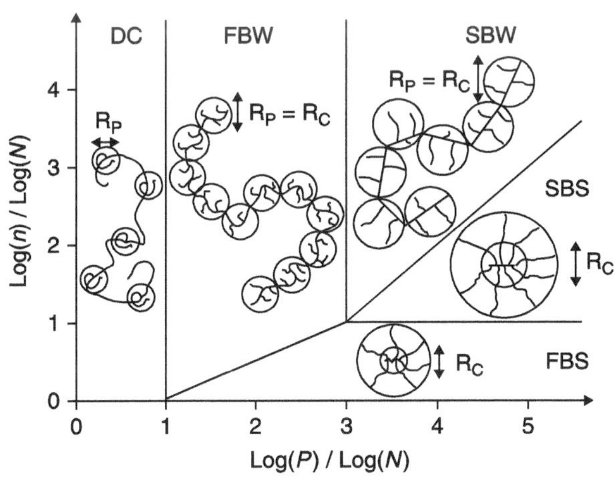

7.20 Phase diagram for comb homopolymers according to Gay and Raphaël (2001). The different domains are: decorated chain (DC), flexible backbone worm (FBW), stretched backbone worm (SBW), stretched backbone star (SBS) and flexible backbone star (FBS). RC denotes core size and Rp is side chain size. Reprinted with permission from Flatt *et al.* (2009a). Copyright 2009, American Chemical Society.

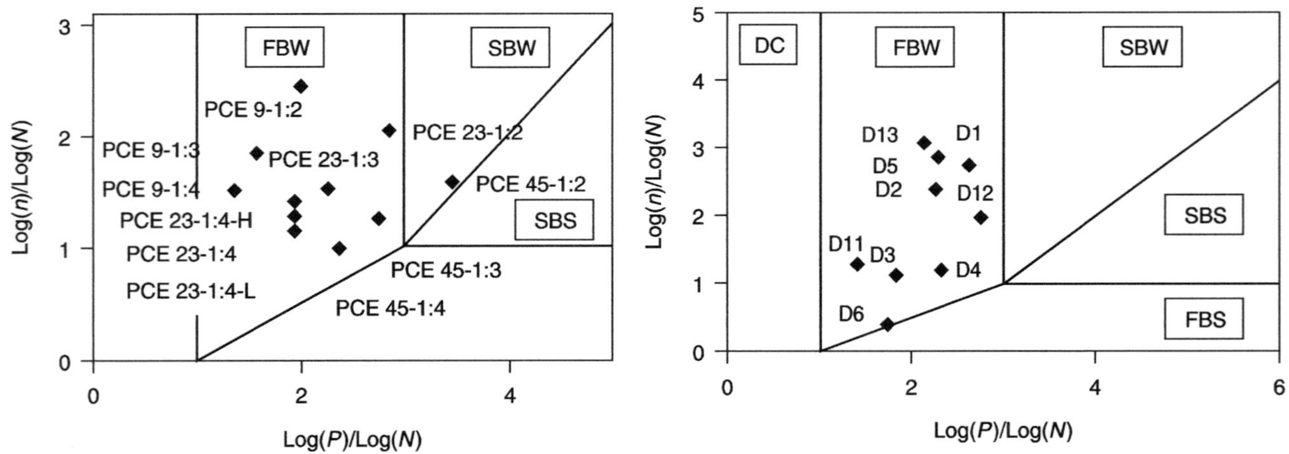

7.21 Representation of the polymers used in two studies on polycarboxylate superplasticizers: left Winnefeld *et al.* (2007) and right Flatt *et al.* (2009a). The different domains are decorated chain (DC), flexible backbone worm (FBW), stretched backbone worm (SBW), stretched backbone star (SBS) and flexible backbone star (FBS).

The conformation of adsorbed polycarboxylate superplasticizers has been examined by atomic force microscopy in various occasions. Early studies on cement (Uchikawa *et al.*, 1997) most probably include artefacts as already reported elsewhere (Flatt and Houst, 2001). In contrast, studies with the same technique using magnesium oxide as substrate, reported values of layer thickness on the order of 2–6 nm, depending on the length of the side chains (Kauppi *et al.*, 2003, 2005). This result emphasizes that these side chains are coiled and not stretched. More recently, the technique has been used with calcium silicate hydrate (Flatt *et al.*, 2009a). This study provides the essential feature of having a realistic substrate at the right pH, with calcium ions and most importantly a substrate with atomically smooth domains (Plassard *et al.*, 2004). This paper showed that the FBW model of the polymer could be extended to account for adsorbed conformation by using a random walk of hemispheres. A schematic representation of how this model would represent the polymer in solution and on the surface is represented in Fig. 7.22.

7.22 Schematic representation of a comb copolymer in solution (chain of spherical cores, each of radius RC) and one adsorbed on a mineral surface (chain of hemispheres, each of radius RAC). Adapted with permission from Flatt *et al.*, (2009a). Copyright 2009 American Chemical Society.

The characteristic dimensions are obtained, first by extending the FBW model for comb homopolymers to comb copolymers. For this, the size of backbone and side chain units are a_N and a_P respectively. The number of side chains in a core of a given FBW is then:

$$n_C^2 = \frac{P}{N}\left(\frac{a_P}{a_N}\right)^2 \quad [7.8]$$

The end-to-end distance of polymer in solution is given by:

$$R = \left(\left(\frac{a_N}{a_P}\right)^2 \frac{(1-2\chi)}{2}\right)^{1/5} a_P\, P^{2/5} N^{1/5} n^{3/5} \quad [7.9]$$

while as for the individual cores it is:

$$R_C = \left(\frac{a_P}{a_N}\frac{(1-2\chi)}{2}\right)^{1/5} a_P\, P^{7/10} N^{-1/10} \quad [7.10]$$

where χ is the Flory parameter. For the polymers of interest in this chapter, we can use $a_P = 0.36$ nm, $a_N = 0.25$ nm and $\chi = 0.37$ at 25°C (Pedersen and Sommer, 2005).

Here it is important to note that the pre-factor in Eq. 7.9, $\left(\left(\frac{a_N}{a_P}\right)^2 \frac{(1-2\chi)}{2}\right)^{1/5} a_P$, is found experimentally by small angle neutron scattering to be 0.26±0.04 nm. The value obtained purely from the theory is 0.21, which is quite close to the range of the experimental value.

On the surface, the number of cores (adsorbed core, AC) n_{AC} is:

$$n_{AC} = \frac{a_P}{a_N}\left(2\frac{P}{N}\right)^{1/2} \quad [7.11]$$

The layer thickness of the adsorbed polymers is then given by:

$$R_{AC} = \left(2\sqrt{2}(1-2\chi)\frac{a_P}{a_N}\right)^{1/5} a_P P^{7/10} N^{-1/10} \quad [7.12]$$

While this gives a good prediction of layer thickness over a broad range of polymer structures, deviations were noted in particular for very short side chains (about ten units) (Flatt *et al.*, 2009a). This can be understood because of their length being too short for coiling. While one could expect that this would be apparent up to much longer chain lengths, the data consistently show that the derived scaling law accounts well for the thickness of the measured layers. The surface occupied by each molecule is given by:

$$S = \frac{\pi}{\sqrt{2}} a_N a_P \left(2\sqrt{2}(1-2\chi)\frac{a_P}{a_N} \right)^{2/5} P^{9/10} N^{3/10} n \qquad [7.13]$$

Results by Flatt *et al.* (2009a) also gave a good correlation for this expression. A comparison of variations in adsorption plateau on a given substrate for different polymers is achieved by rewriting Eq. 7.13 with the molar mass of the polymer written as $n(M_N N + M_P P)$, where M_N and M_P are the molar masses of the backbone and side chain monomers respectively. This gives:

$$m_{sat} \approx \frac{M_N N + M_P P}{P^{9/10} N^{3/10}} \qquad [7.14]$$

For a methacrylate backbone, $M_N = 86$ g/mol and for PEG side chains, $M_P = 44$ g/mol. It can be seen that for side chains on the order of 23 units (1000 g/mol), the contribution of the backbone to the numerator is only minor so that we get:

$$m_{sat} \approx P^{1/10} N^{-3/10} \qquad [7.15]$$

While only an approximation, this equation has the benefit of emphasizing the fact that the amount of polymer adsorbed at saturation plateau does not change much with the polymer structure. For example, an increase in side chain length by a factor of ten would only increase the adsorption plateau by 20% ($10^{1/10}$). Also if it covers the range of values of N in Fig. 7.18, which are from 5 (C/E=4) to 2.5 (C/E = 1.5), the adsorption would increase also by 20%. Both the examples emphasize the fact that, in absence of competitive adsorption and at the saturation plateau, variations in the molecular structure do not have a big effect on the mass of adsorbed polymer. However, it is essential to emphasize that the situation is quite different when the plateau is NOT reached because too low dosages are used. This is the case of the adsorption data in Fig. 7.18 and largely accounts for the strong variation of adsorption of those data when N is changed.

7.4 Superplasticizers and rheology: microscopic behaviour

7.4.1 Dispersion at full surface coverage

The now classical approach for determining interparticle forces consists of an extended DLVO approach in which interparticle forces are assumed to be additive. Force or potential balances are performed and the effect of changes induced by adsorbed polymers is discussed. Given that the particles considered are not Brownian (Flatt, 2004b; Roussel *et al.*, 2010), it is more appropriate to examine interparticle forces than potentials. From this and using basic principles, a relation between yield stress and layer thickness can be derived (Flatt and Bowen, 2006, 2007).

Such calculations typically include a van der Waals force, for which an adequate Hamaker constant is needed (Flatt, 2004a). They also involve an electrostatic force, of which the magnitude decreases strongly with the electrolyte concentration. For cement suspensions, the range of this force ends up being very small (Flatt and Bowen, 2003). The steric hindrance force from the polycarboxylate has been calculated from different forms. The most recent of which was derived from atomic force microscopy (AFM) measurements (Flatt *et al.*, 2009a). The expressions used typically assume a rather high surface coverage by the adsorbed polymer. The result is that the maximum attractive force occurs just about at the start of the overlap of the adsorbed layers. This force is therefore more or less equal to the van der Waals force at a separation distance equal to twice the adsorbed layer thickness. A convenient consequence is also that the details of the expression used for the steric force are not important.

However, these calculations generally assume a complete or at least high surface coverage. This situation is the easiest to calculate as it assumes that interparticle forces are homogeneous all over the particle surfaces. However, this situation can rarely be measured in practice without introducing some artefact or major perturbations. Indeed, owing to their size, cement particles tend to sediment if they are fully dispersed (Flatt, 2004b; Roussel *et al.*, 2010). To avoid sedimentation, it is possible to increase the volume fraction of solids. In order to obtain a system stable at full surface coverage, the limiting volume fraction above which the direct contacts control the response of suspensions is probably reached (Roussel and Coussot, 2005; Roussel *et al.*, 2010). This change in volume fraction could also be achieved by adding finer particles.

This problem can be overcome in consolidation experiments where segregation is used to our advantage. In such experiments, the degree of dispersion is determined based on the degree of compaction of a suspension submitted to centrifugal forces (Bergström *et al.*, 1992). The principle of these experiments is illustrated in Fig. 7.23.

The packing density in such experiments is related to a compressional yield stress. Neglecting a change in system behaviour due to the friction/lubrication regime that must be approached towards the end of the consolidation, this yield stress is related to the shear yield stress. This makes it possible to propose a relationship between the packing density obtained and the layer thickness. In turn this can be related to the polymer structure with the following relationships, previously developed in a paper dealing with the consolidation of MgO suspensions dispersed with superplasticizers of varying structures (Kjeldsen *et al.*, 2006). The compressive yield stress is written according to YODEL (Yield Stress mODEL) (Flatt and Bowen, 2006):

$$\tau_0 \approx f(\phi)F_{max} \qquad [7.16]$$

where $f(\varphi)$ is a function including the dependence on the volume fraction of solids φ. This includes a maximum packing fraction φ^* and a percolation threshold φ_0. As the packing fractions discussed are high, the latter can be neglected giving:

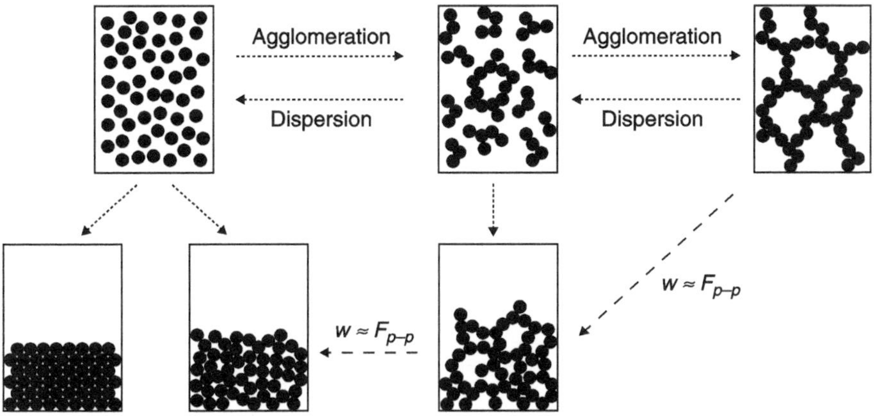

7.23 Schematic illustration of the relation between dispersion degree and consolidation. On the top: suspensions of different degrees of dispersion. The one to the right does not consolidate under its own weight but can when submitted to centrifugal forces. The other suspensions compact to degrees that depend on their initial degree of flocculation. In an ideal case, one could expect a random close pack situation (left), but in most cases one would get a random close packed system (bottom centre).

$$f(\phi) \approx \frac{\phi_3}{\phi^* - \phi} \qquad [7.17]$$

The maximum packing fraction can be estimated based on the following observation. As the side chain length increases, so does the dispersion, but it approaches a plateau value which in this case is about 0.51. The same value is reached in absence of admixtures when the centrifugal speed is increased by a factor 10 (Kjeldsen *et al.*, 2006). Equation 7.17 can therefore be determined for each admixture using the measured degree of consolidation for ϕ.

The maximum attractive force is assumed to come from a non-retarded van der Waals force. It varies with the inverse square of the separation distance, which we can express as a function of the layer thickness of the adsorbed polymers:

$$F_{max} \approx \frac{1}{R_{AC}^2} \qquad [7.18]$$

with Eq. 7.12, this gives:

$$F_{max} \approx P^{-7/5} N^{1/5} \qquad [7.19]$$

The structural parameters of the polymers used in Kjeldsen *et al.* (2006) can be determined from the data given by these authors. The obtained values for P, N and n are shown in Table 7.1. With this in hand we can further use equation 7.16. In particular, we first note that in these consolidation experiments, the stress is the same in all samples at the bottom of the tube. This means that we expect theoretically to

Table 7.1 Structural parameters of superplasticizers used in consolidation experiments by (Kjeldsen *et al.* 2006) as well as maximum attractive force and measured packing density.

Polymer	P	N	n	F_{max}	ϕ
PC 1	23	4.3	16.8	0.494	0.462
PC 2	23	4.2	83.0	0.492	0.459
PC 3	45	7.5	9.5	0.210	0.485
PC 4	114	7.7	10.8	0.058	0.500
PC A	13	2.2	47.3	1.000	0.432
PC B	227	22.5	3.8	0.027	0.501

Note: force is calculated with Eq. 7.19 and normalized by the largest value of the series.

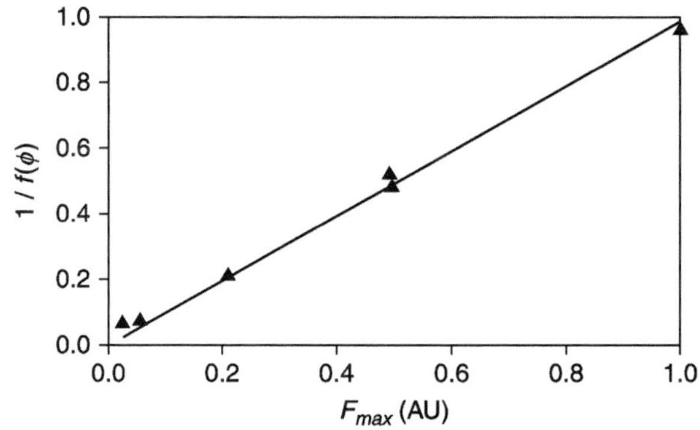

7.24 Plot of maximum attractive interparticle force vs. the inverse of the volume fraction function. Adapted from Kjeldsen *et al.* (2006) using the force calculated with Eq. 7.19 and normalized by the largest value of the series.

find that the maximum attractive force is inversely proportional to the volume fraction function given in equation 7.17. This is indeed what we obtain in Fig. 7.24 is a strong indication that the overall approach is consistent. It indicates that in situations where full surface coverage is obtained, the evaluation of interparticle forces assuming a minimal separation distance equal to twice the adsorbed layer thickness is acceptable. Problems start when full surface coverage is not achieved, as is the case in many practical situations in order to counter segregation by some degree of flocculation.

7.4.2 Dispersion at incomplete surface coverage

Various factors can contribute to a reduced surface coverage. The first is obviously an insufficient dosage. However, even with an *a priori* sufficient dosage, the surface of cement particles may not be well covered by the added superplasticizer, for example the presence of a species that competes for surface sites (or for calcium). Examples of these would include minerals as sulfates or hydroxides or ionic organic compounds as other admixtures present in a formulated product (e.g. additional dispersant or retarders). Ongoing hydration reactions can consume the polymer by adsorption onto freshly created surfaces. In the case of organo-aluminate phases, they may consist in an 'internal' adsorption within the layered structure of hydrocalumnite as explained in Section 7.6.2.

The same study on centrifugal consolidation, when one of the polymers was added at different dosages, demonstrates the difficulty of this problem (Kjeldsen *et al.*, 2006). As the dosage was increased, the packing density increased to approach a maximum characteristic of the dispersion obtainable with that polymer at full surface coverage.

It was hypothesized that the average interparticle force could be determined by taking into account the statistical distribution of contacts between areas with or without polymer:

$$F_{\max} \approx \left(\frac{\theta^2}{\delta_P^2} + 8\frac{\theta(1-\theta)}{(\delta_P+\delta_0)^2} + \frac{(1-\theta)^2}{\delta_0^2} \right) \quad [7.20]$$

where θ is the surface coverage, δ_P is the layer thickness at full surface coverage and δ_0 is an equivalent layer thickness at zero surface coverage, which corresponds to half the minimal separation distance in absence of polymers. To make use of Eq. 7.20, some assumptions need to be made to obtain the missing parameters. First of all we need to estimate the value of δ_0. This is done by extending the line in Fig. 7.24 to the value of $1/f(\phi)$ that corresponds to the packing fraction without admixtures (0.36). From the associated force and 'layer thickness' we get δ_0. The second term we need to obtain is the saturation plateau, as Eq. 7.20 is given in terms of surface coverage $\theta = \frac{m}{m_{sat}}$. To do this we examine the situation at low surface coverage. There, based on the measurements of Perche (2004), the amount of adsorbed polymer can be assumed to vary linearly with dosage. Moreover, at low dosage, Eq. 7.20 simplifies to:

$$F_{\max} \approx (1-\theta)^2 \quad [7.21]$$

We can therefore expect that $\sqrt{1/f(\phi)}$ should decrease linearly with dosage at low dosages. This is indeed the case as shown in Fig. 7.25. It can be shown that the ratio of $\sqrt{1/f(\phi)}$ obtained in absence of polymers with the slope of the regression line in Fig. 7.25 gives m_{sat}, which in this case turns out to be 3.1 mg/g in this specific case.

With this in mind it is possible to estimate surface coverage for dosages lower than m_{sat}. Around and above this value the adsorption isotherm is no more linear and θ can no longer be as easily inferred from dosage. With these assumptions the experimental results from (Kjeldsen *et al.* (2006) can be presented in terms of contributions of contacts between bear surfaces (B–B), between bear surfaces and coated surfaces (P–B) as well as between coated surfaces (P–P):

The results in Table 7.2 show that the first two points, and in particular the one at a dosage of 0.4 mg/g, are largely dominated by contacts between surfaces

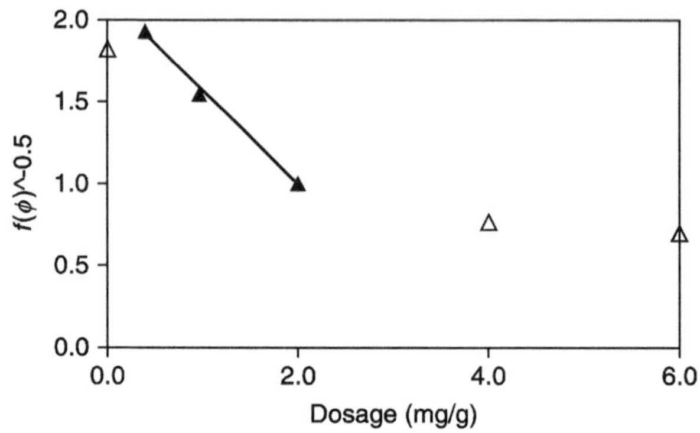

7.25 Plot used to estimate the saturation plateau m_{sat} needed to calculate surface coverage. Details are given in the text.

Table 7.2 Distribution of forces between contacts: polymer–polymer (P–P), polymer–bear surface (P–B) and bear surface–bear surface (B–B)

Dosage	Force fraction			
(mg/g)	P–P	P–B	B–B	Packing
0	0%	0%	100%	0.358
0.4	0%	8%	92%	0.35
1.0	2%	19%	79%	0.38
2.0	13%	40%	47%	0.43
4.0	74%	23%	3%	0.455
6.0	100%	0%	0%	0.462

Note: The line for 4.0 mg/g uses an estimated surface coverage of 0.93.
Source: Kjeldsen *et al.* (2006)

without polymer. This explains why the packing at that dosage is very similar to the one without admixture. This is also reflected by the fact that this point deviates from the extended linear regression in Fig. 7.26.

These data show that the approach relating interparticle forces to yield stress can be extended to incomplete surface coverage. However, the correlations become substantially less good. Moreover various assumptions must be made to estimate the minimum separation distance in absence of admixtures. Extending this approach to normal rheology, would require to performing yield stress measurements at the same solids volume fraction while varying the admixture dosage. This is often not easy because the solids volume fraction must be high to avoid segregation in the paste with the polycarboxylate dispersant. The second limitation is that information is needed concerning the adsorption isotherm. In advantageous cases, this might lead to a linear relation at low dosages between dosage and surface coverage. At high surface coverages, however, a more detailed knowledge of the amount adsorbed is needed and in particular under the

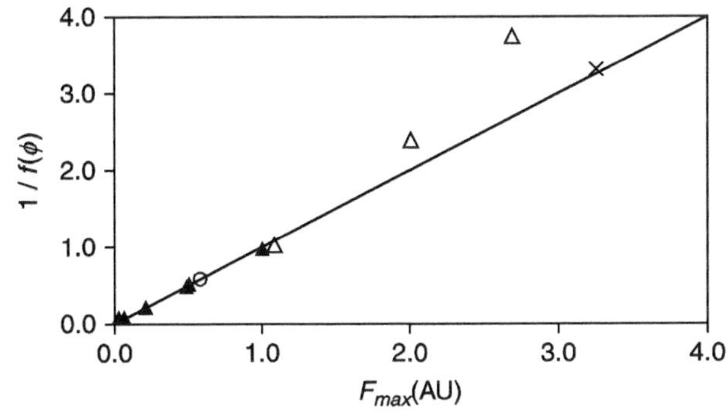

7.26 Extended linear regression from 7.24 (filled triangles). The dosages of PC1 of 0.4, 1.0 and 2.0 mg/g are represented by the open triangles. Their surface saturation is estimated by the linear relation described in the text. The point at 4.0 mg/g dosage uses a surface coverage of 0.93 that fits the point of the linear regression (empty circle). Finally, the cross corresponds to the sample without admixture, for which the force was back-calculated to fit on the line for its measured packing of 0.36. Adapted from Kjeldsen *et al.* (2006).

conditions of the rheological measurement. More work is needed in this area, particularly with cementitious systems.

7.5 Superplasticizers and rheology: macroscopic behaviour

7.5.1 Dosage impact on flow

In general, the higher the dosage of a plasticizer, the higher is the possible water reduction or the flow enhancement, depending on what the user wants to influence. The minimum superplasticizer dosage that shows a dispersing effect can be defined as the critical dosage. Below this dosage, the effects of the superplasticizer are not important, or a least not clearly observable under the specific testing conditions for the mix being evaluated. Above this dosage, the flow will continuously increase up to a certain level at which surface saturation is reached (Fig. 7.27) (Yamada *et al.*, 2000b). Beyond this saturation dosage, a further increase in the dosage has only minimal influence on the flow and a tendency towards segregation, bleeding and delay in hardening can be observed (Aitcin *et al.*, 2001). Segregation and bleeding depend, however, a lot on the volume fraction of fines and their size distribution.

The critical dosage and the increase of flow per dosage unit depend on the type of cement, the W/C and the type of superplasticizer. The saturation dosage has similar dependancies, but is not or much less dependent on W / C. Superplasticizers of the LS, PNS and PMS type have a higher critical dosage, i.e. they need a higher dosage to induce a flow of the paste than superplasticizers of the PCE type.

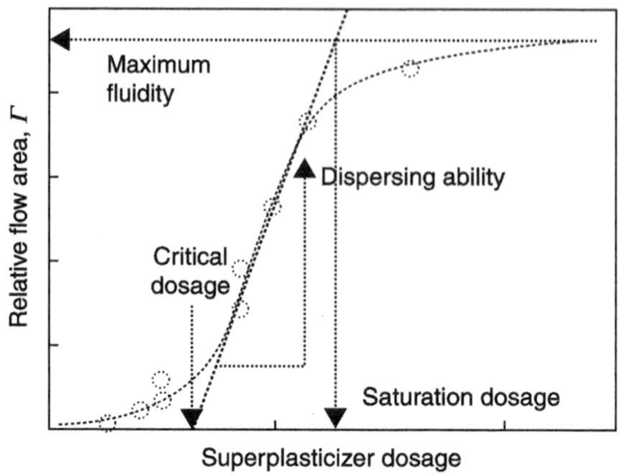

7.27 Critical dosage and dispersing ability. Reprinted from Yamada *et al.* (2000a) with permission from ACI.

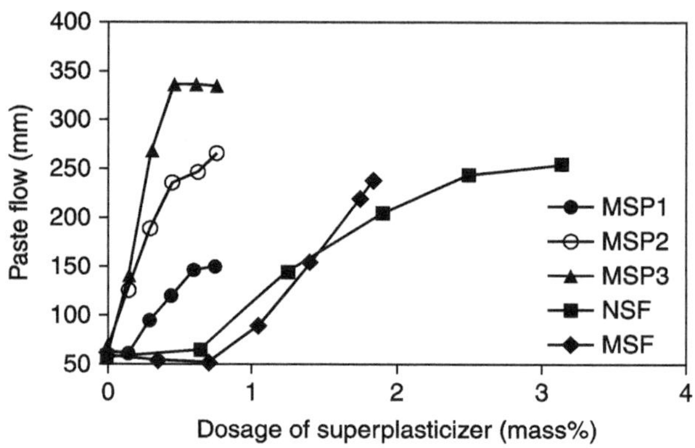

7.28 Relationship between the dosage of superplasticizers and the initial flow of a cement paste. NSF = PNS, MSF = PMS, MSP1-3 are PCEs. Reprinted from Nawa *et al.* (2000) with permission from ACI.

Furthermore, PCE type polymers usually induce a higher increase of the flow per dosed unit (Fig. 7.28).

Another factor that limits the use of some superplasticizers is the packing density of the solids. At high packing density (i.e. at low W/C), the number of interparticle contacts increases substantially. For a given interparticle force, the yield stress diverges as the maximum packing density of the solid is approached (Flatt and Bowen, 2006). To be effective at low values of W/C, superplasticizers must therefore be able to strongly reduce the interparticle forces. At higher W/C an equivalent reduction in interparticle forces can be reached by a low surface coverage of a strong dispersant or by a higher surface coverage of a less effective dispersant. At low W/C, however, one needs a high surface coverage by a good dispersant. Less effective dispersants are no longer capable of providing a sufficient reduction in the average interparticle force at low W/C. This is illustrated by a test described by Kinoshita and Okada (1999): A cement paste of W/C = 0.30 plasticized with PNS or PCE type superplasticizers shows only minimal difference in the performance between different PCE types and the performance of the PNS superplasticizer in this paste is fair. Lowering the W/C to 0.20, however, results in quite different plasticizing effects of the different PCE types and the PNS polymer is ineffective, even at high dosage level as shown in Fig. 7.29.

The different performance of PCEs is due to the variations in their structure i.e. number of ionic groups, length of the side chains, length of the backbone or number of side chains per molecule. These structural features affect, amongst other things, the critical dosage, the efficiency and the saturation level in a cement paste. Figure 7.30 shows the different performance of PCE polymers in two cements. The structures of the tested PCEs are given schematically in Fig. 7.16. PCE polymers with more ionic groups in the backbone, due to lower grafting degree or increased length of the backbone, adsorb more than PCEs with only a few ionic groups. At high volume fraction of fines in suspension, polymers that strongly adsorb and polymers with long side chains can, therefore, better counteract the attractive forces than polymers with lower adsorption and shorter side chains (Schober and Mäder, 2003; Winnefeld *et al.*, 2006).

The critical dosage and efficiency of a superplasticizer are also dependent on the type of cement. In pastes of cements with higher content of aluminates, the critical superplasticizer dosage is higher compared to a paste prepared with a standard cement of comparable specific surface (Fig. 7.30). This can be explained by the ability of organic molecules to interact in early hydrate phases, mainly with the

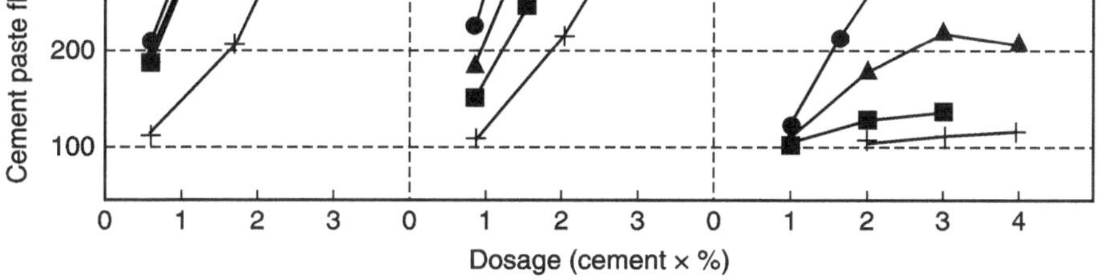

7.29 Fluidity of cement paste at different W/C. NSF = PNS, MSP1-3 are PCEs. Reprinted from Kinoshita and Okada (1999) with permission from RILEM.

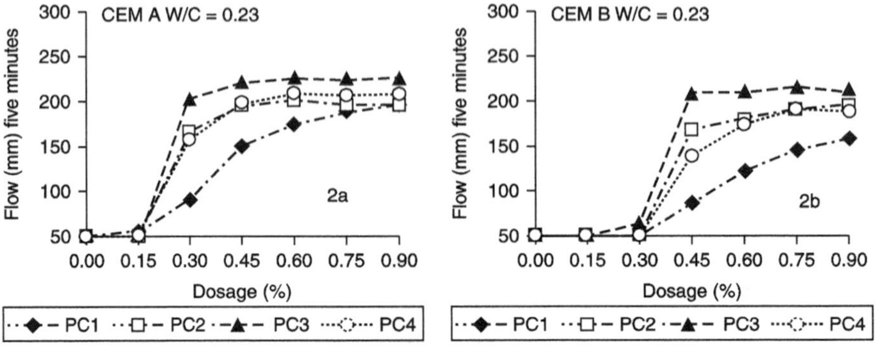

7.30 Initial flow of a cement paste with PCEs with increased dosage. CEM A is a standard NPC (left) and CEM B, a high C3A cement (right). Reprinted from Schober and Mäder (2003) with permission from ACI.

aluminate phases, possibly by intercalation or modification of the specific surface of the newly formed hydrates. Not only small molecules but also polymers like PCEs can intercalate. (Plank *et al.*, 2006b). This intercalation reduces the amount of polymer available for adsorption on the cement surface (Flatt and Houst, 2001). Therefore a higher superplasticizer dosage is needed to compensate for the intercalated amount in order to obtain the desired flow.

The same correlations between efficiency of a superplasticizer and its structure can be drawn from measurements of the yield stress of a cement paste since yield stress and flow of a paste are directly correlated (Flatt *et al.*, 2006). The lower the side chain density (i.e. the higher the number of carboxylic groups in the backbone), the higher is the reduction of the yield stress with dosage (Fig. 7.31).

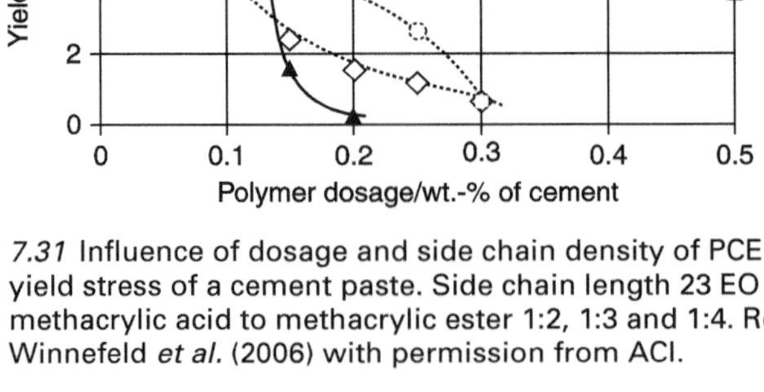

7.31 Influence of dosage and side chain density of PCE polymers on the yield stress of a cement paste. Side chain length 23 EO units, ratio of methacrylic acid to methacrylic ester 1:2, 1:3 and 1:4. Reprinted from Winnefeld *et al.* (2006) with permission from ACI.

Polymers with very high efficiency (i.e. high increase of the flow with only low increase of the dosage) like PCE 23-1:4 in Fig. 7.31, however, will be very sensitive to over dosage. Indeed as shown in figure 7.31, they can show a sharp drop of yield stress over a narrow range of dosages.

7.5.2 Dosage impact on slump life

Fresh concrete is well known to lose its workability with time. This phenomenon is called 'slump loss'. To improve slump retention with PNS or PMS, of which the action is often limited in time, addition of set-retarding water reducers, typically lignosulfonates, was found to be quite adequate. The sometimes fast stiffening of concrete with PNS was explained as an incompatibility with cement and can be overcome to some extent by addition of alkali (see Section 7.2.1 on polynaphthalene sulfonates).

Concretes produced with vinyl copolymers showed extended slump life compared to PNS, PMS or LS superplasticizers. But nowadays vinyl copolymers are mostly replaced by the PCE polymers with higher water reduction and significantly better slump retention. Furthermore, with PCE polymers the slump retention is dependent on structural factors and can more easily be adjusted according to needs than with the linear vinyl copolymers.

Previously we have demonstrated how important the adsorption of a dispersant on the cement surface is. The amount of adsorbed polymer at a certain dosage and the amount of polymer necessary to reach surface saturation depends on the specific structure of the superplasticizer. Figure 7.32 shows the adsorption curves of two PCE polymers, one with high adsorption and high plasticizing power, the

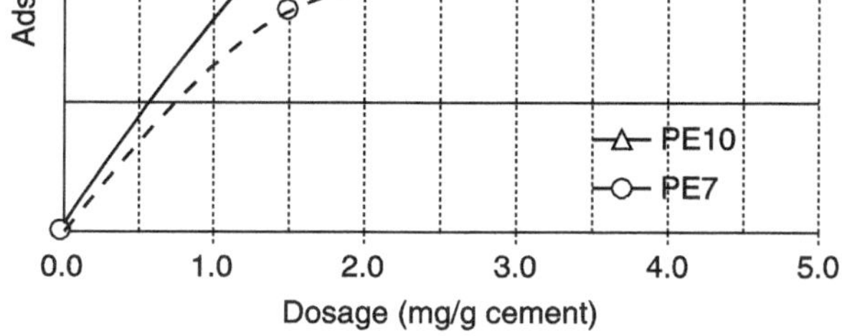

7.32 Adsorption and estimated dosage for surface saturation of two PCE polymers on cement after 30 minutes adsorption time. Reprinted from Schober and Flatt (2006) with permission from ACI.

other with lower adsorption and lower plasticizing power. The dosage estimated to achieve complete surface coverage is indicated by arrows. If the polymer dosage is far below this critical amount, a short slump life of the concrete may result. If the dosage is high enough to reach almost full surface coverage, a good slump life of the concrete can be expected. This was demonstrated by concrete tests performed with dosages below and above full surface coverage (Schober and Flatt, 2006). When PCE PE10 was dosed below its saturation dosage, rapid stiffening of the concrete was the consequence (Fig. 7.33a). In a second concrete batch (Fig. 7.33b), the dosage of PE10 was increased to almost its saturation level and a perfect slump life was the result. Polymer PE7, which has a lower number of carboxylic groups, needed a higher dosage to obtain the desired initial flow. At this dosage the saturation level was almost reached and a fair slump life resulted. The slump life was further improved at higher dosage. PCE types, however, may retard the setting of cement – some strongly, some only marginally. Undesired strong retardation of the setting at higher dosage can limit the use of some types at or beyond their saturation level.

Slump retention was also found to depend on the amount of dispersant remaining in the aqueous phase. For concrete with an extended slump life, PCE polymers with lower initial adsorption (i.e. PCE polymers with higher grafting density) should be preferred to fast and high adsorbing PCE polymers (Vickers *et al.*, 2005).

Another possibility for increasing slump life is to use reactive polymers. These polymers change their structure in the alkaline environment of cement and gradually exhibit dispersion effects. Examples of such polymers are cross-linked polymers in which the cross-linking bonds are cleaved and plasticizer molecules are 'released'. Other examples for reactive polymers are PCE polymers containing structures derived from acrylic esters. These esters are hydrolysed by the alkaline

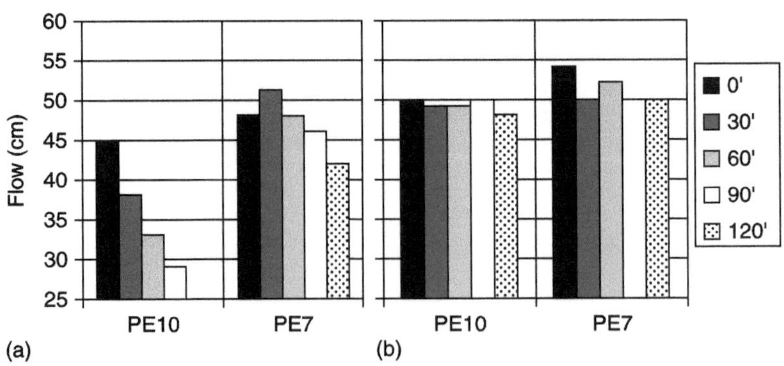

7.33 Flow of concrete with two PCE polymers: (a) W/C=0.39, 0.18% PE10, 0.21% PE7; (b) W/C=0.36, 0.25% PE10, 0.53% PE7. Reprinted from Schober and Flatt (2006) with permission from ACI.

pore solution and thus generate carboxylic groups which drive further adsorption. Since this hydrolysis is time dependent, PCEs with a higher adsorption rate are continuously formed and can adsorb on the newly formed surfaces of cement reaction products. Therefore these types of polymers are excellent slump keepers if used alone or in combination with other superplasticizers.

There are different strategies for the incorporation of such labile ester groups in the PCE polymer. Either all PEG side chains are connected via acrylic ester bonds to the backbone. In this case, PEG chains are released with time and a polymer with reduced or almost no grafting is formed. Other polymers make use of acrylic as well as methacrylic ester bonds to connect the PEG side chains. In this case, the methacrylic esters will not or only to a small extent be hydrolyzed. Here PCE polymers with increased ionic content will be formed with time, but a comb polymer will still remain at the end. New polymer developments have incorporated short chain acrylic esters into the PCE polymers. These esters, mostly ethyl esters, are easily cleaved in the cementitious medium releasing free carboxylic groups. In contrast to the other hydrolyzable PCE polymers, however, in this case the grafting degree (i.e. number of PEG side chains per molecule) will not be changed with time.

7.5.3 Relation between flow and adsorption

Adsorption of the dispersant on the surface of a particle induces repulsive forces when particles come in closer contact. The power of the repulsive force depends on the effective amount of superplasticizer on the surface and its structure in its adsorbed form. The amount of adsorbed polymer is driven by the system's gain in energy (entropic and enthalpic) during adsorption. This again depends on the structural features of the polymer and the surface.

A direct correlation of the adsorbed mass of methacrylate based PCE polymers at a dosage near the saturation dosage and the flow of the cement paste was shown by Schober and Flatt (2006). The adsorbed masses of six PCE polymers with different side chain length, grafting degree and length of the backbone and the flow of the cement pastes could be correlated linearly (Fig. 7.34). A rough correlation of adsorbed mass of PCE polymer with the yield stress of the cement paste can also be drawn from the data given in Winnefeld *et al.* (2006) as shown in Fig. 7.35.

This again demonstrates the importance of adsorption of the dispersant on the cement surface. However, this direct correlation of the mass of adsorbed polymer on the cement with the flow of the paste, independent of the polymer structure, cannot be generalized. The effect of the adsorbed polymer mass on the flow properties of cement or concrete depends, of course, on the type of polymer (PNS, PMS or PCE), on the type of cement and especially on the W/C. In addition, such linear correlation as demonstrated before could not be found for flow table spread measurements of mortar or concrete.

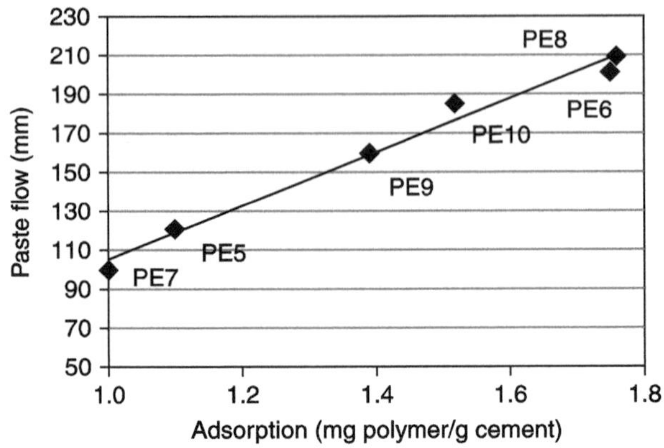

7.34 Correlation of adsorbed polymer with flow of the cement paste (Schober and Flatt, 2006).

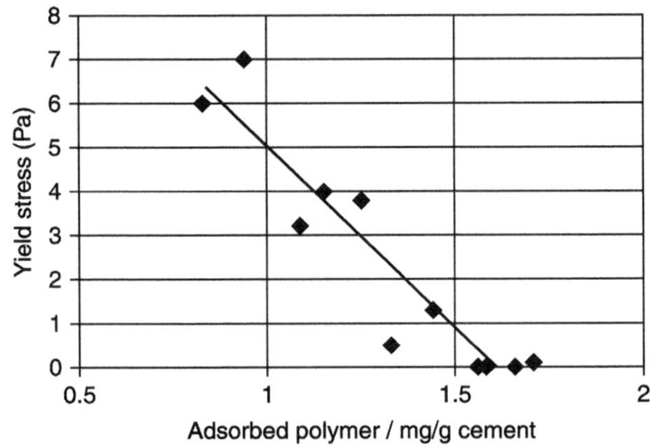

7.35 Correlation of adsorbed polymer with yield stress of a cement paste (Winnefeld *et al.*, 2006).

7.5.4 Effect of addition time and mixing conditions

The performance of superplasticizers is influenced by many factors. Some of them are well-known, others like the influence of soluble alkali sulfates have only been recognized in the last few years. The influence of time of admixture addition has been known for decades but is mostly underestimated in practice when data produced in laboratory are compared with data from the job site. In practice, the cement is first blended with wet aggregates and the admixture is added during the addition of the mixing water. In many tests in the laboratory, the admixture is added with the mixing water to the dry blend of cement and aggregates. This wetting of cement before the addition of the superplasticizer is an important difference.

When lignosulfonate or PNS or PMS superplasticizers are added two minutes after the addition of the mixing water not only is the flow increased but the retardation is also much higher than for immediate addition. A delayed addition also causes a lower reduction of workability with time (Dodson and Farkas, 1964; Chiocchio and Paolini, 1985). The adsorption of PNS and PMS for delayed addition was found to be lower than for immediate addition. This increase in flow, reduction of yield stress, reduction of adsorbed mass of superplasticizer and increased hydration time with delayed addition is more pronounced with PNS superplasticizers than with PCE polymers (Uchikawa *et al.*, 1995; Winnefeld, 2008).

As seen, after delayed addition of a superplasticizer the flow is higher than after immediate addition but the adsorbed mass of the superplasticizer is lower. This contradicts the previously shown correlation of a high adsorption with high flow and vice versa. What is the reason for this contradiction?

What is generally measured by adsorption experiments is the depletion of a superplasticizer in the aqueous phase of a cement paste (Bürge, 1999; Yamada *et al.*, 2001). The difference in the polymer concentration in the mixing water and in the pore solution is generally taken as the amount of adsorbed polymer. But, superplasticizers like PNS or PCE can be intercalated in early cement hydrates (Flatt and Houst, 2001; Plank *et al.*, 2006b) and we cannot distinguish between adsorbed and intercalated admixtures. What we measure is the sum of adsorbed and intercalated polymer. In delayed addition mode, less or no organo-mineral phases are formed and almost no superplasticizer is 'lost' by intercalation. Less dosage is necessary to reach the same mass of 'really' adsorbed polymer on the surface with delayed addition than with immediate addition. This explains the apparent contradiction of lower dispersant adsorption but higher flow of pastes produced by delayed addition. Let us note that some intercalated phases are unstable, so that they may only be transient. This, however, does not alter the possibility of PCEs interacting in early hydration reactions of aluminate phases. As a result one can expect that the specific surface of the system would increase, leading to a higher adsorption requirement and lower overall performance. Whether this surface is coming from layered particles or an ensemble of dispersed particles is only of secondary importance as discussed later.

Overall, it is important to recognize that the same effect can be generated with a delayed addition mode as with an immediate addition of superplasticizers that show slow adsorption. Polymers with initially low ionic charges or polymers that are cross-linked are slow in adsorption. Thus these polymers will not be intercalated, modify aluminate reactivity or be adsorbed on the early hydrates. Therefore their concentration in the aqueous phase will be higher after the initial aluminate reactions than for polymers that are more easily adsorbed. This higher concentration eventually leads to a higher amount of polymer that are able to adsorb on the silicate and hydrated aluminate phases, leading to a higher flow and increased slump retention. Such slow adsorbing or reactive polymers are excellent slump keepers and are used alone or in blends with other polymers. Intercalation of polymers in early cement hydrates and the formation of organo-mineral phases are described in more detail in Section 7.6.2.

Besides the mode of superplasticizer addition, the mixing speed or intensity influences superplasticizer adsorption on cement and thus slump and slump life. The adsorption of PCE superplasticizer in a concrete mix was found to be higher than in a cement paste mixed at 700 rpm with identical mixing time, W/C and dosage. However, the relative differences of adsorbed masses of the different PCEs measured in the cement paste correlated with the results from concrete. Therefore the adsorption results from the cement paste can be used to differentiate their behaviour (Vickers *et al.*, 2005). This higher adsorbed amount of polymer found in the concrete can be explained by the higher shear forces generated during mixing of a concrete. These higher shear forces better disperse the cement particles and might even remove early hydration products from the surface, thus generating a higher surface area that can adsorb superplasticizers. More importantly, however, shear forces developed during mixing seem to change the dissolution behaviour of anhydrous phases (Juilland, 2009) (Julliand *et al.*, 2010). A coherent picture is in fact now emerging on how the formation of etch pits in mixing stage affects later hydration. The sensitivity to mixing appears to be enhanced in presence of admixtures.

The mixing time may also influence the flow behaviour of a cement paste or mortar plasticized with PCE superplasticizer. Besides other changes, prolonged mixing time was shown to be able to reduce the effect of over-fluidification seen with some PCE polymers (Regnaud *et al.*, 2006). This over-fluidification may occur with PCEs having a slow initial adsorption and a higher relative increase of this adsorption over time.

7.5.5 Role of temperature

In Section 7.3.1 we saw that an increase in temperature could enhance the adsorption of at least some PCEs. While these results were obtained on inert suspensions, it is interesting to note that Fernàndez-Altable and Casanova (2006) and Fernàndez-Altable (2007) observed a reduction in yield stress for cement suspensions dispersed with a PCE (Fig. 7.36). Equally interesting is the fact that Fernàndez-Altable observed the opposite effect when using a PNS (Fig. 7.37). Here the yield stress increases with temperature. Moreover, this behaviour is more pronounced in direct than in delayed addition mode. This difference probably comes from the interaction with the aluminate phases and most likely is linked to a higher specific surface of aluminate hydrates. Thereby, they immobilize more PNS, either by adsorption or intercalation, which leads to a lower overall surface coverage and poorer dispersion.

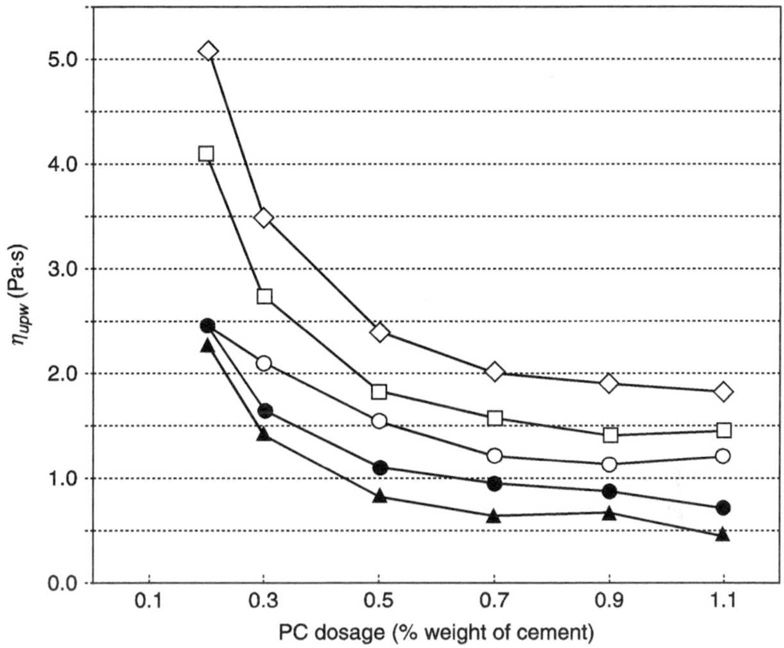

7.36 Yield stress of cement suspensions dispersed with a PCE. Data show values at different temperatures as a function of dosage. η_{upw} is the viscosity obtained from the upward flow curves (Fernàndez-Altable and Casanova, 2006).

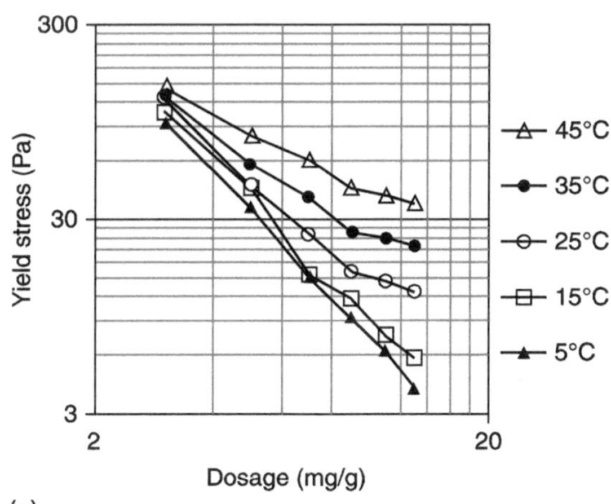

7.37 Yield stress of cement suspensions dispersed with a PNS. Data show values at different temperatures as a function of dosage. Results in (a) direct addition;

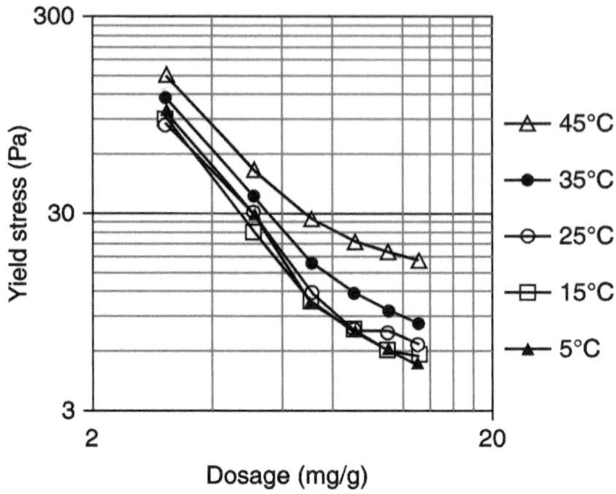

(b)

7.37 Continued (b) delayed addition (Fernàndez-Altable, 2007).

These examples show that temperature effects are not trivial and are very polymer dependent. However, they can in first approximation be rationalized by adsorption behaviour as well as aluminate phase reactivity.

7.6 Superplasticizers and cement chemistry

7.6.1 Sulfate competition

Cement chemistry can affect the dispersion efficiency of superplasticizers in various ways. One of these is through the ionic composition of the aqueous phase and in particular through the amount of soluble sulfates. The milestone paper on this subject comes from Yamada and Hanehara (2001). As explained in Section 7.3.1 on the reversibility of polymer adsorption, they made alternative additions of $CaSO_4$ and Na_2SO_4 that modified the soluble sulfate concentration (Fig. 7.38a) and the flow spread (Fig. 7.38b), as well as the adsorption (Fig. 7.39). The important step in this paper was to plot the relative flow directly with respect to the polymer adsorption (Fig. 7.39b). This emphasizes the reversible nature of the polymer adsorption already discussed above.

As elegant as this experiment may have been, it was only carried out with a single polycarboxylate superplasticizer. In a more recent study, this issue was corrected by conducting the same test with many more polycarboxylates. Results showed that the variability of flow in these tests was very much dependent on the molecular structure parameters of the superplasticizer (Zimmermann *et al.*, 2009).

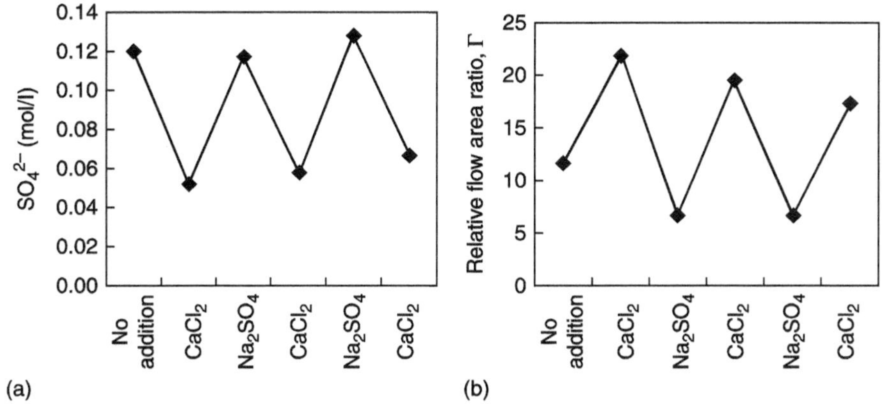

7.38 Variation of soluble sulfate concentration (a) and the relative flow (b) which change with successive additions of $CaCl_2$ and Na_2SO_4. Reprinted from Yamada *et al.* (2001). Copyright 2001, with permission from Elsevier.

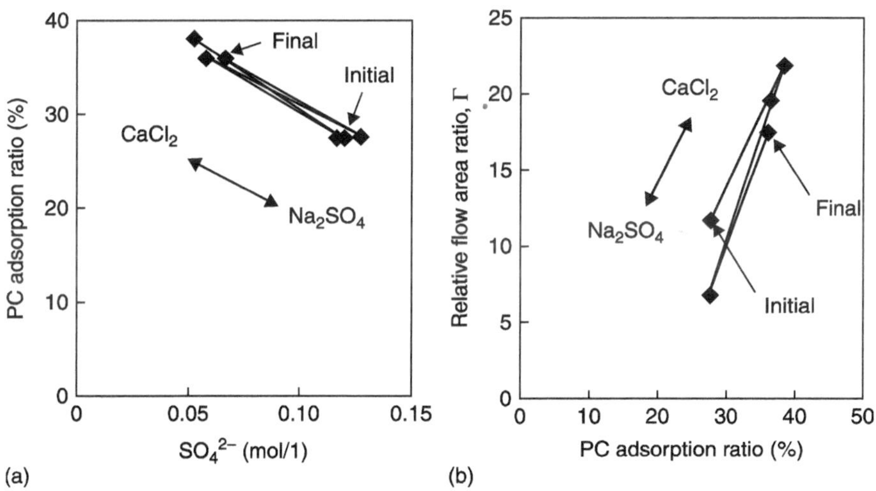

7.39 (a) Variation of polymer adsorption with increase and decrease of soluble sulfate in the solution caused by additions of $CaCl_2$ and Na_2SO_4 respectively and (b) the resulting effect on the flow spread which is dependent on polymer adsorption. Reprinted from Yamada *et al.* (2001). Copyright 2001, with permission from Elsevier.

In this study, the data were analyzed so as to make the different series more directly comparable. Results of flow spread measurements (Fig. 7.40a) were first converted into yield stress (Fig. 7.40b) using an equation that interpolates between the analytical limits of high and low spread taken from Roussel and Coussot (2005) and Roussel *et al.* (2005):

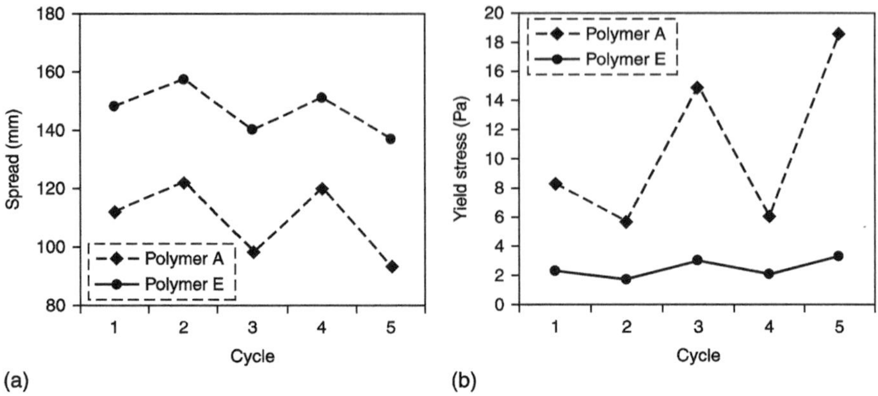

7.40 Example of flow results with two different polymers (a) (adapted from Zimmermann *et al.* (2009)) and the corresponding yield stress values (b).

$$\tau_0 = \frac{225\rho g V^2}{128\pi^2} \frac{R^{-5}}{1 + \frac{225\sqrt{3}}{128\pi} V R^{-3}} \qquad [7.22]$$

where R is the spread radius, V the sample volume, ρ its density and g the gravity constant. For each series of data, the average over all cycles was calculated. The starting point was left out because of a somewhat atypical behaviour that can also be seen in Yamada and Hanehara (2001). Each data series was then normalized with respect to its own average, which makes it easier to compare among admixtures with different performances.

Because the yield stress increased over time for some polymers, a linear regression was then taken, again leaving out the starting point (Fig. 7.41). From this average evolution of the yield stress, the average caused by the salt additions was determined. This was referred to as the sulfate sensitivity parameter (SSP). Low SSP values mean a low sensitivity and high SSP values mean a high sensitivity to variations in the sulfate contents. The most interesting part of this work was to find that the SSP could be related to the molecular structure, with a relation close to P/N^2 (Fig. 7.42a). A theoretical approach was conducted to understand this relation (Flatt *et al.*, 2009b) and came to the result of the SSP being proportional to $P^{4/5}N^{3/5}(N-1)^{-2}$, which gives a slightly better correlation (Fig. 7.42b).

This result demonstrates the possibility of directly relating the structure of a superplasticizer to its macroscopic performance – here its robustness to variations in soluble sulfates. Similar work would be needed concerning other important properties in order to reach a more global capacity of *a priori* selection of admixtures. One of these concerns the interaction with the aluminate phase developed in the next section.

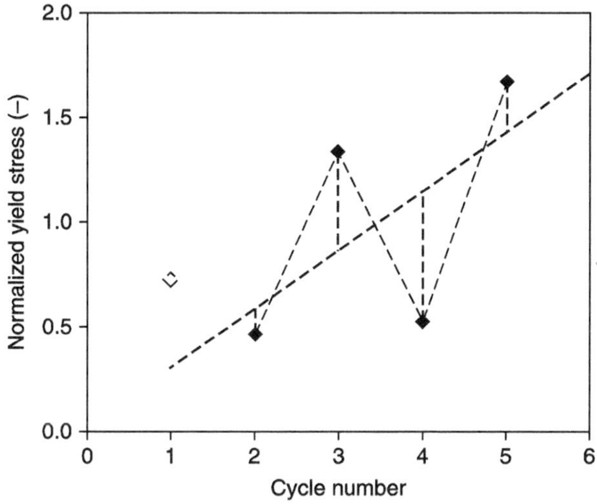

7.41 Example of the linear evolution determined over the cycle and the determination of the departure from this regression (vertical bars). The sulfate sensitivity parameter (SSP) corresponds to the average of these departures. Reprinted from Zimmermann *et al.* (2009) with permission from ACI.

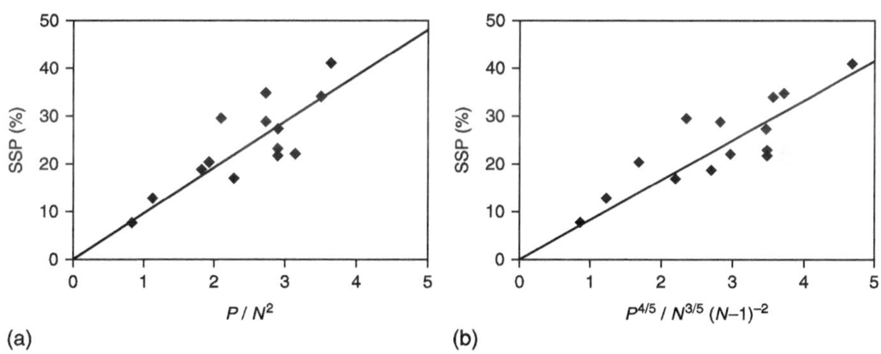

7.42 Relation of the SSP with molecular structure: (a) empirical link to P/N2, (b) theoretical link to $P^{4/5}N^{3/5}(N-1)^{-2}$. Reprinted from Flatt *et al.* (2009b) with permission from ACI.

7.6.2 Organo-mineral formation

General observations and phenomena

The possibility of cement hydration to perturb the action of superplasticizers has been thought about for a long time. In particular, it was well-known that aluminate phases had a big impact (Bonen and Sarkar, 1995). A milestone work on this subject comes from Fernon (1994) and Fernon *et al.* (1997), whose particular target was to understand the reactivity of aluminate in the presence of PNS superplasticizers by carrying out precipitation reactions. The idea was to determine if some organo-mineral compounds were formed in the process. In particular, owing to their layered structure, AFM phases were highly susceptible of being able to intercalate organic admixtures, something that Pöllmann had already shown for small organic acids (Pöllmann, 1992). AFM phases are from the family of layered double hydroxides (LDH), on which many studies have been carried out demonstrating the intercalation of anionic organic compounds (Giraudeau *et al.*, 2009). The main result of the work by Fernon and co-authors was to demonstrate that oligomers of PNS could indeed intercalate (Fernon *et al.*, 1997). However, by X-ray diffraction (XRD), these authors were not able to prove the intercalation of the polymers, although infrared (IR) showed the presence of the admixture in the precipitate.

Some years later, Flatt and Houst presented 'A simplified view on chemical effects perturbing the action of superplasticizers' (Flatt and Houst, 2001). In this review paper, they showed how various 'puzzling' observations of the behaviour of superplasticizers could be explained if an organo-mineral phase formed. This would be the result of the interaction between the admixtures and the highly reactive tricalcium aluminate phase, leading to an 'intercalated, coprecipitated or micellized' compound that could be described as an organo-aluminate. It was in particular argued that this process would take place in the first stages of mixing. The argument was that if the polymer was not present in the mixing water, then the initial aluminate reactions would lead to stable ettringite. In contrast, if admixtures were initially present, then they would be partially 'consumed' by the formation of an organo-aluminate product that would not decompose during typical open working times. The same paper speculated that these processes would not only affect PNS but also PCEs.

The possibility of PCEs being incorporated between hydrocalumnite layers has been clearly demonstrated in recent years by two separate approaches. The first consisted of hydrating C_3A in the presence of PCEs (Plank *et al.*, 2006b, 2006c), while the second followed the precipitation approach of Fernon (Giraudeau *et al.*, 2009; Giraudeau-Lenain, 2009). In the first case, the work was initially performed at high temperature (70°C) in order to favour the layer formation. In the second, the temperature was low (5°C) in order to thermodynamically favour the layered calcium aluminate phase at the expense of the cubic one. Eventually, both groups found that these precautions were not necessary as ambient temperature preparations largely led to the same results (Plank *et al.*, 2006d; Giraudeau *et al.*, 2009). One implication of this is that the presence of PCEs in the system stabilizes the layered phase with respect to the cubic one, leading to stable layered organo-aluminates.

In terms of mechanisms, it was shown that these phases can also form if the polymer is added subsequently to the hydrocalumnite formation (Giraudeau *et al.*, 2009). This happens by a process of dissolution precipitation, in which there is a characteristic induction (nucleation) time that depends on the molecular structure of the PCE.

Finally, an important factor with respect to real cementitious systems is the impact of sulfates. In this case, both groups have concluded that organo-aluminates are not stable with respect to sulfate addition (Plank *et al.*, 2006d). However, as above, a polymer dependent characteristic time exists, which here is concerned with the destabilization of the organo-aluminate (Giraudeau *et al.*, 2009). Under the conditions of these experiments, this was several hours. Transposed to cementitious systems, a similar duration would mean that, if organo-aluminates form (in the first stages of mixing), then they remain stable over the open working time. If they do not remain stable, there conversion would probably lead to a higher specific surface, having a largely similar overall impact on polymer absorption as discussed below. The result is the 'scavenging' of a part of the superplasticizer and a decrease of its effectiveness, as initially proposed by Flatt and Houst (2001).

Structure of organo-aluminates

It appears that the relevance of organo-aluminates to rheology of cementitious systems involves the kinetics of phase formation in the initial stages of mixing as well as sulfate availability. Because of the short time when this process would take place, it is important to emphasize that it is the sulfate availability at C_3A surfaces that is important.

Owing to the sensitivity of this process it is useful to better understand the structure of the phases we are discussing. The studies on this subject that have shown the best layering were generally performed in conditions favouring this and also measured layer spacing on dry or partially dry products. Giraudeau, however, performed XRD measurements at 100% relative humidity (RH) and found the presence of very low angle peaks characteristic of a large interlayer spacing. This layer spacing was found to be about twice the layer spacing predicted by Eq. 7.12. As shown in Fig. 7.43, the slope of R_{AC} versus layer spacing is equal to two and the ordinate at the origin is the interlayer spacing in absence of PCEs. However, this very nice result poses a fundamental problem in terms of the stability of such phases. Indeed, with such interlayer spacing, the residual attractive forces between the layers should be very small. It is therefore surprising that an organized layered structure should form. Nevertheless, this statement is not inconsistent with Giraudeau *et al.* (2009) who reported a stacking distance of only about four layers.

As these experiments were conducted under very clean conditions and because sample preparation might have favoured some alignment, it can be reasonably assumed that such phases would be highly disorganized in cementitious systems. How close they are to being even individual platelets is an open question. The question has an important practical relevance in regard to the additional observation that crystallite sizes are small, on the order of 2 nm (Giraudeau *et al.*, 2009). Indeed, in an extreme case, one could imagine a collection of small well dispersed platelets being formed. If the conversion to ettringite then takes place rapidly, most of these platelets would probably convert to ettringite nuclei,

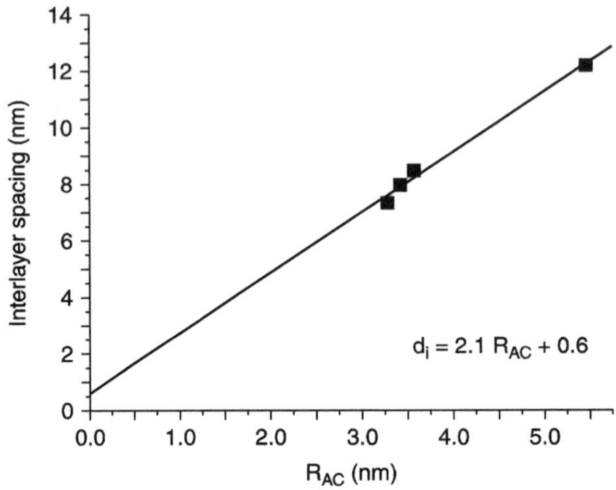

7.43 Interlayer spacing of organo-aluminate phases measured at 100%RH plotted against the theoretical layer thickness from Eq. 7.12 (Giraudeau *et al.*, 2009).

increasing the number of ettringite crystals, decreasing their size and most of all increasing their specific surface. Such a process would lead to an increase in specific surface, whereby the admixture would be mobilized on ettringite rather than on AFM. The overall result in terms of dispersion efficiency would, however, be the same as less polymer would be available for the rest of the cement surfaces.

While much progress has been made on the subject of organo-aluminates, important questions therefore remain to be answered in terms of their impact within real cementitious systems.

7.7 Conclusions and outlook

Polycarboxylate polymers, more than all other superplasticizers, have brought about a major change in concrete manufacturing. This comes largely from the flexibility they offer to vary their molecular structure to achieve the requirements of quite different applications. This flexibility, together with the traditional approach of formulation, has established polycarboxylates as an essential superplasticizer type throughout the markets and across borders.

These superplasticizers work through their adsorbed fraction which induces steric hindrance. As the dosage of these admixtures is increased, the surface coverage increases until saturation is reached. Since these polymers are good dispersants, one often reaches a high fluidity before this. Because cement particles are large enough to be affected by gravity forces, this means that, depending on the solids volume fraction, segregation may occur. In practice the use of superplasticizers is always faced with this balance between enhancing dispersion while trying to avoid segregation. In this regime the rheological response of cementitious systems is much more governed by the amount of adsorbed superplasticizer than by the conformation of the adsorbed polymers.

In this range of dosage, the molecular structure of superplasticizers affects their performance in as much as this affects their adsorption. Another consequence is that, again in this dosage range, any factor affecting adsorption will affect dispersion directly. One of the challenges for future research is to rationalize these effects and their dependence on the molecular structure of superplasticizers. Preliminary work on the role of sulfates has been presented and demonstrates promising perspectives.

Among the factors affecting adsorption, there is also the interaction with the aluminate phases. It has been demonstrated that AFM can intercalate superplasticizers, but that the stability of the resulting organo-aluminate phase is limited. Issues of kinetics are therefore probably crucial in determining whether this does or not affect the ultimate dispersion. On a simpler level, however, the first question to answer is whether these superplasticizers modify the specific surface of the early hydrates. If they increase it, more admixtures will be consumed and, for a given dosage, dispersion will be reduced. This will hold whether one forms well-intercalated products or not. More research is needed to address these specific questions.

Understanding how the dispersing effect of superplasticizers can be perturbed provides very useful information for the design of robust products. Moreover, with the increasing diversity of blended cements and their growing role on the market, we can only expect that the requirements for robust admixtures to increase. We hope that this review will contribute to highlighting the important aspects that need to be taken into account to achieve this.

7.8 References

Aignesberger A and Michaud H (1966), 'Anorganisch-organischer Baustoff', DE Patent 1671017.

Aitcin P-C, Jiang S, Kim B-G, Nikinabubanzi P-C and Petrov N (2001), 'Cement/superplasticizer interaction. The case of polysulfonates', *Bulletin des laboratoires des ponts et chaussées*, 233, 89–99.

Amaya T, Ikeda A, Imamura J, Kobayashi A, Kaname S, *et al.* (1999), 'Cement dispersant and concrete composition containing the dispersant', EP Patent 1184353.

Arfaei A and Lambert L (1995), 'Air controlling superplasticizer', US Patent 57063174.

Arfaei A, Darwin D C, Gartner E M, Chun B-W, Koyata H and Kuo L (1995), 'Improved cement admixture product having improved rheological properties and process of forming same', WO Patent 95/09821.

Banfill P F G (1979), 'A discussion of the papers 'Rheological properties of cement mixes' by M. Daimon and D.M. Roy', *Cement and Concrete Research*, 9, 795–796.

Banfill P F G, Bowen P, Flatt R J, Galmiche L, Houst Y F, *et al.* (2007), 'Improved superplasticisers for high performance concrete: the SUPERPLAST Project', 12th International Congress on the Chemistry of Cement – ICCC Montreal, Canada, RILEM.

Bergström L, Schilling C H and Aksay I A (1992), 'Consolidation behaviour of flocculated alumina suspensions', *Journal of the American Ceramic Society*, 75, 3305–3314.

Bonen D and Sarkar S L (1995), 'The superplasticizer adsorption capacity of cement paste, pore solution composition, and parameters affecting flow loss', *Cement and Concrete Research*, 25, 1423–1434.

Borget P, Galmiche L, Le Meins J-F and Lafuma F (2005), 'Microstructural characterisation and behaviour in different salt solutions of sodium polymethacrylate-g-PEO comb copolymers', *Colloids and Surfaces A: Physicochemical and Engineering Aspects*, 260, 173–182.

Bradeley G and Szymanski C D (1985), 'Cementiferous compositions', EP Patent 0097513.

Bürge T (1999), 'Multi component polymer concrete admixture', in Skarendahl A and Petersson Ö (ed.), Proceedings 1st International RILEM Symposium on Self-Compacting Concrete, Stockholm, RILEM, pp. 411–424.

Bürge T, Widmer J, Meyer T and Sulser U (1981), 'Verfahren zur Herstellung von hochkonzentrierten, niedrigviskosen, wässrigen Lösungen von N-modifizierten Melamin/Aldehydharzen', CH Patent 644876.

Bürge T, Schober I, Huber A, Widmer J and Sulser U (1990), 'Hydrosoluble polymers, process for their preparation and their use as dispersing agents in suspensions of solid materials, EP Patent 0402563.

Cheung J, Jeknavorian A, Roberts L and Silva D (2011), 'Impact of admixtures on the hydration kinetics of Portland cement', *Cement and Concrete Research*, in press.

Chiocchio G and Paolini A E (1985), 'Optimum time for adding superplasticizers to portland cement pastes', *Cement and Concrete Research*, 15, 901–908.

Collepardi M (1984), 'Water Reducers/Retarders', Ramachandran V S (ed.) in *Concrete Admixtures Handbook*, Noyes Publications, Park Ridge, NJ, USA.

Comparet C, Nonat A, Pourchet S, Guicquero J P, Gartner E and Mosquet M (2000), 'Chemical interaction of di-phosphonate terminated monofunctional polyoxyethylene superplasticizer with hydrating tricalcium silicate', in Malhotra V M (ed.), Proceedings 6th CANMET/ACI Internatonal Conference on Superplasticizers and Other Chemical Admixtures in Concrete, Nice, ACI, SP-195, pp. 61–74.

Daimon M and Roy D M (1978), 'Rheological properties of cement mixes: I. Methods, Preliminary experiments, and adsorption studies', *Cement and Concrete Research*, 8, 753–764.

Danzinger W M, Saitoh K, Jost P and Maeder U (2003), 'New generation of admixture polymers and its role in concrete technology', in Grieve G and Owens G (ed.), Proceedings 11th International Congress on the Chemistry of Cement (ICCC), Durban, pp. 2271–2275.

De Gennes P-G (1987), 'Polymers at an interface: a simplified view', *Advances in Colloid and Interface Science*, 27, 189–209.

Dodson V H and Farkas E (1964), 'Delayed additions of set retarding admixtures to Portland cement concrete', *ASTM Proceedings*, 64, 816–829.

Fernàndez-Altable V (2007), *Rheological performance of superplasticisers in cement-based pastes: influence of temperature, mixing procedure, and mineral additives*, PhD thesis, Universitat Polytechnica of Catalunya, Spain.

Fernàndez-Altable V and Casanova I (2006), 'Influence of mixing sequence and superplasticizer dosage on the rheological response of cement pastes at different temperatures', *Cement and Concrete Research*, 36, 1222–1230.

Fernon V (1994), *Etude de nouveaux solides lamellaires obtenus par coprécipitation d'hydrate alumino calcique et de sulfonates aromatiques*, PhD thesis, Univerisité d'Orléans, Orléans, France.

Fernon V, Vichot A, Goanvic N L, Colombet P, Corazza F and Costa U (1997), 'Interaction between portland cement hydrates and polynaphthalene sulfonates', in Malhotra, V. M., ed., Proceedings 5th CANMET/ACI International Conference on Superplasticizers and Other Chemical Admixtures in Concrete, Rome, ACI, SP-173, pp. 225–235.

Flatt R J (2001), 'Polymeric dispersants in concrete', Hackley V A, Somasundaran P and Lewis J A (ed.), in *Polymers in particulate systems – properties and applications*, Marcel Dekker, New York.

Flatt R J (2004a), 'Dispersion forces in cement suspensions', *Cement and Concrete Research*, 34, 399–408.

Flatt R J (2004b), 'Towards a prediction of superplasticized concrete rheology', *Materials and Structures*, 37, 289–300.

Flatt R J and Houst Y F (2001), 'A simplified view on chemical effects perturbing the action of superplasticizers', *Cement and Concrete Research*, 31, 1169–1176.

Flatt R J and Ferraris C (2002), 'Acoustophoresis of cement suspensions', *Materials and Structures*, 35, 541–549.

Flatt R J and Bowen P (2003), 'Electrostatic repulsion between particles in cement suspensions: domain of validity of linearized Poisson–Boltzmann equation for nonideal electrolytes', *Cement and Concrete Research*, 33, 781–791.

Flatt R J and Bowen P (2006), 'Yodel: a yield stress model for suspensions', *Journal of the American Ceramic Society*, 89, 1244–1256.

Flatt R J and Bowen P (2007), 'Yield stress of multimodal powder suspensions: an extension of the YODEL (Yield Stress mODEL)', *Journal of the American Ceramic Society*, 90, 1038–1044.

Flatt R J, Houst Y F, Bowen P, Hofmann H, Widmer J, *et al.* (1997), 'Interaction of superplasticizers with model powders in a highly alkaline medium', in Malhotra V M (ed.), Proceedings 5th CANMET/ACI International Conference on Superplasticizers and Other Chemical Admixtures in Concrete, Rome, ACI, SP-173, pp. 743–762.

Flatt R J, Houst Y F, Oesch R, Hofmann H, Widmer J, *et al.* (1998), 'Analysis of superplasticizers in concrete', *Analusis*, 26, M28=NM35.

Flatt R J, Houst Y F, Bowen P and Hofmann H (2000), 'Electrosteric repulsion induced by superplasticizers between cement particles: an overlooked mechanism?', in Malhotra V M (ed.) Proceedings 6th CANMET/ACI International Conference on Superplasticizers and Other Chemical Admixtures in Concrete, Nice, ACI, SP-195, pp. 29–42.

Flatt R J, Larosa D and Roussel N (2006), 'Linking yield stress measurements: spread test versus Viscomat', *Cement and Concrete Research*, 36, 99–109.

Flatt R J, Schober I, Raphaël E, Plassard C and Lesniewska E (2009a), 'Conformation of adsorbed comb copolymer dispersants', *Langmuir*, 25, 845–855.

Flatt R J, Zimmermann J, Hampel C, Kurz C, Schober I, *et al.* (2009b), 'The role of adsorption energy in the sulfate-polycarboxylate competition', in Holland T C, Gupta P R and Malhotra V M (ed.), Proceedings 9th ACI International Conference on Superplasticizers and Other Chemical Admixtures in Concrete, Seville, ACI, SP-262, pp. 153–164.

Gay C and Raphaël E (2001), 'Comb-like polymers inside nanoscale pores', *Advances in Colloid and Interface Science*, 94, 229–236.

Giraudeau-Lenain C (2009), *Interactions organo-aluminates dans les ciments. Interaction de polymethacrylates-g-PEO dans l'hydrocalumite*, PhD thesis, University Pierre et Marie Curie, Paris.

Giraudeau C, d'Espinose De Lacaillerie J-B, Souguir Z, Nonat A and Flatt R J (2009), 'Surface and intercalation chemistry of polycarboxylate copolymers in cementitious systems', *Journal of the American Ceramic Society*, 92, 2471–2488.

Guicquero J-P, Mosquet M, Chevallier Y and Perchec P L (1999), 'Thinners for aqueous suspensions of mineral particles and hydraulic binder pastes', US Patent 5879445.

Hamada D (2005), 'Phosphoric ester polymer', EP Patent 1767504.

Hamada D, Yamato F and Shimoda M (2003), 'Admixture for a hydraulic composition', EP Patent 1344754.

Hermida G, Moranville M and Flatt R (2009), 'The role of paste volume on concrete performance ', in Holland T C, Gupta P R and Malhotra V M (ed.), Proceedings 10th ACI International Conference on Recent Advances in Concrete Technology and Sustainability Issues, Seville, ACI, SP-261, pp. 201–214.

Jiang S, Kim B-G and Aitcin P-C (1999), 'Importance of adequate soluble alkali content to ensure cement/superplasticizer compatibility', *Cement and Concrete Research*, 29, 71–77.

Jolicoeur J, Nkinamubanzi P C, Simard M A and Piotte M (1994), 'Progress in understanding the functional properties of superplasticizers in fresh concrete', in Malhotra V M (ed.), Proceedings 4th CANMET/ACI International Conference on Superplasticizers and Other Chemical Admixtures in Concrete, Montreal, ACI, SP-148, pp. 63–88.

Juilland P (2009), *Early hydration of cementitious systems*, PhD thesis, EPFL, Lausanne, Switzerland.

Karl W, Lohaus L and Pott J-U (2009), 'Synthese und Eignungsprüfung neuer Betonzusatumittel unter Verwendung nachwachsender Rohstoffe', Tagung der GDCh-Fachgruppe Bauchemie, Freiberg, GDCH, 41, pp. 89–94.

Kauppi A, Banfill P F G, Bowen P, Galmiche L, Houst Y F, *et al.* (2003), 'Improved superplasticizers for high performance concrete', in Grieve G and Owens G (ed.), Proceedings 11th International Congress on the Chemistry of Cement, Durban, pp. 528–537.

Kauppi A, Andersson K M and Bergström L (2005), 'Probing the effect of superplasticizer adsorption on the surface forces using the colloidal probe AFM technique', *Cement and Concrete Research*, 35, 133–140.

Kim B-G, Jiang S and Aitcin P-C (1999), 'Influence of molecular weight on PNS superplasticizers on the properties of cement pastes containing different alkali contents', in Proceedings International Conference on the Role of Admixtures in High Performance Concrete, RILEM, pp. 97–111

Kim B-G, Jiang S and Aitcin P-C (2000), 'Slump improvement mechanism of alkalies in PNS superplasticized cement pastes', *Materials and Structures*, 33, 363–369

Kinoshita M and Okada K (1999), 'Performance of methacrylic water-soluble graft polymers with different types of cement', in Cabrera J G and Rivera-Villarreal R (ed.), Proceedings International Conference on the Role of Admixtures in High Performance Concrete, RILEM, pp. 34–47.

Kjeldsen A M, Flatt R J and Bergström L (2006), 'Relating the molecular structure of comb-type superplasticizers to the compression rheology of MgO suspensions', *Cement and Concrete Research*, 36, 1231–1239.

Lewis J A, Matsuyama H, Kirby G, Morisette S and Young J F (2000), 'Polyelectrolyte effects on the rheological properties of concentrated cement suspensions', *Journal of the American Ceramic Society*, 83, 1905–1913.

Lorenz K, Kraus A, Wimmer B, Wagner P, Scholz C, Huebsch C, *et al.* (2010), 'Dynamic copolymers for workability retention of cementitious composition', WO Patent 2010029117.

Mäder U and Schober I (2003), 'Performance of blends of polycarboxylate polymers in different cements', in Grieve G and Owens G (ed.), Proceedings 11th International Conference on the Chemistry of Cement (ICCC), Durban, pp. 504–513.

Mäder U, Kusterle W and Grass G (1999), 'The rheological behaviour of cementitious materials with chemically different superplasticizers', in Proceedings International Symposium on the Role of Admixtures in High Performance Concrete, Monterrey, RILEM, pp. 357–376.

Mosquet M, Chvalier Y, Brunel S, Guicquero J P and Perche P L (1997), 'Polyoxyethylene di-phosphonates as efficient dispersing polymers for aqueous suspensions', *Journal of Applied Polymer Science*, 65, 2545–2555.

Mosquet M, Maitrasse P and Guicquero J P (2003), 'Ethoxylated di-phosphonate: An extreme molecule for extreme applications', in Malhotra V M (ed.), Proceedings 7th CANMET/ACI International Conference on Superplasticizers and Other Chemical Admixtures in Concrete, Berlin, ACI, SP-217, pp. 161–176.

Nakajima Y and Yamada K (2004), 'The effect of the kind of calcium sulfate in cements on the dispersing ability of poly -naphthalene sulfonate condensate superplasticizer', *Cement and Concrete Research*, 34, 839–844.

Nawa T, Ichiboji H and Kinoshita M (2000), 'Influence of temperature on fluidity of cement paste containing superplasticizer with polyethylene oxide graft chains', in Malhotra V M (ed.), Proceedings 6th CANMET/ACI International Conference on Superplasticizers and Other Chemical Admixtures in Concrete, Nice, ACI, SP-195, pp. 195–210.

Ohta A, Sugiyama T and Tanaka Y (1997), 'Fluidizing mechanisms and application of polycarboxylate-based superplasticizers', in Malhotra V M (ed.), Proceedings 5th CANMET/ACI International Conference on Superplasticizers and Other Chemical Admixtures in Concrete, Rome, ACI, SP-173, pp. 359–378.

Pedersen J S and Sommer C (2005), 'Temperature dependence of the virial coefficients and the chi parameter in semi-dilute solutions of PEG', *Progress in Colloid and Polymer Science*, 130, 70–78.

Perche F (2004), *Adsorption de polycarboxylates et de lignosulfonates sur poudre modèle et ciments*, PhD thesis, EPLF, Lausanne, Switzerland.

Perche F, Houst Y F, Bowen P and Hofmann H (2003), 'Adsorption of lignosulfonates and polycarboxylates – depletion and electroacoustic methods', in Malhotra V M (ed.), Proceedings 7th CANMET/ACI International Conference on Superplasticizers and Other Chemical Admixtures in Concrete, Berlin, ACI, Supplementary Papers, pp. 1–16.

Pfefferkorn E., Carroy A and Varoqui R (1985), 'Dynamic behavior of flexible polymers at a solid/liquid interface', *Journal of Polymer Science: Polymer Physics Edition*, 23(10), 1997–2008.

Pieh S (1987), 'Polymere Dispergiermittel I', *Die Angewandte Makromolekulare Chemie*, 154, 145–159.

Piotte M (1993), *Charactérisation du poly(naphtalènesulfonate) – Influence de son contre-ion et de sa masse moléculaire sur son interaction avec le ciment*, PhD thesis, University of Sherbrooke, Canada.

Plank J and Winter C (2008), 'Competitive adsorption between superplasticizer and retarder molecules on mineral binder surface', *Cement and Concrete Research*, 38, 599–605.

Plank J, Bassioni G, Dai Z, Keller H, Sachsenhauser B and Zouaoui N (2006a), 'Neues zur Wechselwirkung zwischen Zementen und Polycarboxylat-Fließmitteln', in Stark J (ed.), 16. Internationale Baustofftagung – ibausil, F.A. Finger-Institut für Baustoffkunde, Bauhaus Universität Weimar, Volume 1, pp. 579–598.

Plank J, Keller H, Andres P R and Zhimin D (2006b), 'Novel organo-mineral phases obtained by intercalation of maleic anhydride-allyl ether copolymers into layered calcium aluminate hydrates', *Inorganic Chimica Acta*, 359, 4901–4908.

Plank J, Dai Z and Andres P R (2006c), 'Preparation and characterization of new Ca-Al-polycarboxylate layered double hydroxides', *Materials Letters*, 60, 3614–3617.

Plank J, Dai Z, Zouaoui N and Vlad D (2006d), 'Intercalation of polycarboxylate superplasticizers into C3A hydrate phases', in Malhotra V M (ed.), Proceedings 8th CANMET/ACI International Conference on Superplasticizers and Other Chemical Admixtures in Concrete, Sorrento, ACI, SP-239, pp. 201–213.

Plank J, Pöllmann K, Zouaoui N, Andres P R and Schaefer C (2008), 'Synthesis and performance of methacrylic ester based polycarboxylate superplasticizers possessing hydroxy terminated poly(ethylene glycol) side chains', *Cement and Concrete Research*, 38, 1210–1216.

Plank J, Sachsenhauser B and Reese J D (2010a), 'Experimental determination of the thermodynamic parameters affecting the adsorption behaviour and dispersion effectiveness of PCE superplasticizers', *Cement and Concrete Research*, 40, 699–709.

Plank J, Recalde-Lummer N and Dugonjic-Bilic F (2010b), 'Competitive adsorption between AMPS-based fluid loss polymer and welan gum biopolymer in oil well cement', *Journal of Applied Polymer Science*, 116, 2913–2919.

Plassard C, Lesniewska E, Pochard I and Nonat A (2004), 'Investigation of the surface structure and elastic properties of calcium silicate hydrates at the nanoscale', *Ultramicroscopy*, 100, 331–338.

Plassard C, Lesniewska E, Pochard I and Nonat A (2005), 'Nanoscale experimental investigation of particle interactions at the origin of the cohesion of cement', *Langmuir*, 21, 7263–7270.

Platel D (2005), *Impact de l'architecture macromoléculaire des polymères sur les propriétés physico-chimiques des coulis de ciement*, PhD thesis, University Paris VI, Paris, France.

Platel D (2009), 'Impact of polymer architecture on superplasticizer efficiency', in Holland T C, Gupta P R and Malhotra V M (ed.), Proceedings 9th ACI International Conference on Superplasticizers and Other Chemical Admixtures in Concrete, Seville, ACI, SP-262, pp. 381–394.

Pöllmann H (1992), 'Carboxylic acids anions: the reaction mechanisms and products with aluminate phase of cement', in Proceedings 9th International Congress on the Chemistry of Cement, New Dehli, pp. 198–204.

Ramachandran V S (1984), 'Accelerators', in Ramachandran V S (ed.), *Concrete Admixtures Handbook*, Noyes Publications, Park Ridge, NJ, USA.

Regnaud L, Nonat A, Pourchet S, Pellerin B, Maitrasse P, *et al.* (2006), 'Changes in cement paste and mortar fluidity after mixing induced by PCP: a parametric study', in Malhotra V M (ed.), Proceedings 8th CANMET/ACI International Conference on Superplasticizers and Other Chemical Admixtures in Concrete, Sorrento, SP-239, pp. 389–407.

Reknes K and Gustafsson J (2000), 'Effect of modifications of lignosulfonate on adsorption on cement and fresh concrete properties', in Malhotra V M (ed.) Proceedings 6th CANMET/ACI International Conference on Superplasticizers and Other Chemical Admixtures in Concrete, Nice, ACI, SP-195, pp. 127–141.

Reknes K and Petersen B G (2003), 'Novel lignosulfonate with superplasticizer performance', in Malhotra V M (ed.), Proceedings 7th CANMET/ACI International Conference on Superplasticizers and Other Chemical Admixtures in Concrete, Berlin, ACI, Supplementary Papers, pp. 285–299.

Roussel N and Coussot P (2005), 'Fifty-cent rheometer for yield stress measurements: from slump to spreading flow', *Journal of Rheology* 49, 705–718.

Roussel N, Stefani C and Leroy R (2005), 'From mini-cone test to Abrams cone test: measurement of cement-based materials yield stress using slump tests', *Cement and Concrete Research*, 35, 817–822.

Roussel N, Lemaître A, Flatt R J and Coussot P (2010), 'Steady state flow of cement suspensions: a micromechanical state of the art', *Cement and Concrete Research*, 40, 77–84.

Schober I and Mäder U (2003), 'Compatibility of polycarboxylate superplasticizers with cements and cementitious blends', in Malhotra V M (ed.), Proceedings 7th CANMET/ACI International Conference on Superplasticizers and Other Chemical Admixtures in Concrete, Berlin, ACI, SP-217, pp. 453–468.

Schober I and Flatt R F (2006), 'Optimizing polycarboxylate polymers', in Malhotra V M (ed.), Proceedings 8th CANMET/ACI International Conference on Superplasticizers and Other Chemical Admixtures in Concrete, Sorrento, ACI, SP-239, pp. 169–184.

Spiratos N, Pagé M, Mailvaganam N P, Malhotra V M and Jolicoeur C (2003), '*Superplasticizers for Concrete*', Supplementary Cementing Materials for Sustainable Development Inc., Ottawa.

Tsubakimoto T, Hosoido M and Tahara H (1984), 'Copolymer and method for manufacture thereof', US Patent 4471100.

Tsuyoshi H, Tsutomu Y, Koichiro N, Hirokatsu K, Hirata T, *et al.* (2000), 'Cement additive', EP Patent 1041053.

Uchikawa H, Sawaki D and Hanehara S (1995), 'Influence of kind and added timing of organic admixture on the composition, structure and property of fresh cement paste', *Cement and Concrete Research*, 25, 353–364.

Uchikawa H, Hanehara S and Sawaki D (1997), 'The role of steric repulsive force in the dispersion of cement particles in fresh paste prepared with organic admixture', *Cement and Concrete Research*, 27, 37–50.

Velten U, Schober I, Sulser U and Mäder U (2001), 'Blends of polycarboxylate-type superplasticizers in use for concrete admixtures', in Proceedings 2nd International Symposium on Self-Compacting Concrete, Tokyo, pp. 187–194.

Vickers T M, Farrington S A, Bury J R and Brower L E (2005), 'Influence of dispersant structure and mixing speed on concrete slump retention', *Cement and Concrete Research*, 35, 1882–1890.

Winnefeld F (2008), 'Influence of cement ageing and addition time on the performance of superplasticizers', *ZKG International*, 61, 68–77.

Winnefeld F, Becker S, Pakusch J and Götz T (2006), 'Polymer structure/concrete property relations of HRWR', in Proceedings 8th CANMET/ACI International Conference on Recent Advances in Concrete Technology, Sorrento, ACI, Supplementary Papers, pp. 159–177.

Winnefeld F, Becker S, Pakusch J and Götz T (2007), 'Effects of the molecular architecture of comb-shaped superplasticizers on their performance in cementitious systems', *Cement and Concrete Composites*, 29, 251–262.

Witt J and Plank J (2009), 'Synthese und Charakterisierung von silylierten Polycarboxylat-Flie*ß*mitteln und deren Wechselwirkung mit Portlandzement', Tagung der GDCh-Fachgruppe Bauchemie, GDCh, 41, 173–175.

Yamada K and Hanehara S (2001), 'Interaction mechanism of cement and superplasticizers – the roles of polymer adsorption and ionic conditions of aqueous phase', *Concrete Science and Engineering*, 3, 135–145.

Yamada K, Takahashi T, Hanehara S and Matushisa M (2000a), 'Effects of the chemical structure on the properties of polycarboxylate-type superplasticizer', *Cement and Concrete Research*, 30, 197–207.

Yamada K, Ogawa S and Hanehara S (2000b), 'Working mechanism of poly-beta-naphthalene sulfonate and polycarboxylate superplasticizer types from point of cement paste characteristics', in Malhotra V M (ed.), Proceedings 6th CANMET/ACI International Conference on Superplasticizers and Other Chemical Admixtures in Concrete, Nice, ACI, SP 195–23, pp. 367–382.

Yamada K, Ogawa S and Hanehara S (2001), 'Controlling of the adsorption and dispersing force of polycarboxylate-type superplasticizer by sulfate ion concentration in aqueous phase', *Cement and Concrete Research*, 31, 375–383.

Yamashita A, Hirata T, Yuasa T and Niwa H (1999), 'Cement admixture and cement composition', EP Patent 0931776.

Zhang X, Pagé M, Dénommé Y and Jolicoeur C (2006), 'Novel polymer blends as concrete superplasticizers', in Malhotra V M (ed.), Proceedings 8th CANMET/ACI International Conference on Recent Advances in Concrete Technology, Montreal, ACI, SP-235, pp. 217–240.

Zimmermann J, Hampel C, Kurz C, Frunz L and Flatt R (2009), 'Effect of polymer structure on the sulfate-polycarboxylate competition', in Holland T C, Gupta P R and Malhotra V M (ed.), Proceedings 9th ACI International Conference on Superplasticizers and Other Chemical Admixtures in Concrete, Seville, ACI, SP-262, pp. 165–176.

8
Viscosity-enhancing admixtures and the rheology of concrete

K.H. KHAYAT, Université de Sherbrooke, Canada and Missouri University of Science and Technology, USA and N. MIKANOVIC, HeidelbergCement Technology Center GmbH Germany

Abstract: Viscosity-enhancing admixtures (VEAs) are water-soluble polymers that increase viscosity and cohesion of cement-based materials. Such enhancement is essential in highly flowable concrete, including self-consolidating concrete (SCC), to control the risk of segregation and to maintain a homogenous suspension in the plastic stage until the onset of hardening. For a given mixture composition, the VEA's ability to enhance stability can vary widely with the type and dosage rate of the admixture as well as that of the high-range water reducer often needed to maintain the targeted fluidity. This chapter discusses the mode of action of various types of VEAs and their effect on the rheological properties and workability of various types of cement-based materials.

Key words: bleeding, chemical admixtures, polysaccharide, rheology, robustness, self-consolidating concrete, segregation, stability, viscosity-enhancing admixture, workability.

8.1 Introduction

Today's concrete has to fulfill a wide range of requirements in both plastic and hardened states. The pressing need for higher quality concrete in terms of strength and durability inevitably points towards low porosity materials; these imply minimum water content and the use of ultrafine supplementary materials (fillers or pozzolans). In other recent developments, requirements in terms of the properties of flowing concrete have been raised to achieve 'self-compacting' and 'self-levelling' abilities. Flowable concrete is commonly used for casting congested structural members and in restricted areas where access for placement and consolidation is limited. The self-compacting requirement calls for highly fluid mixtures, which are inherently prone to separation effects, i.e. bleeding, sedimentation and segregation of aggregates. In order to minimize these phenomena, a new class of chemical admixtures has been designed, the main function of which is to increase concrete viscosity while retaining the desirable features of free-flowing, self-consolidating concrete. They are commonly known as viscosity-enhancing admixtures (VEAs), but they are also referred to as viscosity-modifying admixtures, water-retaining admixtures and colloidal admixtures. Since VEAs increase the cohesion of cement-based systems in general, they can also function as anti-washout admixtures (for underwater concreting), or to reduce rebound and sagging in shotcrete applications.

In general, VEAs are water-soluble polymers that increase the yield stress and/or the plastic viscosity of cement-based materials. Such enhancement of the rheology of the liquid phase in cement-based materials is essential to reduce the rate of separation of material constituents, which is critical in flowable systems. When proper dosage rate of the VEA is combined with a high-range water reducer (HRWR), highly flowable yet cohesive cement-based material can be produced with high segregation resistance. The improved homogeneity of the concrete can enhance bond strength to reinforcement, and aggregate and enhance the overall performance of such systems. The information offered in this chapter is complementary to the comprehensive discussion on the effect of VEAs on fresh and hardened properties of cement-based materials in the literature.[1] This chapter deals mainly with the mode of action of various types of VEAs and their influence on the rheological performance and workability of cement-based materials.

8.2 Chemical nature, classification and mode of action of viscosity-enhancing admixtures

8.2.1 Classification

Materials that have been used as VEAs in cement-based materials are mainly hydrophilic, water-soluble polymers with a high molecular weight. The main groups of polymers currently used in such application can be classified as follows:[2]

1. Natural polymers including starch, welan gum, xanthan gum and other natural gums, as well as plant proteins;
2. Semi-synthetic polymers that include modified starch and its derivatives, cellulose-ether derivatives such as hydroxypropyl methyl cellulose (HPMC) and carboxy methyl cellulose (CMC), sodium alginate and propylene glycol alginate;
3. Synthetic polymers such as polyethylene oxide, polyacrylamide, polyacrylate and polyvinyl alcohol.

The VEAs most frequently used in cement-based materials are cellulose-ether derivatives and polysaccharides from microbial sources such as welan gum and diutan gum. The chemical structure of these molecules is shown in Fig. 8.1. Basically, these admixtures are high-molecular-weight polysaccharides composed of one or a few different mono-saccharide repeating units joined together by glucosidic bonds. The multiple hydroxyl groups in the structure of these molecules allow hydrogen bonding between VEA molecules, a mechanism that is responsible for building viscosity in solution.

Cellulose ethers are the compounds produced by replacing the hydrogen atoms in the glucose units of cellulose with alkyl or substituted alkyl groups (Fig. 8.1(a)). The important properties of cellulose ethers are determined by the molecular

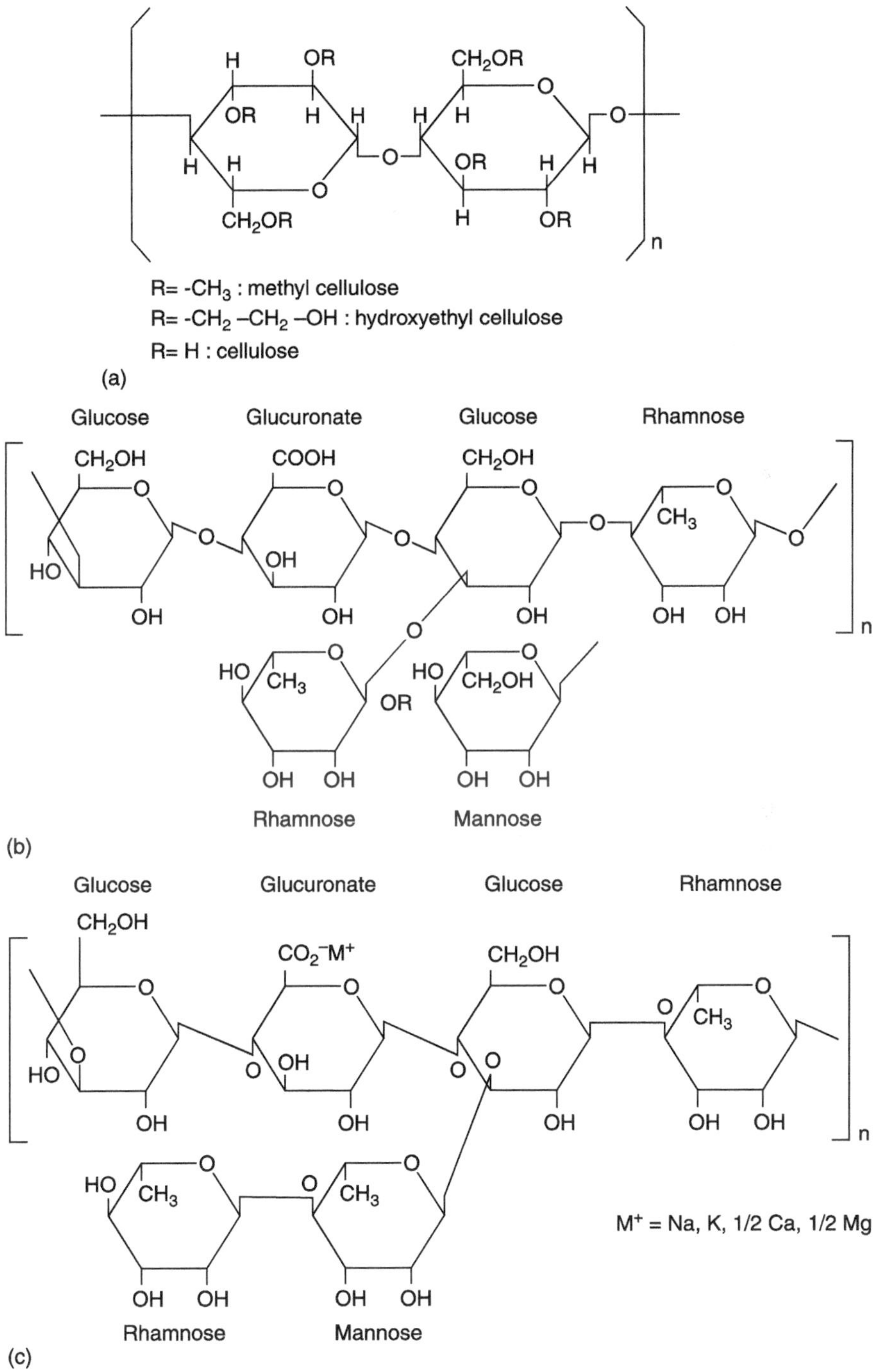

8.1 Molecular structure of common colloidal admixtures: (a) cellulose ethers; (b) welan gum; (c) diutan gum.

weight of the cellulose used, the chemical structure of the substituent group and the degree of substitution. These properties generally include solubility, viscosity in solution, surface activity and stability against biodegradation, heat, hydrolysis and oxidation. Viscosity in solution varies directly with molecular weight. Typically, cellulose-based VEAs have a molecular weight between 10^5 and 10^6 g/mol, a range which is determined by the capacity of viscosity enhancement and appropriate solubility of cellulose ether molecules in solution.

Welan gum and diutan gum are high-molecular weight biopolymers (microbial polysaccharides) produced by aerobic fermentation. The molecular structure of welan gum is illustrated in Fig. 8.1(b). The welan gum backbone is composed of D-glucose, D-glucuronic acid, D-glucose and L-rhamnose units, while the side chain consists of a single L-mannose or L-rhamnose unit. A significant structural difference between diutan and welan gum is the longer side chain, i.e. diutan has a two unit rhamnose side chain (Fig. 8.1(c)). Also, the molecular weight of diutan gum is up to three times higher than that of welan gum (~1×10^6 g/mol). It is believed that these properties contribute significantly to their performance relative to each other.[3]

Beside these molecules, new low-cost VEAs such as modified starch, precipitated silica and new biopolymer molecules are actively being developed and introduced to the concrete industry.[4–6] The stability of highly fluid cementitious systems may also be improved by incorporating materials that contain colloidal-size particles such as flyash, finely divided calcium carbonate, blast furnace slag and diatomaceous earth, or materials having a high surface area or unusual surface properties such as fine clays, silica fume, colloidal silica and milled asbestos.[7] These fine materials have water-retaining ability, and they often affect the void structure by acting as fillers of capillary pores in the hydrated cement paste.

8.2.2 Adsorption

As described above, the VEA molecules are hydrophilic, high molecular weight polymers containing multiple hydroxyl groups per monomeric unit of structure. Therefore, it is not surprising that VEAs of all types adsorb more or less significantly on hydrating cement particles either by hydrogen bonding or by specific interaction involving the hydroxyl groups of VEAs and calcium ions on the surface of cement particles.[8] The adsorption is typically studied by adding the admixture to the cement paste, and after a period of time, filtering the system. The filtrate is analysed to give the quantity of admixture left in solution, and the amount adsorbed is calculated by difference.

Figure 8.2 shows adsorption isotherms for three commonly used VEA molecules. Cellulose-based VEAs are seen to adsorb slightly on hydrating cement particles and their binding affinity depends mainly on the type and degree of substitution.[9] The greater adsorption of welan and diutan gum can be explained by their anionic character, i.e. by the presence of the carboxylic acid functions on

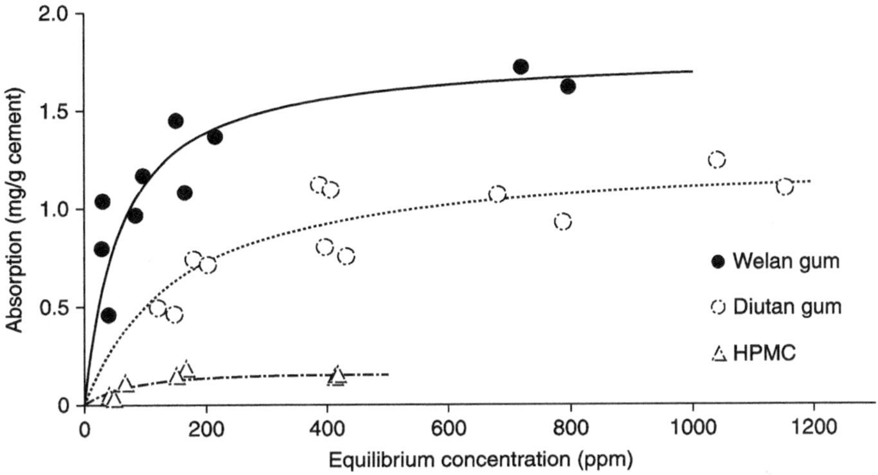

8.2 Adsorption of commonly used VEAs onto cement particles (w/c=10) (unpublished results from Université de sherbrooke).

their molecular backbone, which are fully ionized at high pH. On the other hand, the higher adsorption of welan relative to diutan gum is attributed to the effectiveness of side chains of these two molecules to sterically shield carboxylate groups.[3]

In addition to viscosity enhancement effects in solution, long-chain VEAs may adsorb simultaneously onto neighboring cement particles, forming a bridged structure.[10,11] This effect results in enhanced particle–particle interactions and the formation of flocks in which free water may be entrapped. In other words, the VEA may behave as a flocculating agent and oppose the dispersion function of HRWR used in highly fluid cement-based materials. When used together, the VEA and HRWR molecules compete for the same surface-binding sites on cement particles.[10] Depending on the composition (hydrophobic/hydrophilic character, ionic charge), molecular weight and configuration in interstitial solution, the VEA thus may interfere more or less with the function of HRWR. In this sense, the required dosage of the latter may increase with increasing dosage of the VEA.[10,12]

8.2.3 Mode of action

The incorporation of VEA increases the cohesiveness and stability of cement-based materials through a combination of several physico-chemical phenomena, depending on the type and concentration of the polymer used. These phenomena are illustrated in Fig. 8.3 and described as follows:[13]

1. Water retention: The long-chain hydrophilic VEA molecules adsorb and fix free water molecules. In doing so their apparent volume increases by swelling, which increases the viscosity of mix water.

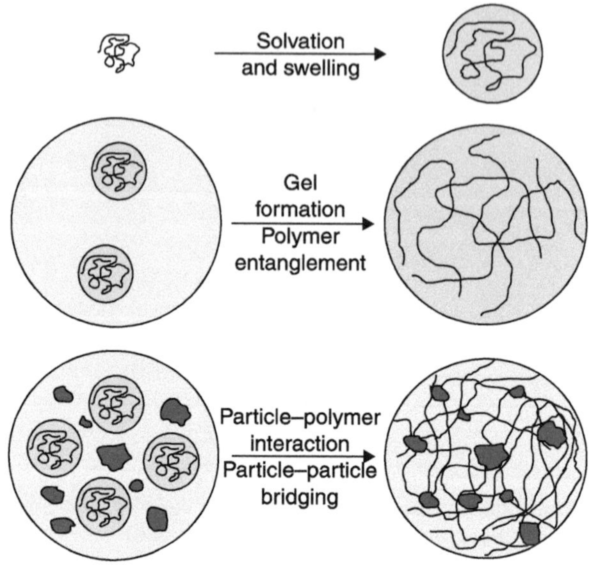

8.3 Schematic illustration of mechanisms contributing to the function of VEAs.[13]

2. Polymer–polymer interaction and entanglement: Adjacent VEA polymer chains can develop attractive forces, resulting in the formation of a gel-like network, thus further blocking the motion of water and increasing the viscosity of the whole system. At higher concentrations, the VEA polymer molecules can entangle, resulting in an increase in the apparent viscosity.
3. Polymer–particle interaction: Adsorption of polymer onto particles leads to an increased particle size and an increase in drag. Furthermore, higher polymer concentrations lead to particle–particle bridging producing a relatively rigid network.[10,11]

The particle bridging effect results in the formation of large flocs in which free water may be entrapped. Since this flocculation effect opposes the dispersion function of the superplasticizers used in flowing concrete, the use of non-adsorptive colloidal agents is generally preferable.[10]

8.3 Effect of viscosity-enhancing admixtures on rheology of water–cement systems

8.3.1 Rheological effects in solution

The main function of VEAs is to enhance the cohesion and viscosity of highly fluid cementitious material at low shear rate, while maintaining a relatively low resistance to flow at high shear rates, such as those encountered during the mixing,

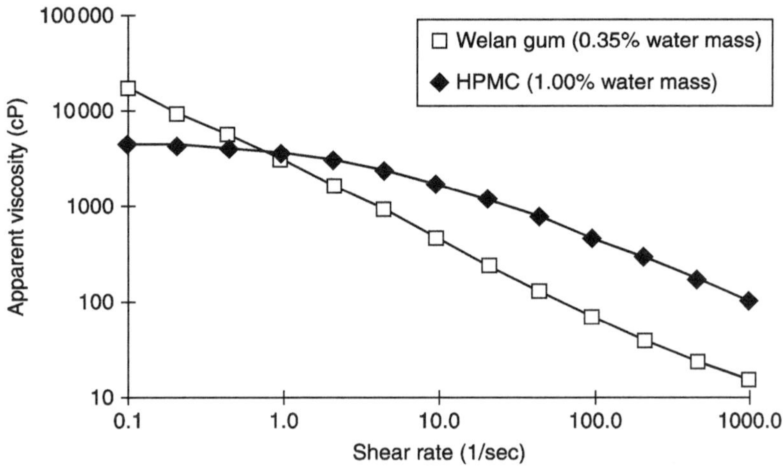

8.4 Variations of apparent viscosity with shear rate of aqueous solutions made with VEAs.[14]

pumping and casting of cement-based systems. In rheological studies, it has been shown that aqueous solutions of VEAs exhibit shear-thinning (or pseudoplastic) behavior where the apparent viscosity decreases with increasing shear rate. The importance of this effect depends on the type and dosage of the VEA polymer. As shown in Fig. 8.4, a shear-thinning behavior is more pronounced in aqueous solution containing welan gum than in that containing cellulose derivative under the same conditions.[14,15] The flow curve observed for the HPMC shows a Newtonian plateau in the low shear rate region, followed by exponential decrease of the apparent viscosity at higher shear rates. The Newtonian plateau viscosity and the critical shear rate (shear rate at which the solution becomes shear-thinning) depend on the polymer concentration and molecular weight.

This kind of rheological behavior is typical for 'real' polymer solutions. The rheological properties of such solutions are essentially governed by water retention through hydrogen bonding involving polymer hydroxyl groups, or by the entanglement of polymer molecules at higher concentrations. On the other hand, more pronounced shear-thinning behavior of welan (and diutan) gum solutions is attributed to the formation of a 'weak' gel-like structure between polymer molecules, similar to that found in xanthan gum solutions.[16,17] Pseudoplastic behaviour is the result of disentanglement or breaking of the network structure between VEA molecules and their alignment in the direction of flow at higher shear rates.

An important feature of these polymeric admixtures is their stability and performance under high pH conditions and in the presence of calcium ions, or at high temperature. Figure 8.5 shows the variation of apparent viscosity of 1%

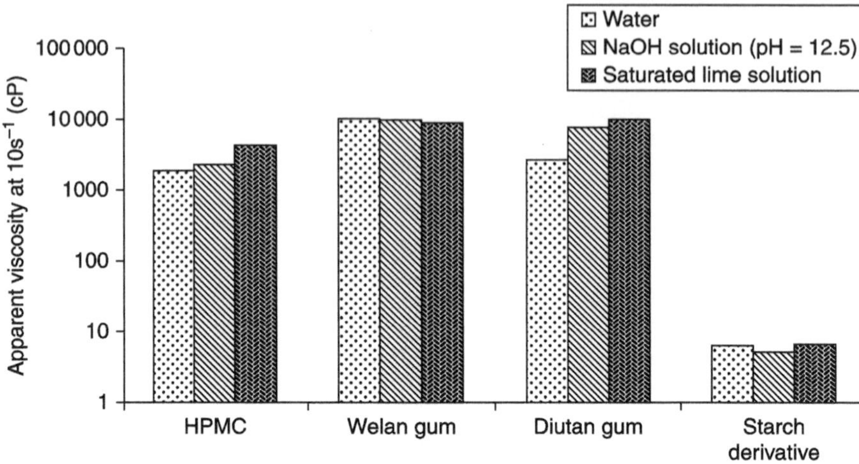

8.5 Variations of apparent viscosity of water solution made with 1% VEA, sodium hydroxide solution and saturated lime solution. Polymer molecules were allowed to swell over 24 hours before rheological measurements (unpublished results from Université de Sherbrooke).

HPMC, welan and diutan gum solutions in three different media: water, saturated lime (pH 12.5) and NaOH solution (pH 12.5). The concentration of 1% is a typical concentration of these molecules in interstitial solution (free water) in most concrete applications. The apparent viscosity of welan gum solution does not seem to be affected by the calcium ion concentration and pH; however, the viscosities of cellulose-based derivative (HPMC) and diutan gum are higher under alkaline conditions and in the presence of calcium ions. The performance of these molecules under these conditions is attributed to intermolecular hydrogen bonding between the polymer molecules and the sterical shielding of the carboxylic acid function on the main chain of welan and diutan gum by molecules' side groups.[3,16] In contrast, there is no evidence that modified starches increase the viscosity of aqueous solution under considered experimental conditions. Therefore, an alternative mechanism other than increase of the viscosity of aqueous solution is responsible for improving the cohesiveness of cement-based materials by starch derivatives.

8.3.2 Rheological effects in cement paste

The detailed rheological influence of VEAs can be described from studies of their shear stress vs. shear rate curve (or stress amplitude vs rate) using specially designed rheometers. The VEAs are found to increase yield stress value, plastic viscosity and apparent viscosity of cement pastes at all shear rates.[12,13,18–20] The effect of VEAs on the yield stress and flow characteristics of cement-based

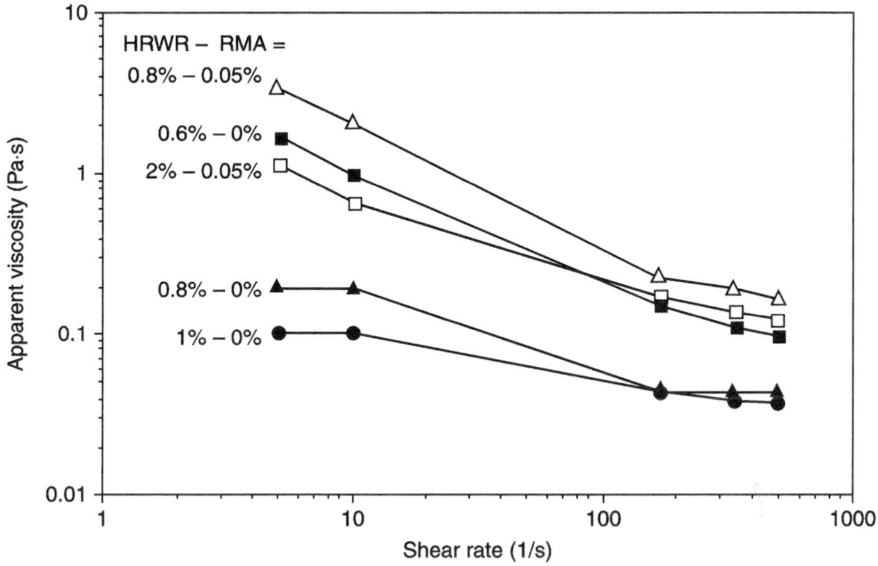

8.6 Variations in apparent viscosity with shear rate for cement pastes (w/c = 0.4) containing PNS-based HRWR and 0 and 0.5% of welan gum.[12]

materials can be substantial, especially when a high dosage of colloidal admixture is applied. In practice, highly flowable cement-based materials with adequate resistance to bleeding, segregation, settlement, etc, may only be obtained by use of both a VEA and a HRWR.

As shown in Fig. 8.6 for cement paste made with a water-to-cement ratio (w/c) of 0.4, cement pastes containing welan gum as the VEA exhibit high apparent viscosities at low shear rates and significantly lower viscosities at high shear ratio, regardless of the content of HRWR based on polynaphthalene sulphonate (PNS). A high apparent viscosity at low shear rates is essential to reduce the rate of sedimentation of solid particles. On the other hand, because of the shear-thinning behavior at higher shear rates associated with mixing, pumping and consolidation, the paste exhibits a lower apparent viscosity that can facilitate the deformability of the mixture. With an increase in the HRWR dosage, the shear-thinning character of the paste decreases, i.e. the apparent viscosity at low shear rate decreases more dramatically than at high shear rates. This effect is more important in pastes containing no VEA than in pastes with VEA. In view of the partially opposing effects of VEA and HRWR, only the combined use of proper contents of these admixtures helps secure a highly fluid yet cohesive paste.

Cement pastes containing a VEA can also exhibit a thixotropic behavior.[18,21] Such behavior associated with a structural build-up of the yield stress (and in some cases of the plastic viscosity) after a given period of rest can enhance the resistance of concrete to surface settlement, segregation and bleeding.[21] A fast build-up of viscosity can be due to thixotropy whereby the cohesiveness increases with the elapsed time of rest. This can be due to physical and/or chemical effects associated with cement hydration. The physical explanation relates to a build-up of inter-particle friction and cohesion between the various cement particles and admixture molecules.[21] Cellulose-based VEA and microbial polysaccharides, such as welan gum, function by occupying volume in the liquid phase resulting in a thickening of that phase. On the other hand, thixotropic admixtures (such as propylene carbonate) function by inducing a network structure in the fluid phase through increased interactions of the solids particles. The thixotropic nature of some thixotropy-enhancing admixtures, such as propylene carbonate, can also be the result of extensive intermolecular association through hydrogen bonding that forms a three-dimensional structure at rest.[21]

Thixotropic behavior depends on the VEA and HRWR types and concentrations, as well as the w/c. For example, in cement grouts made with 0.4, 0.5 and 0.55 w/c, the apparent viscosities at low shear rates were reported to increase over an extended period of time after mixing. However, such increase was easily recovered following some mixing, which means that the observed process appears to be a physical one rather than a result of cement hydration. The thixotropic character was especially high in cement paste containing an HRWR together with VEA (Fig. 8.7); the degree of thixotropy was reported to increase with the concentration of the HRWR.[18]

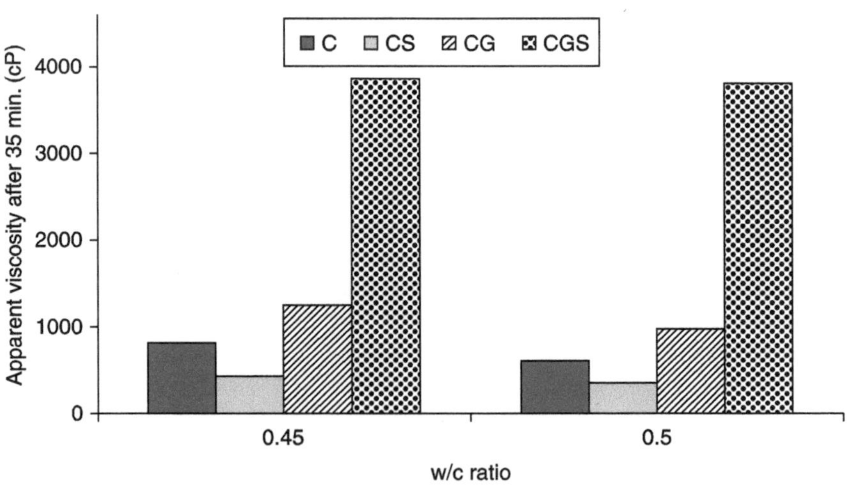

8.7 Apparent viscosity at low shear rate (8 rpm) after 35 min. of mixing for cement pastes having two different w/c; C: plain cement paste, CS: cement paste containing 0.4% PNS-based HRWR, CG: cement paste containing 0.1% VEA (welan gum), CGS: cement paste containing both HRWR and VEA.[18]

8.3.3 Effect of viscosity-enhancing admixtures (VEA) type and concentration on rheological properties of concrete-equivalent mortar made with fixed HRWR dosage

The increase in VEA concentration typically requires an increase in HRWR demand in order to maintain the targeted fluidity level. In the case of a well-dispersed system with a fixed HRWR, which is necessary to ensure self-consolidating consistency in the absence of VEA, an increase in VEA dosage can have an adverse effect on fluidity. This is illustrated in Fig. 8.8 for concrete-equivalent mortar based on a self-consolidating concrete (SCC) mixture of 0.57 water-to-powder ratio (w/p).[22] The reference mortar made without any VEA was proportioned with a spread of 240 mm corresponding to a slump flow consistency of 665 mm. Two types of HRWR were used: a polycarboxylate ether-based HRWR (PC) and a polynaphthalene sulphonate-based HRWR (PNS). The dosage rates of these admixtures were set at 0.3 and 1.5 l/m^3, respectively, for the remaining VEA mixtures. In total, 22 concrete-equivalent mortars were prepared with welan gum, diutan gum, HPMC and modified starch of different concentrations. The concrete incorporated a 30% replacement of the total powder content by limestone filler. The degree of drop in slump flow consistency of the mortar with the increase in VEA content was especially high for the bio-gums. On the other hand, the cellulose-based VEA had a rather limited effect on reducing fluidity, despite the high addition rate.

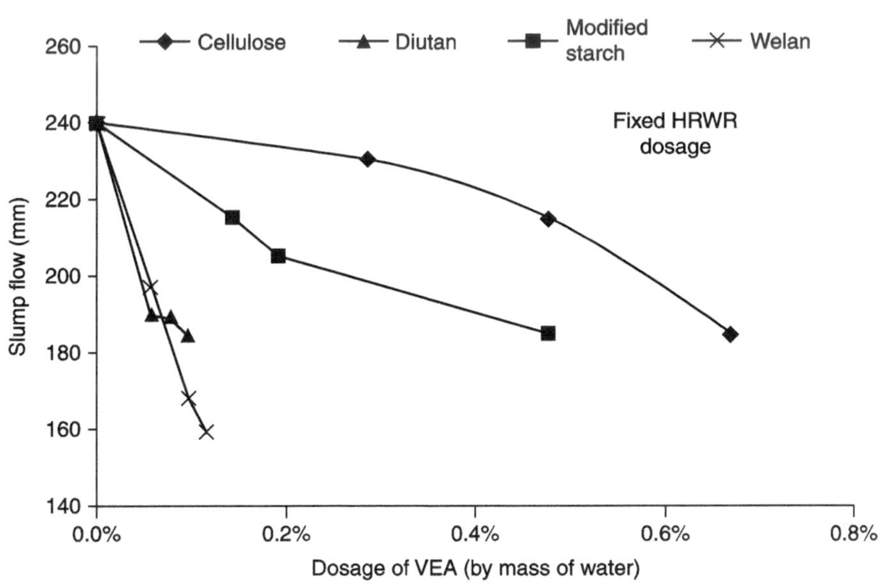

8.8 Effect of VEA concentration on slump flow of concrete-equivalent mortar of SCC consistency made with fixed dosage of HRWR.[22]

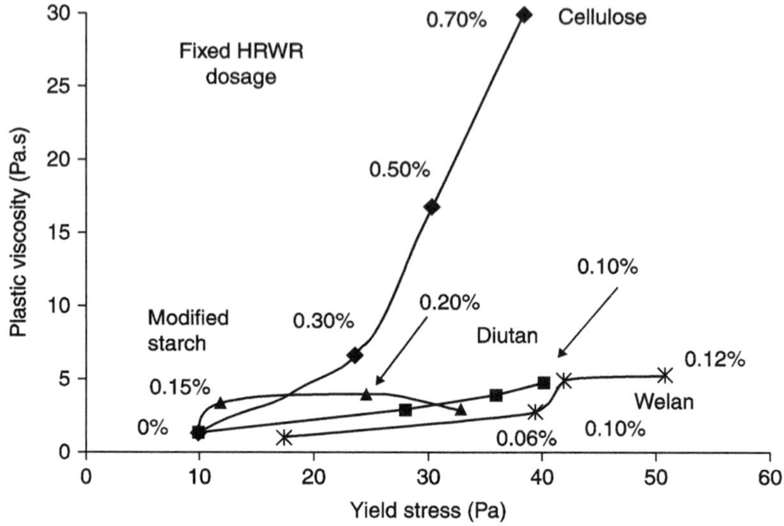

8.9 Effect of VEA type and concentration on rheology of concrete-equivalent mortar of SCC consistency made with constant HRWR dosage.[22]

As illustrated in Fig. 8.9, the increase in VEA content for a fixed dosage of HRWR leads to an increase in rheological properties.[22] The highest increase in plastic viscosity was observed in the mortar made with the cellulose-based VEA. Plastic viscosity values of 6, 16 and 30 Pa.s were obtained in mixtures made with cellulose-based VEA concentrations of 0.30, 0.50 and 0.70%, respectively, by mass of water. The increase in the dosage of modified starch from 0 to 0.5% had a similar effect on yield stress as the cellulose VEA, but practically no effect on plastic viscosity. As expected from the results of the drop in slump flow shown in Fig. 8.8, the increase in the dosage of welan gum and diutan gum incorporated to concentrations of 0.12% and 0.10%, respectively, led to considerable increase in yield stress. Limited gain in viscosity was observed. An increase in 0.02% of either bio-gum resulted in an approximately 30% increase in plastic viscosity. The apparent viscosity at low shear rate was also found to be highly affected by the increase in VEA content; this effect was greater than that of the apparent viscosity at high shear rate.[22] As mentioned earlier, the apparent viscosity is greater at low shear rate given the higher degree of VEA polymer entanglement and association of adjacent polymer chains.

As in the case of yield stress and plastic viscosity, Fig. 8.10 shows that the increase in VEA content in the mortar proportioned with a fixed dosage of HRWR leads to a net increase in thixotropy.[22] Thixotropy was determined by integrating the area between the initial shear stresses and equilibrium shear stress values determined at four shear rates varying between 0.3 and 0.9 s^{-1}. The mortar was subjected to five minutes of rest prior to each measurement. Mortars made without any VEA and

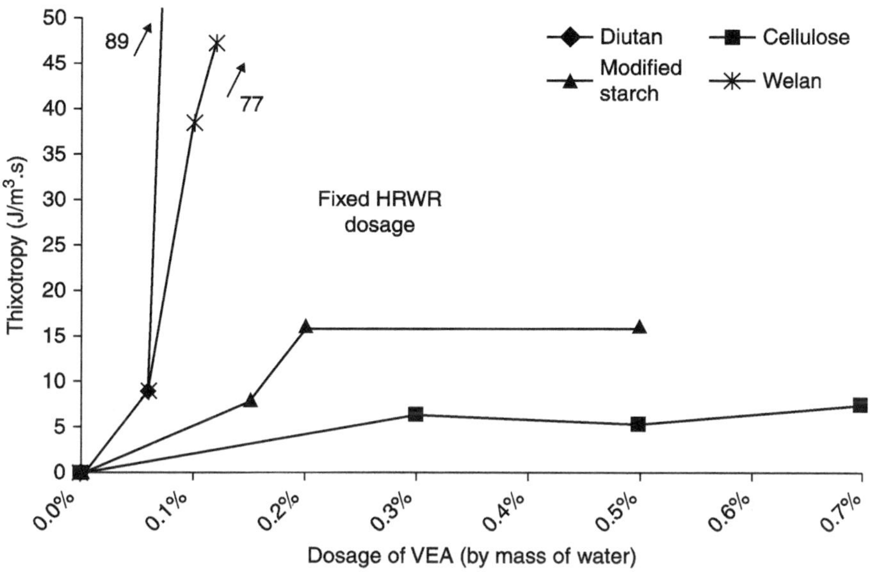

8.10 Effect of VEA type and concentration on thixotropy of concrete-equivalent mortar of SCC consistency made with fixed dosage of HRWR.[22]

with either PNS- or PC-based HRWR had no thixotropic properties (0 J/m^3.s) given the relatively high water content of the mixtures (w/p of 0.57). On the other hand, mixtures made with welan and diutan gums exhibited considerable degrees of increase in thixotropy with the increase in VEA dosage. For example, at the highest dosage of diutan gum of 0.1%, a thixotropic index of 89 J/m^3.s was obtained. Such increase in thixotropy (structural breakdown area) can lead to greater static stability of the mortar. Despite the increase in viscosity and dynamic yield stress of the mortar with the increase in cellulose-based VEA, the mortar exhibited a limited change in thixotropy (from 0 to 8 J/m^3.s at the high concentration of 0.7%).

8.3.4 Effect of VEA on rheological properties of self-consolidating concrete

As described earlier, the increase in VEA concentration requires greater HRWR demand to maintain a given fluidity. In the case of SCC of a given slump flow consistency, the effect of an increase in VEA concentration is shown to depend on the type of VEA and HRWR in use, as indicated in Fig. 8.11.[22] In total, 13 self-consolidating concrete (SCC) mixtures with a w/p of 0.57 were prepared with welan gum, diutan gum, HPMC and modified starch incorporated at various concentrations along with two types of compatible HRWRs. The HRWR dosage was adjusted to maintain an initial slump flow of 665±15 mm. The concrete

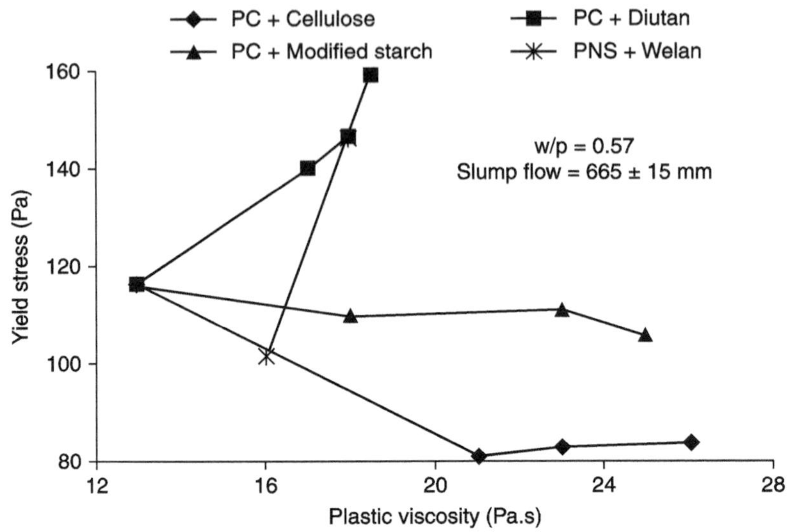

8.11 Effect of VEA type and concentration on yield stress and plastic viscosity of SCC of constant slump flow.[22]

incorporated a 30% replacement of the total powder content by limestone filler. A modified Tattersall MKIII rheometer operating in a co-axial mode was used.

The results show that both yield stress and plastic viscosity are affected by the dosage and type of VEA. In the case of the SCC made with diutan gum and PC, both yield stress and plastic viscosity are shown to increase with the VEA dosage, despite the increase in HRWR dosage needed to maintain a constant slump flow of 665±15 mm. For the welan gum and PNS mixture, a large increase in yield stress was observed. For the SCC made with the modified starch, VEA and PC, the increase in these admixtures led to a net increase in plastic viscosity with limited effect on yield stress. A similar increase was observed with the cellulose-based VEA and PC, except that the VEA mixtures exhibited lower yield stress than the non-VEA concrete, which had lower HRWR demand.[22] Cellulose-based and modified-starch and PC mixtures exhibited greater plastic viscosity at higher VEA concentrations than those with welan gum or diutan gum. The increase in welan gum led to a significant increase in PNS demand; the increase of welan gum dosage from 0 to 0.08%, by mass of water, necessitated approximately 0.6% to 0.9% PNS, by active mass of the HRWR, by mass of cement and limestone filler (powder). A similar effect was observed with the diutan gum and PC concrete where the increase in diutan gum dosage from 0 to 0.05%, by mass of water, necessitated approximately 0.08% to 0.18% PC. Such increase in PC demand was limited to 0.11% for the modified starch and PC concrete and the cellulose-based VEA and PC mixture.[22]

8.4 Effect of viscosity-enhancing admixtures on stability of cement-based systems

8.4.1 Effect of VEA–HRWR combinations on stability of cement paste

As discussed earlier, highly fluid cement-based materials are prone to bleeding, sedimentation and segregation phenomena due to relative migration of liquid and solid phases under the force of gravity before setting occurs. The incorporation of VEA molecules can adsorb some of the free water in the system, thus reducing bleeding and surface settlement. The enhanced yield stress and plastic viscosity, as well as the structural build-up of the static yield stress with rest due to the addition of VEA, can improve the capacity of the liquid phase to suspend solid particles and decrease the rate of sedimentation (Stokes' law).

The performance of the VEA correlates with its ability to alter the rheology of paste and mortar. The relation between the stability and mini-slump spread for cement pastes incorporating various dosages of HPMC and welan gum together with either PNS- or PC-type HRWR is presented in Fig. 8.12. The stability is expressed in terms of equilibrium stability index, which characterizes the overall homogeneity of the sample at the end of sedimentation experiment; the studied system is as much prone to bleeding–sedimentation–segregation phenomena as its value deviates from one.[13] Both VEAs improve stability while reducing the fluidity of cement pastes as VEA dosage increases. However, the data in Fig. 8.12 suggest that the presence of VEA lowers the mini-slump spread significantly (increased yield stress and plastic viscosity), while the stability is only slightly improved for PNS/HPMC and PC/welan gum combinations. In view of opposing effects of VEA and HRWR, we can declare these admixtures incompatible. On the

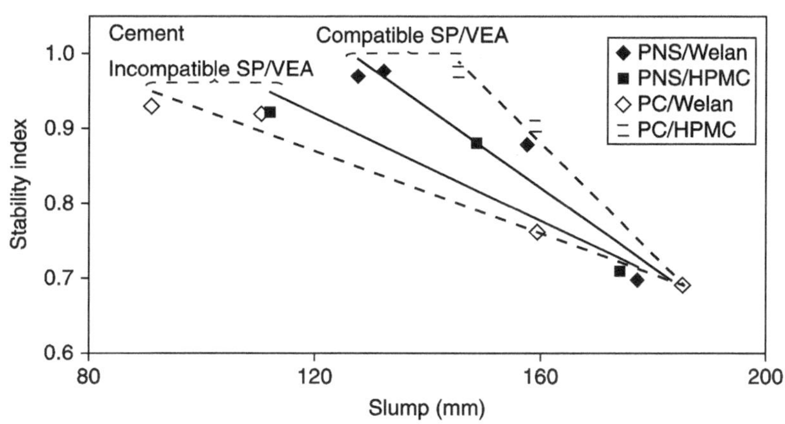

8.12 Stability of cement pastes as a function of paste fluidity (w/c = 0.65, constant dosage of HRWR).[13]

other hand, a significant improvement of the overall stability and a small slump spread (fluidity) reduction with an increase in VEA dosage should be observed in the case of a compatible HRWR/VEA combination (PC/HPMC and PNS/welan gum). The incompatibility between cellulose-based VEAs and PNS-type HRWR has also been reported elsewhere.[23,24] Studies carried out on welan gum showed no apparent incompatibility with either melamine-based or naphthalene-based HRWR;[25] however, some researchers reported its incompatibility with certain PC HRWR.[13,22]

8.4.2 Effect of VEA on stability of self-consolidating concrete

Several investigators have reported that the incorporation of VEA can enhance the stability of SCC.[4,22,26–30] The incorporation of the proper concentration of VEA, along with enough HRWR to secure the targeted deformability and passing ability of SCC, can greatly enhance the resistance of the concrete to bleeding, segregation and surface settlement. This is especially the case of SCC mixtures made with relatively low binder content (425 kg/m^3) and/or water-to-cementitious materials ratio (w/cm) greater than 0.40.[31] An example of the benefit of using VEA to enhance segregation resistance is illustrated in Fig. 8.13. The investigated SCC is designated for cast-in-place commercial construction and is proportioned with a total of 280 kg/m^3 of ordinary Portland cement and 120 kg/m^3 of limestone

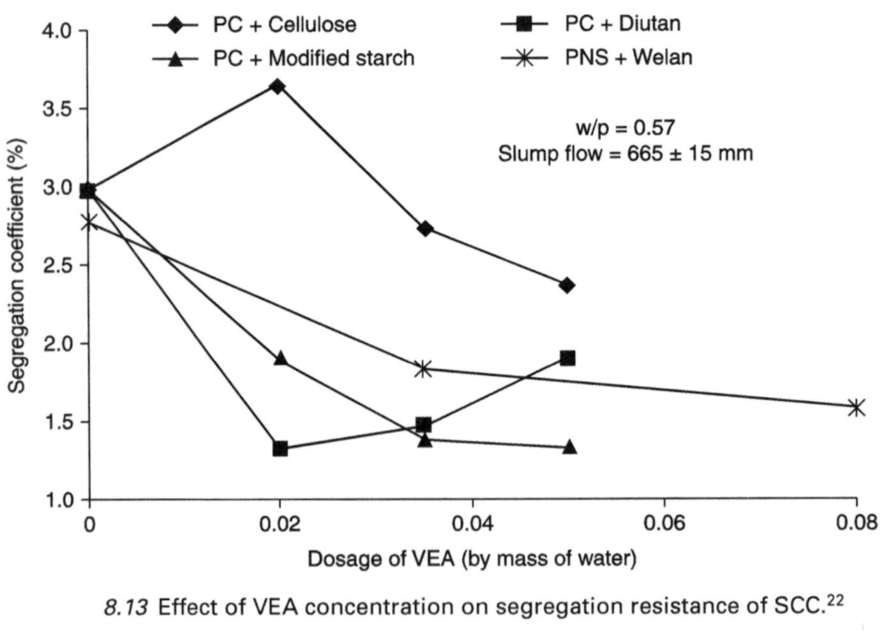

8.13 Effect of VEA concentration on segregation resistance of SCC.[22]

powder having a mean diameter of 10 microns.[22] The w/p is set to 0.57. As indicated in Section 8.3.4, the SCC mixtures were proportioned with four types of VEA and two compatible HRWR types added to secure an initial slump flow of 665±15 mm.

The results indicate that the increase in VEA content considerably decreases the segregation. Segregation is determined by casting concrete in a column measuring 660 mm in height and 200 mm in diameter, and then determining the variation in the relative concentration of coarse aggregate at four sections along the concrete sample. The concrete was consolidated five times using a 20-mm diameter rodding bar and left to rest for 15 minutes before testing. The coefficient of variation of the aggregate distribution along the column is used to quantify segregation. The increase in VEA is shown to reduce segregation, especially in the case of SCC made with diutan gum or modified starch along with the PC-based HRWR. The incorporation of the cellulose-based VEA did not secure high resistance to segregation. In fact, a small addition of that VEA resulted in greater segregation, possibly due to the higher demand of HRWR which led to a considerable reduction in yield stress (Fig. 8.11).

8.4.3 Use of VEA to enhance robustness of self-consolidating concrete

Given the sensitivity of SCC to small variations in raw material characteristics and daily changes in production conditions, care should be taken in the selection of constituent materials and for the mixture proportioning of such concrete to secure adequate robustness. Control of robustness is an essential step in the production of reliable concrete since variations of a single mixture parameter could adversely change the flow properties of the concrete, including segregation and blocking, if the concrete is too fluid or insufficient fluidity to ensure proper filling of the formwork in the absence of mechanical consolidation. The incorporation of VEA can enhance the robustness of SCC even in mixtures where the VEA is not required to ensure stability (i.e. mixtures with low w/cm).

For SCC proportioned with approximately 550 kg/m^3 of powder (of which 39% consists of sand dust) and a high w/p of 0.53, the incorporation of welan gum was reported to substantially reduce the sensitivity of the concrete to changes in concrete temperature (10 to 30°C), fineness modulus of sand (2.08 to 3.06) and Blaine fineness of the cement (3180 to 3420 cm^2/g) compared to a similar SCC made without any VEA.[32] Similar findings were reported regarding changes in sand humidity and HRWR dosage.[30,33] The incorporation of VEA can reduce variations in workability of SCC due to changes in sand humidity (±1% of saturated surface-dry condition) and HRWR content (±10% of the targeted value).[30] The investigated SCC was prepared with 0.37 w/c and 470 kg/m^3 of Type I/II (MH) cement. Both PNS- and PC-based HRWRs were used along with different types of VEAs (welan gum, two diutan gums of different particle fineness, HPMC and modified starch). SCC made with PNS-based HRWR and no VEA was found to be more robust than similar concrete with PCE-based HRWR and no VEA, as well as SCC made with PCE-based HRWR and HPMC. For SCC with PC-based HRWR, the incorporation of modified starch led to greater robustness than using the coarsely ground diutan gum or the HPMC VEA. For SCC with PNS-based HRWR, the finely ground diutan gum was more effective in enhancing robustness than the welan gum. This difference was attributed to the high molecular weight and higher fineness of the former VEA; the latter can affect the rate of hydration of the water-soluble polymer in water and its ability to fix free water from the concrete mixture.

8.5 References

1. Khayat, K.H., Use of viscosity-enhancing admixtures in cement-based systems – An overview, *Cement and Concrete Composites*, 1998, Vol. 20, pp. 171–188.
2. Kawai, T. *Non-dispersible underwater concrete using polymers. Marine concrete.* International Congress on Polymers in Concrete, Brighton, UK, 1987, Chapter 11.5.
3. Sakata, N., Yanai, S., Yoshizaki, M. and Phyfferoen, A. *Evaluation of S–657 biopolymer as a new viscosity-modifying admixture for self-compacting concrete.* Proceedings of 2nd International Symposium on Self-Compacting Concrete, Tokyo, Japan, 2001, pp. 229–236.
4. Rols, S., Ambroise, J. and Pera, J. Effects of different viscosity agents on the properties of self-leveling concrete. *Cement and Concrete Research*, 1999, Vol. 29, pp. 261–266.
5. Pera, J., Ambroise, J., Rols, S. and Chabannet, M. *Influence of starch on the engineering properties of mortars and concretes.* Proceedings of 20th International Conference on Cement Microscopy, Lyon, 1998, pp. 120–128.
6. Sari, M., Prat, E. and Labastire, J.-F. High strength self-compacting concrete – original solutions associating organic and inorganic admixtures. *Cement and Concrete Research*, 1999, Vol. 29, pp. 813–818.
7. Mailvaganam, N.P. *Miscellaneous admixtures. Concrete Admixture Handbook.* Ramachandran, V.S. ed., Noyes Publications, 2nd edition, 1995, pp. 939–1024.
8. Liu, Q, Zhang, Y. and Laskowski, J.S., The adsorption of polysaccharides onto mineral surfaces: an acid/base interaction, *International Journal of Mineral Processing*, 2000, Vol. 60, pp. 229–245.
9. Pourchez, J., Grosseau, P. and Ruot, B., Changes in C_3S hydration in the presence of cellulose ethers, *Cement and Concrete Research*, 2010, Vol. 40, pp. 179–188.
10. Yammamuro, H., Izumi, T. and Mizunuma, T. *Study of non-adsorptive viscosity agents applied to self-compacting concrete.* Proceedings of 5th CANMET/ACI International Conference on Superplasticizers and Other Chemical Admixtures in Concrete, Malhotra, V.M. ed., ACI SP–173, Rome, 1997, pp. 427–444.
11. Nawa, T., Izumi, T. and Edamatsu, Y. *State-of-the-art report on materials and design of self-compacting concrete.* Proceedings of International Workshop on Self-Compacting Concrete, Japan, 1998, pp. 160–190.
12. Khayat, K.H. and Yahia, A. Effects of welan gum – high range water reducer combination on rheology of cement grout, *ACI Materials Journal*, 1997, Vol. 94, pp. 365–372.
13. Mikanovic, N., Sharman, J., Jolicoeur, C., Khayat, K. and Pagé, M. *Compatibility of viscosity-enhancing agents and superplasticizers in cementitious and model systems:*

rheology, bleeding and segregation. Proceeding of 9th CANMET/ACI International Conference on Superplasticizers and Other Chemical Admixtures in Concrete, Malhotra, V.M. ed., ACI SP262–5, Seville, Spain, 2009, pp. 67–84.

14. Khayat, K.H. Effects of anti-washout admixtures on fresh concrete properties. *ACI Materials Journal*, 1995, Vol. 92, pp. 164–171.
15. Skaggs, C.B., Rakitsky, W.G. and Whitaker, S.F. *Application of rheological modifiers and superplasticizers in cementitious systems*. Proceeding of 4th CANMET/ACI International Conference on Superplasticizers and Other Chemical Admixtures in Concrete, Malhotra, V.M. ed., ACI SP–148, Montréal, 1994, pp. 189–207.
16. Lee, E.J. and Chandrasekaran, R. X-Ray and computer modeling studies on gellan-related polymers: Molecular structures of welan, S–657, and rhamsan. *Carbohydrate Research*, 1991, Vol. 214, pp. 11–24.
17. Morris, E.R., Gothard, M.G.E., Hember, M.W.N., Manning, C.E. and Robinson, G. Conformational and rheological transitions of welan, rhamsan and acylated gellan. *Carbohydrate Polymers*, 1996, Vol. 30, pp. 165–175.
18. Ghio, V.A., Monteiro, P.J.M. and Demsetz, L.A. The rheology of fresh cement paste containing polysaccharide gums, *Cement and Concrete Research*, 1994, Vol. 24, pp. 243–249.
19. Lachemi, M., Hossain, K.M.A., Lmbros, Y., Nkinamumbanzi, P.-C. and Bouzoubaa, N., Performance of new viscosity-modifying admixtures in enhancing the rheological properties of cement paste, *Cement and Concrete Research*, 2004, Vol. 34, pp. 185–193.
20. Kwasny, J., Sonebi, M., Taylor, S.E., Bai, Y., Basheer, P.A.M., *et al.* The influence of different viscosity-modifying admixtures on the fresh properties of superplasticized cement-based grouts. Proceedings of 3rd International RILEM Symposium on Rheology of Cement Suspensions such as Fresh Concrete, Reykjavik, Iceland, 2009, pp. 208–218.
21. Khayat, K.H., Saric-Coric, M. and Liotta, F., Influence of thixotropy on stability characteristics of cement grout and concrete, *ACI Materials Journal*, 2002, Vol. 99, No. 3, pp. 234–241.
22. Mechaymech, A. *Effect of viscosity-enhancing admixtures on rheology and stability of flowable cementitious systems* [in French]. PhD thesis, Université de Sherbrooke, 2010.
23. Saucier, K.L. and Neeley, B.D. Antiwashout admixtures for underwater concrete. *Concrete International*, 1987, Vol. 9, pp. 42–47.
24. Kawai, T. *Non-dispersible underwater concrete using polymers. Marine concrete.* International Congress on Polymers in Concrete, Brighton, UK, 1987, Chapter 11.5.
25. Kawai, T. and Okada, T. Effect of superplasticizer and viscosity-increasing admixture on properties of lightweight aggregate concrete, in *Superplasticizers and Other Chemical Admixtures in Concrete*, ACI-SP–119, 1989, pp. 583–604.
26. Khayat, K.H. Workability, testing, and performance of self-consolidating concrete. *ACI Materials Journal*, 1999, Vol. 96, No. 3, 1999, pp. 346–353.
27. Petrov, N., Khayat, K.H. and Tagnit-Hanou, A. *Effect of stability of self-consolidating concrete on the distribution of steel corrosion characteristics along experimental wall elements*. Proceedings of 2nd International Symposium on Self-Compacting Concrete, Ozawa, K. and Ouchi, M., ed., Tokyo, Japan, October 2001, pp. 441–450.
28. Long, W. J., Khayat, K.H. and Xing, F. Statistical models to predict fresh properties of self-consolidating concrete, *Advanced Materials Research*, 2010, Vol. 129–131, pp. 853–856.

29. Khayat, K.H., Hwang, S.-D. and Belaid, K. Performance of cast-in-place self-consolidating concrete made with various types of viscosity-enhancing admixtures. *ACI Materials Journal*, 2010, Vol. 107, No. 4, pp. 403–412.
30. Naji, S., Hwang, S.-D. and Khayat, K.H. Robustness of self-consolidating concrete incorporating different viscosity-enhancing admixtures, *ACI Materials Journal*, 2011, Vol. 108, No. 4, pp. 432–438.
31. Khayat, K.H. and Mitchell, D. *Self-consolidating concrete for precast, prestressed concrete bridge elements*. National cooperative highway research program (NCHRP). NCHRP Report 628, Transportation Research Board of the National Academies, Washington DC, 2009.
32. Yurugi, M., Sakai, G. and Sakata, N. *Viscosity agent and mineral admixtures for highly fluidized concrete*. Proceedings of International Conference on Concrete under Severe Conditions: Environment and Loading, Sakai, K., Banthia, N. and Gjorv, O., ed., Vol. 2, 1995, pp. 995–1004.
33. Billberg, P. and Khayat, K. *Use of viscosity-modifying admixtures to enhance robustness*. Proceedings of 3rd North American Conference on the Design and Use of Self-Consolidating Concrete, November 2008, Chicago.

9
Fibre reinforcement and the rheology of concrete

S. GRÜNEWALD, Delft University of Technology, The Netherlands

Abstract: The addition of fibres can improve the performance of cementitious materials in the hardened state. With an optimized mixture composition and controlled rheology and quality, fibres can become more effective. Significant progress has been made during the past years on the field of flow simulations and the rheology of fibre suspensions. The main difficulties are related to the non-Newtonian behaviour of fibre suspensions (i.e. shear-thinning due to fibre orientation and local flow-induced structures) and difficulties in predicting and measuring the different contributions of, for example, hydrodynamic effects and mechanical interaction. Fibre rheology and flow simulation are excellent tools to optimize fibre suspensions and form the basis for predictions of structural performance.

Key words: fibres, cementitious materials, rheology, flow simulation, fibre orientation.

9.1 Introduction

This chapter addresses aspects related to fibres and rheology. Rheology and the simulation of flow are important to understand and to optimize the structural behaviour of fibre-reinforced cementitious materials. Rheologically speaking, fibre suspensions are non-Newtonian fluids and can display normal stress differences. Fibre simulation is a branch of mechanics that deals with modelling the dynamics and rheology of particles of large aspect ratio, or fibres, with the goal of predicting the properties of fibre suspensions and networks.

Many studies have demonstrated the benefits of adding fibres to cementitious materials. Fibres can improve the ductility, increase tensile as well as shear strength, are able to enhance fatigue strength and impact resistance, can counteract shrinkage cracking and prevent spalling in case of fire loading. Fibres have been applied in cementitious materials as reinforcement to replace rebars, to improve the structural performance and for crack width control. Innovative, slender structures can be designed with, for example, engineered cementitious composites (Li, 2002) and ultra high performance concrete (Resplendino, 2011), taking into account the specific tensile behaviour.

The effect of fibres on workability has been recognized and studied for many years due to the need to produce structures with a 'workable' concrete. Recent developments in material technology show that fibre dosages can be significantly increased with adjusted mixture compositions and production techniques, i.e. concrete with self-compacting ability. In contrast to traditionally vibrated concrete, highly flowable/workable concrete requires little or no vibration energy to be placed. This implies that not only the mixture composition but also the production process have to be 'tailored' to the devised application, finally aiming at an optimized structural performance. The performance of flowable cementitious materials depends on the material behaviour, the production method and influences related to the structure (Fig. 9.1).

Fibres can be more efficient in the hardened state with an optimized mixture composition and a controlled rheological behaviour. This is achieved by minimizing mechanical interaction and interlocking of fibres during the flow and the production stage. Due to the flow or strain of the concrete, fibres are able to orient; this makes the prediction of the structural behaviour of fibre-reinforced concrete more complex, but it also offers the potential for an improved structural performance. Synergetic effects were observed with flowable concrete on the micro-level (matrix-fibre interaction: Markovic *et al.*, 2002), the macro-level (preferred fibre orientation: Ferrara *et al.*, 2010 and Laranjeira *et al.*, 2011) and the structural level (i.e. due a lower variation of fibre orientation at increasing alignment of the fibres: Laranjeira *et al.*, 2010). With flowable concrete, special attention is required to obtain the desired performance.

Rheological characteristics are essential input parameters for flow simulations of both vibrated and flowable fibre concretes. The position and the orientation of fibres in a structure can differ from the assumed isotropic orientation and distribution due to dynamic/static segregation or floating of fibres, blocking during the flow and/or the orientation of fibres caused by the flow and induced by walls. As well as for cementitious materials, the rheology of fibre suspensions is

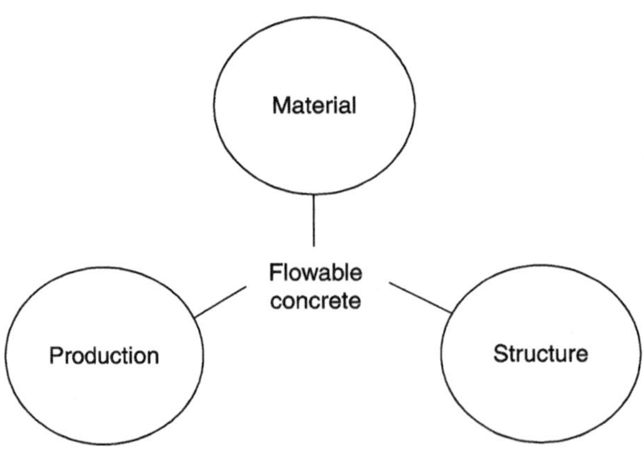

9.1 Structural performance as a result of three components: material behaviour, production effects and structural boundaries (Grünewald *et al.*, 2010).

relevant to other industrial suspensions like polymers, which contain rod-like particles (i.e. Kevlar, nylon, glass or wollastonite).

9.2 Fibres in cementitious materials

9.2.1 Introduction to fibres used in cementitious materials

Fibres are long, slender particles with a high aspect ratio r ($r = L_f/d_f$: fibre length divided by the fibre diameter) that can cause difficulties during production and processing. In order to optimize cementitious materials with fibres, aspects such as workability and rheology have to be addressed. Due to the elongated shape of the fibres, their effect depends on the position and the orientation within a structure and relative to principal stresses. Below fibres in cementitious materials are discussed, particularly fibre types and their effect on mixture characteristics. The physical characterisation of fibres and their influence on packing density is also addressed.

9.2.2 Fibre types

Steel and plastic fibres are the most common fibre types applied by the building industry; other fibre types like glass and carbon fibres contribute a smaller share to the market. Table 9.1 lists fibre types and material characteristics.

Fibres affect the characteristics of concrete in the fresh state. They are needle-like particles that increase the resistance to flow and contribute to the formation of an internal structure of aggregate grains and fibres. The effect of fibres on the

Table 9.1 Fibre types and material characteristics (Holschemacher and Dehn)

Fibre type	Typical fibre diameter (μm)	Typical fibre length (mm)	Density (g/cm3)	E-modulus (kN/mm2)	Tensile strength (N/mm2)	Elongation at break (%)
Steel fibre						
– hooked end	500–1300	30–60	7.85	160–210	>1000	3–4
– crimped	400	26–32	7.85	210	980	
AR-glass fibre	3–30	3–25	2.68–2.70	72–75	1500–1700	1.5–2.4
Polypropylene fibre						
– monofilament	18–22	6–18	0.91	4–18	320–560	8–20
– fibrillated	50–100	6–19	0.91	3.5–10	320–400	5–15
Polyacrylnitrile fibre	18–104	4–24	1.18	15–20	330–530	6–20
Carbon fibre	5–10	6	1.6–2.0	150–450	2600–6300	0.4–1.6

workability is mainly due to four reasons: First, the shape of the fibres is more elongated compared to the aggregates; the surface area at the same volume is larger. Second, stiff fibres change the structure of the granular skeleton, whereas flexible fibres fill the space between them. Stiff fibres push apart particles that are relatively large compared to the fibre length, which increases the porosity of the granular skeleton. Third, the surface characteristics of fibres differ from that of cement and aggregates, i.e. plastic fibres might be hydrophilic or hydrophobic. Finally, fibres can be deformed (i.e. have hooked ends, be crimped or wave-shaped) to improve the anchorage between them and the surrounding matrix.

With long and stiff steel fibres (aspect ratio 45–80, fibre length 30–60 mm), the mechanical interaction of fibres and aggregates dominates the flow behaviour; the surface area of these fibre types is comparable to coarser sand fractions. In contrast, thin plastic fibres and fibres at a very high aspect ratio can have a much higher surface area. Plastic fibres mainly affect the rheological behaviour of the paste. Due to their flexibility, the mechanical interaction with aggregates is much less pronounced but they can still form a network with aggregates and other fibres. The flexibility of fibres has two effects: 'the deformation of fibres' and 'the formation of entangled structures' (Yamanoi *et al.*, 2010). The shear viscosity increased at increasing flexibility (Keshtar *et al.*, 2009). The presence of steel fibres affects the yield stress of concrete but alters the paste characteristics to a minor degree. A structure produced with a fibre-reinforced concrete still can have a smooth and faultless surface, even at a relatively small slump flow (550–600 mm) or medium slump (100–150 mm).

The stiffness of fibres can be calculated using Eq. 9.1 (Martinie *et al.*, 2010). The lower the ratio f/L_f (f: flexibility, L_f: fibre length) the stiffer the fibre behaves. The ratio f/L_f was 0.03% for a stiff steel fibre (r = 50, E = 190000 MPa, self-compacting concrete: τ_0 = 50 Pa), whereas it is 66% for a flexible carbon fibre (r = 500, E = 210000 MPa, traditional vibrated concrete: τ_0 = 1000 Pa).

$$\frac{f}{L_f} \cong \frac{\tau_0 \cdot r^3}{E} \qquad [9.1]$$

where f = fibre flexibility (mm^4), E = E-modulus (MPa) and τ_0 = yield value (Pa).

In order to optimize the performance of an individual fibre, fibres need to be homogenously distributed in a matrix and clustering of fibres has to be counteracted. The size of the fibres relative to the aggregates determines their distribution (Johnston, 1996). To be effective in the hardened state it is recommended to choose fibres not shorter than the maximum aggregate size (Johnston, 1996; Vandewalle, 1993). Usually, the fibre length is 2–4 times that of the maximum aggregate size.

9.2.3 Packing density

According to Yu and Zou (1998), the initial packing density of irregular particles depends on the shape and size of the particles and the applied compaction energy.

The packing density of fibres (when randomly placed) is much lower compared to a single grain fraction. The packing density of fibres is also affected by the stiffness of the fibres. Flexible fibres (i.e. plastic fibres) can be compressed and, at commonly applied fibres dosages, have little effect on the packing density of the aggregates. Placing fibres in an orderly (aligned) manner, which cannot be realised by mixing cementitious materials, can result in a much higher packing density. Zou and Yu (1996) proposed a formula (Eq. 9.2) to approximate the experimental (dense and randomly placed) packing density of cylinders. The 'sphericity' (Eq. 9.3) is the ratio of the surface area of the sphere, having the same volume as the particle and its actual surface area; the 'volume diameter' is the diameter of a sphere having the same volume as the particle (Yu *et al.*, 1993). The comparison of predicted values obtained with Eq. 9.2 and experimental results on the packing density carried out with a variety of (stiff) steel fibres showed a good agreement (Grünewald, 2004). Crimped and hooked ends fibres were applied in this study.

$$\ln \varepsilon_{Od,cyl} = \Psi^{6.74} \cdot \exp[8.00 \cdot (1-\Psi)] \cdot \ln 0.36 \qquad [9.2]$$

where $\varepsilon_{Od,cyl}$ = dense (experimental) packing density of fibres and ψ = sphericity, with Yu *et al.* (1993):

$$\psi = 2.621 \cdot \frac{(L_f / d_f)^{2/3}}{1 + 2 \cdot (L_f / d_f)} \qquad [9.3]$$

where d_f = fibre diameter (mm) and L_f = fibre length (mm).

Philipse (1996) showed that, for high aspect ratio bodies, the packing density can be calculated with: $\phi_c = \alpha_c/r$ (loose packing) and $\phi_m = \alpha_m/r$ (dense packing). In the aspect ratio range of 50–100, the parameters α_c and α_m were found to be 3.2 and 4.0, respectively (Martinie *et al.*, 2010). The relative size of fibres and aggregates also affects the packing density. Figure 9.2 shows results of packing experiments (average of three measurements) with (1.5 vol.-%) and without steel fibres for different sand contents (sand: 0–4 mm, coarse aggregates 4–16 mm). The actual fibre content in concrete at 1.5 vol.-% of the granular skeleton is lower since concrete also contains paste: assuming a paste content of 38 vol.-% in self-compacting concrete, the content of steel fibres equals 0.93 vol.-%. The higher the aspect ratio and the lower the content of sand, the more pronounced is the effect on packing density. At a sand content of 75 vol.-%, about the same packing density was found for different types of steel fibres. The maximum packing density decreases after the addition of steel fibres and shifts towards higher sand contents. To compensate for the effect of the fibres, the mixture composition has to be adjusted by increasing the content of grains that are relatively small compared to the fibre length. It was not possible to test 3.0 vol.-% of fibres having an aspect ratio of 60 or higher since fibre clustering counteracted a homogenous distribution of the fibres.

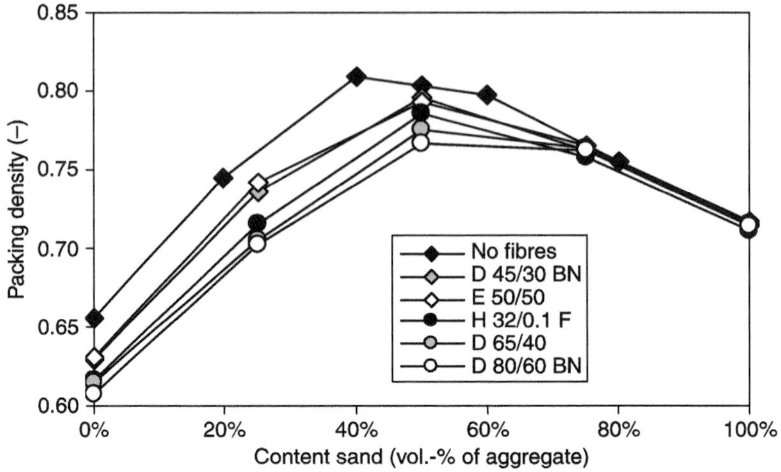

9.2 Effect of the sand content and the type of the steel fibres (at 1.5 vol.-%) on packing density (Grünewald, 2004).

Stiff fibres can be implemented in numerical calculations to determine the packing density (i.e. with the 'compressible packing model (CPM)'; de Larrard, 1999 (cf. Chapter 7)) by applying one of the following approaches:

Concept of the 'perturbed zone'

The perturbed zone is the disturbed area (with a locally decreased packing density) close to a wall, in this case the fibre (de Larrard, 1999). The mean packing density can be calculated using Eq. 9.4. The perturbed volume v_P can be determined according to Table 9.2, depending on the size and shape of the fibres.

$$\bar{\alpha} = (1 - \phi_f - N_{sf} \cdot v_p) \cdot \alpha \qquad [9.4]$$

where $\bar{\alpha}$ = mean packing density (affected by the container size) (–), ϕ_f = percentage of fibres of the granular skeleton (–), N_{sf} = number of steel fibres (–), v_p = perturbed volume in a container (vol.–%) and α = unperturbed packing density (–).

Concept of the 'equivalent packing diameter'

Yu *et al.* (1993) proposed the concept of the 'equivalent packing diameter' to include non-spherical particles in numerical simulations. In their approach, the shape and the dimension of a non-spherical particle were related to the diameter of a fictitious sphere having an equivalent diameter that did not result in a change of the packing density when combined with a spherical grain of the same diameter. Equation 9.5 was proposed to calculate the equivalent packing diameter for a particle with a cylindrical shape. The volume diameter d_v can be calculated according to Eq. 9.6:

Table 9.2 Calculation of the perturbed volume

Cross-section fibre	Perturbed volume	*R*	L	A	B
Round with diameter d_f	$\pi \cdot R^2 \cdot L$	$0.5 \cdot (d_f + k_f \cdot d)$	$L_f \cdot d/2$	–	–
Rectangular with sides $a_f + b_f$	A·B·L	–	$L_f d/2$	$a_f + k_f \cdot d$	$b_f + k_f \cdot d$

Note: $k_f = 0.065$, d = grain diameter

$$d_p = (3.1781 - 3.6821 \cdot \frac{1}{\psi} + 1.5040 \cdot \frac{1}{\psi^2}) \cdot d_v \qquad [9.5]$$

where d_p = equivalent packing diameter (mm) and ψ = sphericity (–).

$$d_V = 1.145 \cdot (L_f / d_f)^{1/3} \cdot d_f \qquad [9.6]$$

where d_f = fibre diameter (mm) and L_f = fibre length (mm).

9.2.4 Mix design principles for fibre-reinforced concrete

The size, shape and content of the coarse aggregates as well as the geometry and volume fraction of steel fibres affect the workability of concrete (Swamy, 1975). The maximum fibre content is the critical fibre dosage at which the compactability/workability drastically decreases and/or the content of the fibres beyond which fibre balling takes place. Figures 9.3 to 9.5 show flow patterns (slump flow test) that indicate that the maximum fibre dosage is exceeded (Grünewald, 2004). The type of flow pattern depends on the fibre type (Fig. 9.3: fibres with a high surface area, Fig. 9.4: long stiff fibres and Fig. 9.5: intermediate aspect ratio, stiff fibres).

The relative fibre to coarse aggregate volume and the 'balling up' phenomenon govern the maximum possible content of steel fibres (Swamy and Mangat, 1974). Depending on the mixture composition, the maximum concentration of steel fibres was reached at a fibre factor ($V_f \cdot L_f / d_f$) between 0.3 and 1.9 (Grünewald, 2004). Mix design methods for fibre-reinforced concrete have been proposed by Rossi (1990), Grünewald (2004), Ferrara *et al.* (2007) and Martinie and Roussel (2010).

9.3 Fibre rheology

9.3.1 Introduction to fibre rheology

Cementitious materials with fibres are often highly concentrated suspensions. The interaction of fibres and fibres with grains (fillers, cement and aggregates) determines the kind of network they form. In flowing suspensions, three kinds of forces coexist to various degrees (Barnes *et al.*, 1989):

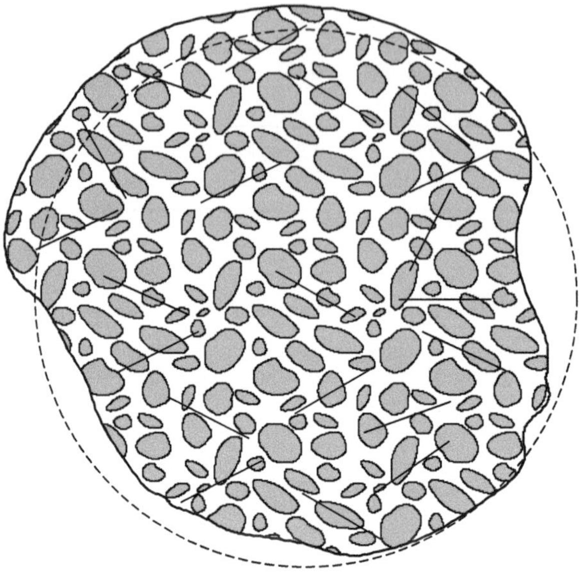

9.3 Flow pattern A, decreased flowability due to a high fibre surface area.

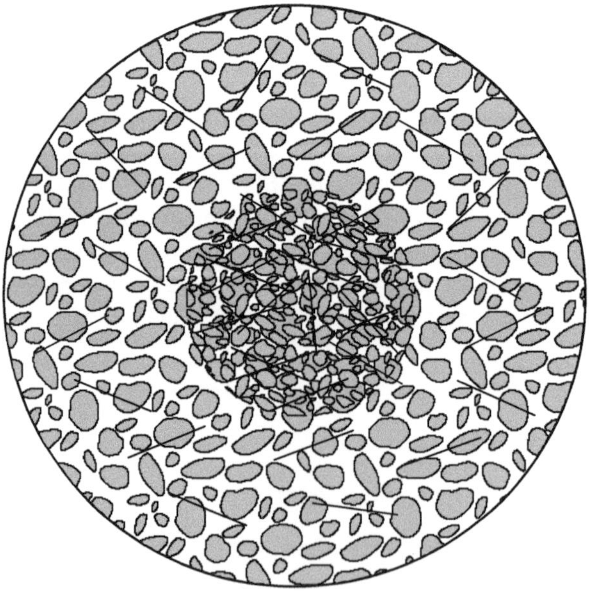

9.4 Flow pattern B, caused by unstable mix and/or high dosage of long fibres.

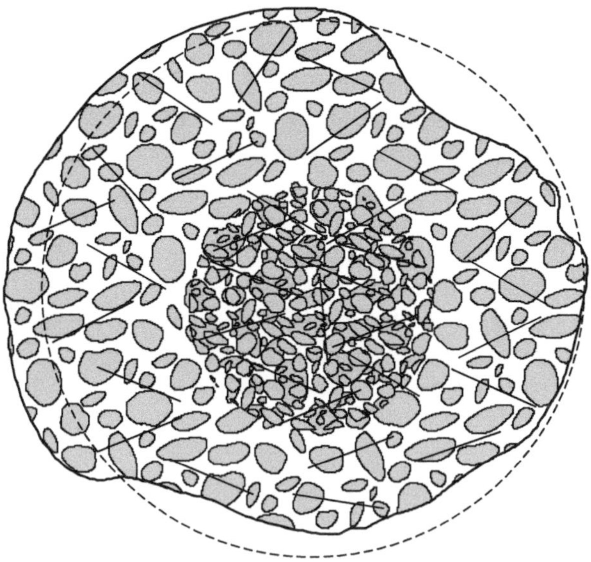

9.5 Flow pattern C, Flow patterns A and B combined, obstructed flow and fibre clustering.

- forces of colloidal origins (interactions between particles result in repulsion or attraction);
- Brownian randomising force, which ensures that particles (mainly smaller than 1 μm) are in constant movement;
- viscous forces acting on particles, which are proportional to the local velocity difference between the particle and the surrounding fluid.

Fibre–fibre interactions include hydrodynamic effects as well as mechanical contacts. The phase volume of solids (Φ) suspended in a liquid determines the extent to which hydrodynamic forces act on the surface of the particles. Resistance to flow arises because particles have to move out of each other's way. The slender shape of fibres causes them to rotate and to orientate during the flow. This motion is affected and/or counteracted by neighbouring particles/fibres. Constitutive equations and fibre orientation evolution equations can be applied to model the behaviour of fibre suspensions. Direct simulations can be executed with the input of acting forces to predict fibre motion and stresses in a suspension.

9.3.2 Influence of fibres on viscosity and yield value

Concerning their rheological effect, fibre suspensions can be divided in four regimes: dilute, semi-dilute, semi-concentrated and concentrated Table 9.3 (Fan *et al.*, 1998). Fibres are able to rotate freely in the diluted regime; at higher concentrations (semi-dilute and semi-concentrated regimes), the free rotation of the fibres is counteracted to some degree by mechanical contacts.

Fibres interact only via hydrodynamic effects in the semi-dilute state, whereas direct contact governs the semi-concentrated state. The fibre–fluid interaction governs the dilute and semi-dilute regimes, whereas with concentrated suspensions (and long fibres), the fibre–fibre interaction is dominating (Thomasset *et al.*, 1997). Increasing the fibre dosage also increases possible fibre interactions. In the concentrated state, the maximum packing density is reached. Martinie *et al.* (2010) defined three regions with different degrees of fibre interaction (Table 9.4).

Einstein (1906, 1911) predicted the viscosity of dilute dispersed suspensions (phase volume ≤ 10%) with Eq. 9.7. Batchelor (1977) extended Eq. 9.7 and included the effect of particles of other sizes by adding higher order terms.

$$\eta = \eta_S \cdot (1 + 2.5 \cdot \phi) \qquad [9.7]$$

where ϕ = volume fraction of solids (vol.-%), η = viscosity of suspension (Pa·s) and η_s = viscosity of suspending fluid (Pa·s).

The ratio between the phase volume and the packing density has to be considered in order to predict the flow behaviour of concentrated suspensions like cementitious materials (Sedran, 1999). When the ratio of phase volume to packing density equals one, the flow is counteracted by a solid contact network. The packing concept was applied for the development of ultra high performance concrete: by increasing the packing density a high flowability was maintained and the strength increased. Krieger and Dougherty (1959) applied the packing of particles to predict the viscosity of concentrated suspensions (Eq. 9.8). The intrinsic viscosity (η) depends on the rate of strain, the Péclet number and on the geometry of the suspended particles (Petrie, 1999).

Table 9.3 Rigid fibre suspension classification

$\Phi \cdot r^2$	$\Phi \cdot r$	Φ	Regime
$\Phi \cdot r^2 << 1$	$\Phi \cdot r << 1$	$\Phi << 1$	Dilute
$\Phi \cdot r^2 >> 1$	$\Phi \cdot r << 1$	$\Phi << 1$	Semi-dilute
$\Phi \cdot r^2 >> 1$	$\Phi \cdot r >> 1$	$i << 1$	Semi-concentrated
$\Phi \cdot r^2 >> 1$	$\Phi \cdot r >> 1$	$\Phi \cong 1$	Concentrated

Source: Martinie *et al.* (2009)

Table 9.4 Fibre interaction dependent on the packing density

Regime	Fibre volume	Influence
1	$\Phi_{fc} \leq 3.2/i$	Low influence
2	$\Phi_{fc} > 3.2/r$; $\Phi_{fm} \leq 4.0/r$	Dominated by contact network
3	$\Phi_{fm} > 4.0/r$	No flow, ball formation and air entrained

Note: Φ_{fc}: random loose packing; Φ_{fm}: random dense packing

$$\eta = \eta_s \cdot (1 - \frac{\phi}{\phi_m})^{-[\eta]\cdot\phi_m} \quad [9.8]$$

where ϕ_m = solid volume (dense packing) (vol.-%), η = intrinsic viscosity (–) and η_s = viscosity of suspending fluid (Pa·s).

The packing density of aggregates (without fibres) can be increased by widening the particle-size distribution (cf. Chapter 7). The interaction between stiff fibres and aggregates increases with larger aggregates and, as a consequence, the packing density might decrease when adding larger aggregates with the result that this mixture adjustment not necessarily results in a decrease of the viscosity. Non-spherical particles lead to poorer space-filling (Barnes *et al.*, 1989). Clarke (1967) determined the effect of the shape of particles on viscosity (Fig. 9.6). Dependent on the deviation of fibres from the spherical shape (as a measure, Eq. 9.3 defines 'sphericity'), the viscosity increased. A relative increase of the viscosity at increasing aspect ratio was found by Giesekus (1983). Particle flocculation and local concentrations of particles (like fibre balling) further decrease the packing density and increase the viscosity.

In particular, stiff fibres decrease the packing density and, consequently, increase the viscosity at the same phase volume compared to grains. Two equations have been proposed to predict the critical volume fraction of fibres (ϕ_{cf}) in a suspension: Eq. 9.9 (Kitano *et al.*, 1981) was obtained for polymer melts with average aspect ratios of 6–27, whereas Eq. 9.10 (Pabst *et al.*, 2006) was proposed

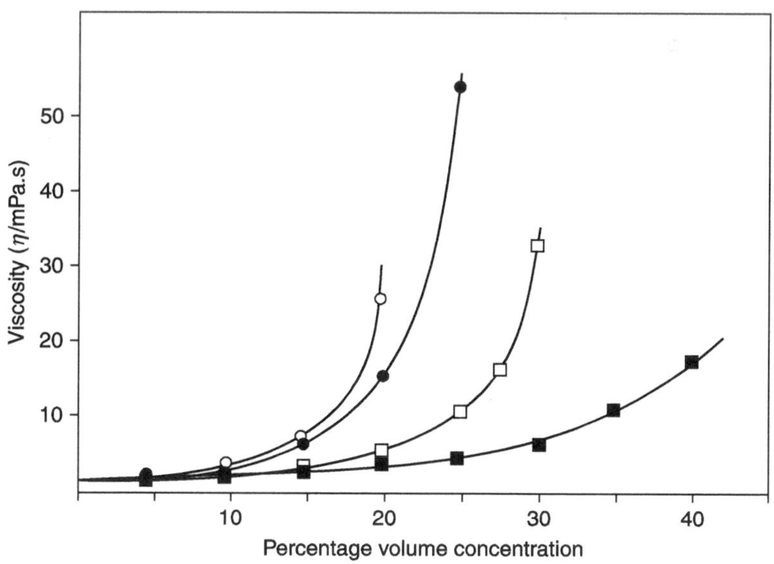

9.6 Dependence of the viscosity of differently shaped particles in water on the concentration at a shear rate of 300 1/s; (■) spheres, (□) grains, (●) plates, (○) rods (Clarke, 1967).

for wollastonite suspensions ($r \leq 22$). Possible reasons for differences in the critical volume fraction are the agglomeration and the presence of high aspect ratio wollastonite particles.

$$\phi_{cf} = 0{,}54 - 0{,}0125 \cdot r \qquad [9.9]$$

$$\phi_{cf} = 0{,}51 - 0{,}0223 \cdot r \qquad [9.10]$$

9.3.3 Shear-thinning/shear-thickening

When properly placed in one direction, the packing density of fibres can be higher than 0.80 (Martinie *et al.*, 2010). This orderly state cannot be obtained when randomly adding and mixing fibres. However, the packing density of fibres in concrete can be increased after mixing by orienting fibres (i.e. during the flow along walls, testing a suspension with a viscometer, or casting by pumping). The flow brings about a more favourable arrangement (Barnes *et al.*, 1989) and rheological characteristics can become 'anisotropic' with a possible favourable rearrangement of fibres in one or two directions. The floc formation of flexible fibres in a suspension is the result of mechanical contact and coherence due to interlocking by the elastic bending of the fibres with shear-thinning being the consequence (Sepehr *et al.*, 2004).

All concentrated suspensions will show shear-thickening at the correct conditions; the circumstances and severity depend on the phase volume, the particle size distribution and the continuous phase viscosity (Barnes *et al.*, 1989). Above a critical shear stress that decreases at increasing phase volume, shear thickening takes place.

9.3.4 Modelling fibre rheology

Fibre suspensions have many degrees of freedom due to the large number and complex geometry of the components, the complexity of the flow and non-linear material behaviour. Most modelling of fibre suspensions has been executed for dilute suspensions and at a defined aspect ratio. The prediction of the flow behaviour of concentrated fibre suspensions is more complex and requires the use of numerical tools in order to take into account the interaction of many particles. Advani and Ait-Kadi (1997) discussed the difficulty in obtaining and comparing experimental results with simulations. The coupling of rheology and fibre orientation dynamics complicates predictions since it also affects the fibre motion. Petrie (1999) reviewed the literature on the rheology of fibre suspensions and paid special attention to modelling. According to Petrie, three steps are required to construct a rheological model:

- Model the motion of an individual fibre in a homogenous, possibly time-dependent, flow field.

- Model the evolution of the distribution of the orientation of many such fibres.
- Calculate the contribution to the bulk stress due to the fibres in the terms of bulk flow.

It has been experimentally confirmed that a rod-like particle rotates in a tumbling motion (Petrie, 1999). The orbit which a fibre describes is independent of time and depends on the individual orientation of the particle and a steady state cannot be obtained without particle interaction. Particle–particle interactions, the polydispersity and non-uniformity of particles mainly affect the rotation of the fibres (Zirnsak *et al.*, 1994). Jeffrey's equation (Jeffrey, 1922; Eq. 9.11) provides a basic mathematical solution to how a single ellipsoid flows and rotates in a viscous Newtonian fluid and has been applied to describe the flow of fibre suspensions. The motion is referred to as Jeffrey's Orbit. The following differential equation describes the evolution of orientation for an ellipsoid, represented by its unit vector p. W is the vorticity tensor and D the rate of strain tensor for the flow unperturbed by the fibre. λ is a parameter dependent only on the slenderness r of the fibres. In the absence of interactions and in concentrated suspensions, fibres spend a large portion of time nearly aligned in the flow direction.

$$\dot{p} = -W \cdot p + \lambda \cdot (D \cdot p - D : ppp) \qquad [9.11]$$

The position and the velocity of particles was re-evaluated at different time steps with non-equilibrium molecular dynamics (NEMD) after providing an initial set of both parameters (Barnes *et al.*, 1989). The output of the calculation is a stress (pressure) tensor with which the viscosity and normal stress differences can be calculated. Simulations showed the formation of two-dimensional layers (a flow-induced structure) for spherical particles. The tendency of a material to form two-dimensional structures causes shear-thinning.

Simulations with the particle simulation method (PSM) were executed to predict the evolution of fibre orientation and the interaction of fibre structures and rheological characteristics for rigid glass fibres (Yamanoi and Maia, 2010) and flexible nylon fibres (Yamanoi *et al.*, 2010). Each fibre was modelled as an array of beads. The translational and the rotational movements of each bead of a fibre were calculated. Hydrodynamic interactions were considered for intra-fibre and lubrication forces to determine inter-fibre interaction; repulsive forces were also taken into account by the PSM method. The calculation according to the stress expression method requires solving constitutive equations. The stress tensor τ, the stresslet S_i and the orientation tensor 'a' are the basic input parameters for the software. The considered concentration region was $1 \leq n{\cdot}d_f{\cdot}L_f \leq 5$ (Yamanoi *et al.*, 2010); n being the number of fibres per unit volume, d_f the diameter of the fibres and L_f the length of the fibres. Rheological properties were determined by parallel plate rheometry. Afterwards, results of simulations were compared with experimental results. The agreement for rigid fibres was good up to the onset of the concentrated regime ($n{\cdot}d_f{\cdot}L_f \leq 3$); the shear viscosity and the elasticity were overestimated in the concentrated regime. The simulations were less accurate with flexible fibres; the discrepancy probably being caused by the fact that hydrodynamic interactions become more important for flexible fibre suspensions. Yamamoto and Matsuoka (1993, 1995a, 1995b, 1996) executed direct simulations with a similar approach and modelled fibres as an array of spheres, which allowed them to predict the flow of rigid as well as flexible fibres.

To simulate the flow of cementitious fibre suspensions, a variety of particles (especially aggregates) concerning size and shape has to be considered. Ausias *et al.* (2006) simulated the flow of fibre suspensions taking into account a distribution of fibre aspect ratios (average r: 19.2 and σ: 3.05) and studied the transient (first stage of flow) and steady state (almost all fibres are oriented) behaviour under simple shear. Hydrodynamic interaction via lubrication forces, contact forces and hydrodynamic forces was taken into account in order to predict the motion and the orientation of the fibres. The effective stresses of fibre suspensions were calculated and showed qualitative agreement with experimental results.

9.3.5 Fibre orientation

By simulating the flow and the orientation evolution of fibres, their position and alignment relative to the direction of stress in a structure can be predicted. This information is essential for the design of structures and the micro- and macro-mechanical analysis of the performance of fibre-reinforced concrete. The assumption of a homogenous fibre distribution has to be verified and can be achieved by adequate mix design and an adjusted production technique. Dependent on the rheological characteristics and the mixture composition, a concrete with fibres can be either ‘rolling’ or ‘flow as a plug’. The former is the case for a low yield value self-compacting concrete at a low fibre loading, whereas the latter is obtained when the concrete is less flowable and/or the fibres form a network, which influences the flowability and the yield value. Consequently, the kind of flow pattern determines whether fibres orient or not. In shear flows, fibres orient in the direction of flow, whereas in extensional flow they tend to a stable orientation perpendicular or parallel to the flow according to the sign of the elongational rate (Vincent *et al.* 1997).

The most common way to calculate the orientation distribution involves a structure tensor (defined as an average of the second moment of the fibre orientation vector) and to obtain a complete set of equations – a closure approximation (Petrie, 1999). Several parameters influence the orientation of fibres (Martinie and Roussel, 2010): the slenderness, the interaction with other particles, the phase concentration and the location of a fibre (proximity to a wall, location within or outside the plug-flow area). The closer to a wall a fibre starts, the quicker it aligns in the direction of the flow. Martinie and Roussel applied Eq. 9.12 to calculate the average orientation factor deduced from calculations with Eq. 9.11. For a two-dimensional flow, the orientation factor relates the behaviour of N fibres relative to the direction d and is calculated as the ratio between the total projection of each unit vector p on this direction and all the projections.

$$\alpha_{\vec{d}} = \frac{\sum_{k=1}^{N} p_{\vec{d}}^{k}}{\sum_{k=1}^{N} (p_1^k + p_2^k)} \qquad [9.12]$$

Barnett *et al.* (2010) studied the orientation and the distribution of steel fibres in ultra high performance fibre-reinforced concrete with X-ray computed tomography (CT) imaging and the electrical sensitivity resistance technique. The pouring method was varied and the results indicated that the fibres aligned perpendicular to the direction of the flow but parallel to the wall of the test specimens. A preferred fibre orientation in the flow direction was observed by Groth and Nemegeer (1999) applying the X-ray technique on hardened concrete slices. Steel fibres in self-consolidating fibre-reinforced concrete (SCFRC) oriented during the flow through the pipe of the truck mixer, whereas in the absence of walls (in a tunnel segment), they remained aligned and were mainly oriented perpendicular to the flow direction (Grünewald, 2004). The orientation of the fibres depended on the position and the direction of tested cylinders in the segment. The flow of concrete in the tunnel segment was more or less comparable with a plug flow in a pipe, which was the result of the rheological characteristics of the applied SCFRC. How fibres are oriented in a structure depends on the rheological characteristics, the mixture composition and the flow 'history' during production.

9.4 Rheology of fibre concrete

9.4.1 Introduction to theology of fibre concrete

With the development of self-compacting concrete, ultra high performance concrete and engineered cementitious composites, the volume and number of applications produced with fibre-reinforced concrete has increased. Surprisingly high fibre dosages can be added to a flowable matrix, promoted by a decreased content of coarse aggregates to prevent blocking and/or to obtain a higher strength and an increased paste content to improve the flowability as well as the robustness. In the following, the rheology of fibre-reinforced concrete is discussed, measurement techniques are listed, affecting parameters on rheological parameters are presented and models to predict the yield value and the viscosity are addressed.

9.4.2 Rheological measurements

There is considerable evidence that the rheology of cement-based materials conforms to the Bingham model (Eq. 9.13, Banfill, 2003). The fibres contribute to the formation of an internal network in concrete and increase the yield stress. A Bingham material behaves like a solid below the yield value; when the stress exceeds the yield value it starts to flow. During the flow, mechanical contacts between fibres and aggregates increase the resistance to flow and the plastic viscosity.

$$\tau = \tau_0 + \mu \cdot \dot{\gamma} \qquad [9.13]$$

where $\dot{\gamma}$ = shear rate (1/s), τ = shear stress (Pa), τ_0 = yield value (Pa) and μ = plastic viscosity Bingham model (Pa·s).

Standard workability tests have been applied for traditional vibrated fibre-reinforced concrete, although only the V-B test was at all successful (Tattersall and Banfill, 1983). Due to the internal structure, fibre-reinforced concrete often appears stiffer; with some vibration this drawback can be overcome. Test methods that require little or no vibration are also applicable for flowable concretes with fibres. Table 9.5 lists rheological instruments reported in the literature for the study of the rheological effect of fibres on cementitious materials.

Designed for conventional cementitious materials, the ratio of the gap width to particle size of a viscometer decreases since a fibre is usually the largest component of a mixture. Fibres can align due to the effect of walls and during the movement of a shearing body in a rheometer and, as a consequence, the rheological characteristics of fibre-reinforced concrete are shear-rate and shear strain-dependent. Fibre orientation is counteracted with the MKIII two-point workability test due to the planetary motion of the impeller (Tattersall and Banfill, 1983). Martinie *et al.* (2010) measured the (static) yield value of fibre-reinforced cementitious materials, assuming that orientation had not yet taken place at this flow stage.

Table 9.5 Rheological measurements with fibre-reinforced cementitious materials

Type of viscometer	Name of test instrument	Reference
Impeller with planetary motion	MK III two point workability tester	Tattersall and Banfill, 1983
	MK III two point workability tester	Tattersall, 1991
	UBC rheometer	Beaupré, 1994
	IBB rheometer	Khayat and Roussel, 1999
Coaxial cylinder viscometer	BML viscometer	Grünewald, 2004
	Viskomat NT	Banfill *et al.*, 2006
	BML viscometer	Krage and Wallevik, 2007
	BML viscometer	Kristjánsson *et al.*, 2009
Vane	Haake viscotester VT550	Martinie *et al.*, 2010
Parallel plate/vane	Own development	Laskar and Talukdar, 2008
	Own development	Kuder *et al.*, 2007

9.4.3 Effect of fibres on rheological characteristics

Few studies are reported in literature on the effect of fibres on rheological characteristics of cementitious materials. Tattersall (1991) studied the effect of steel fibres on the rheological characteristics of concrete and found that the addition of the fibres increased both the yield value as well as the plastic viscosity compared to a reference concrete without fibres. Tattersall and Banfill (1983) found that the parameters g and h (correlated with yield value and viscosity, respectively) increased with increasing fibre content, whereas increasing the fibre length mainly increased g but had little effect on h. Laskar and Talukdar (2008) studied the influence of the fibre volume (same fibre type), the aspect ratio (same diameter and fibre dosage) and the fibre diameter (same fibre dosage and aspect ratio) on rheological characteristics. In most cases, the yield value and the plastic viscosity increased at increasing fibre volume and with a higher aspect ratio. However, at the lowest fibre dosage (0.25 vol.-%), the yield stress and the plastic viscosity were lower in one case, respectively, indicating that the fibres are not yet dominating the rheological behaviour. An effect of the fibre diameter was observed only on the plastic viscosity; the dosage steel fibres (0.125 vol.-%) was very low in these tests. Banfill *et al.* (2006) determined the effect of carbon fibres (V_f: 0.15–0.5 vol.-% and L_f: 3–12 mm) on the rheological characteristics of a mortar. The yield value and the plastic viscosity increased at increasing fibre content, increasing fibre length and decreasing water/cement ratio. Krage and Wallevik (2007) studied the effect of polymeric synthetic fibres (L_f = 40 mm, monofilament fibre type, up to 0.75 vol.-%) on concrete rheology with a BML viscometer. The fibres increased the yield value as well as the plastic viscosity. With a fibre dosage of 0.75 vol.-%, the increase in both characteristics was more than proportional (yield value > 70 Pa); these mixtures were not considered self-compacting. Figure 9.7 shows the effect of the addition of three types of steel fibres (the fibres are characterized by aspect ratio/fibre length) in self-compacting concrete. The slump flow of the reference mixture was 728 mm (yield value: 31 Pa; plastic viscosity: 81 Pa·s). The volume of the aggregates was replaced by the same volume of the fibres and the ratio of sand to total aggregate was kept constant. The higher dosage for each fibre type exceeded the maximum fibre dosage (Grünewald, 2004).

An important aspect to consider for the design of workable/flowable fibre-reinforced concrete is maintaining the fibre volume below the critical dosage. Figure 9.8 shows that mixture components other than the fibres also determine the degree to which the plastic viscosity is affected (Grünewald, 2004). In particular, the composition, dosage and type of aggregates are affecting parameters. The highest plastic viscosity of self-compacting fibre-reinforced concrete was 267 Pa·s (not all mixtures of Fig. 9.8 were considered self-compacting); the plastic viscosity of the corresponding reference mixture without fibres was in the range 56–98 Pa·s.

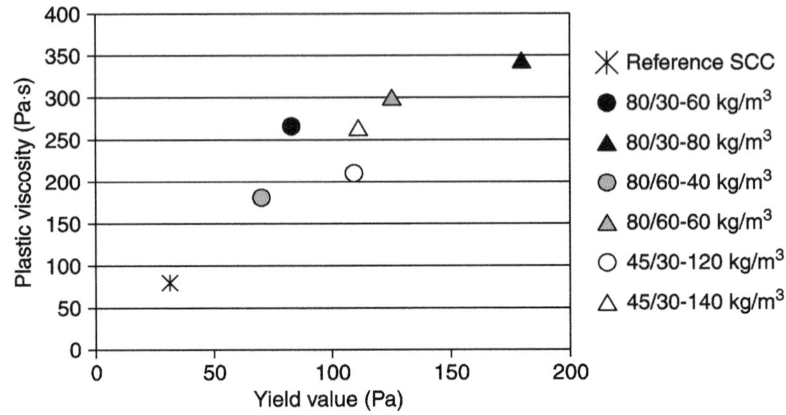

9.7 Increase of yield value and plastic viscosity of SCC due to the addition of steel fibres.

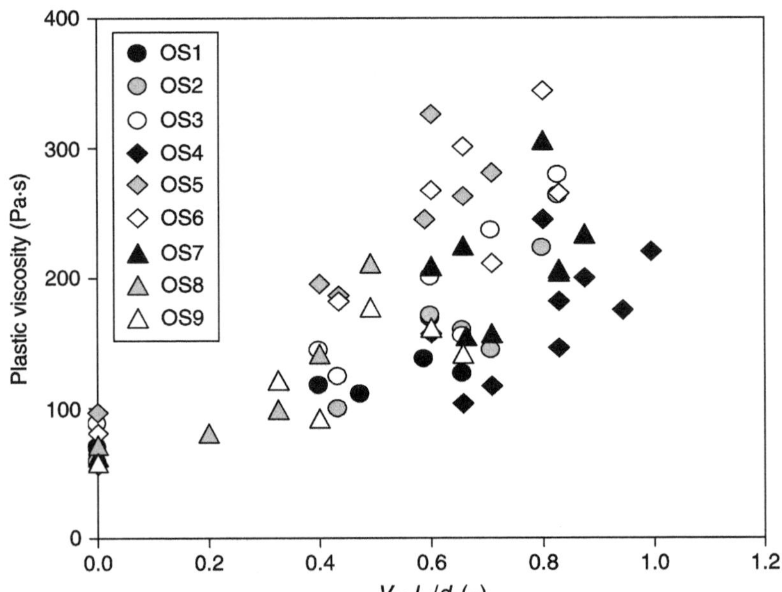

9.8 Effect of fibre factor and mixture composition on the plastic viscosity.

At the lower limit of flowability (550–600 mm), the yield value was in a wide range at a given slump flow (Fig. 9.9). The yield value was in few cases below zero, which is not possible in technical sense. The yield value was not directly measured but interpolated with the Bingham model as the axis intercept τ_0 (Grünewald, 2004). The lower yield values originate from mixtures with less/

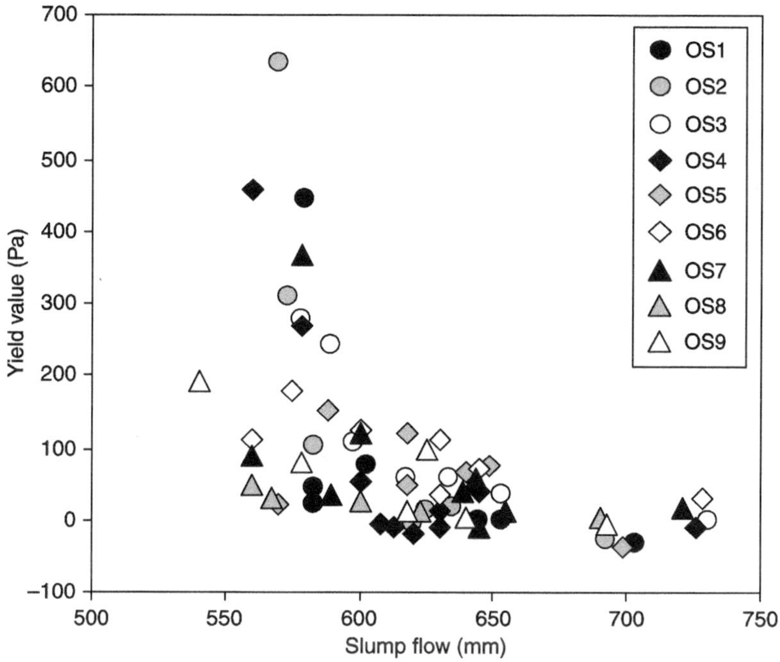

9.9 Rheological measurements on self-compacting fibre-reinforced concrete: slump flow vs. yield value.

weaker mechanical contacts caused by the fibres (i.e. shorter fibres or with a lower aspect ratio), resulting in a homogenous flow pattern but a decreased flowability (Fig. 9.3). The higher yield values at a given slump flow indicate significant mechanical interlocking of fibres during rheological testing; the flow was less homogenous (Fig. 9.4 and Fig. 9.5).

Prediction of the yield value

Martinie and Roussel (2010) obtained a relationship between the relative yield stress (relative to the yield stress of the suspending paste) and the relative packing fraction (Fig. 9.10). The relative packing fraction is $\Phi_f \cdot r/4 + \Phi_s/\Phi_{ms}$ for a mixture of fibres and aggregates. A mixture with a relative packing fraction between 0.8 and 1.0 can be considered to be optimized, whereas for a high flowability the relative packing fraction should be limited to 0.80.

Prediction of the plastic viscosity

Ghanbari and Karihaloo (2009) predicted the effective viscosity of self-compacting fibre-reinforced concrete using Eq. 9.14 and validated it using experimental

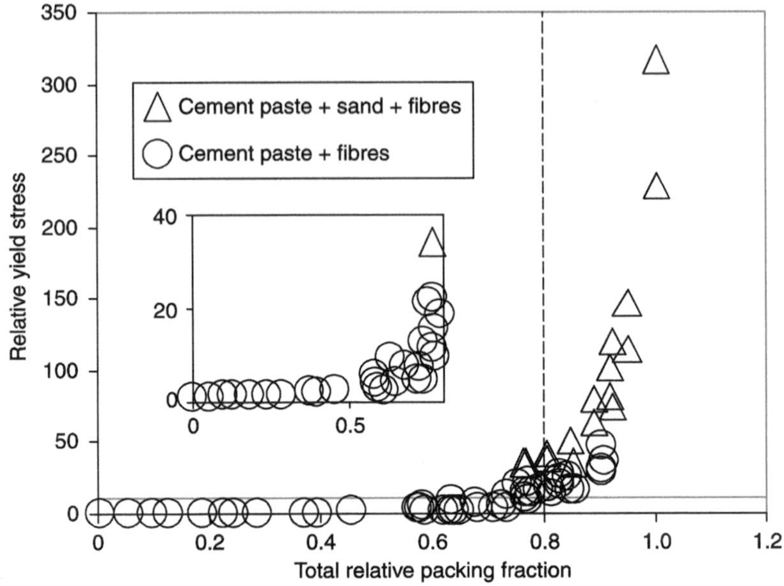

9.10 Relative yield stress as a function of the total packing fraction. The dashed line corresponds to the theoretical random loose packing (Martinie and Roussel, 2010).

results from the literature. Equation 9.14 can be applied to calculate the viscosity of suspensions containing spherical rigid particles in large concentrations with or without needle-shaped rigid steel fibres ($r \leq 85$, $V_f \leq 2.0$ vol.-%).

$$\eta_e = \eta \cdot \left\{ (1-\phi) + \frac{\pi \cdot \phi \cdot r^2}{3 \cdot \ln(2 \cdot r)} \right\} \qquad [9.14]$$

where ϕ = volume fraction of fibres (vol.-%), η_e = plastic viscosity of mix with fibres (Pa·s) and η = viscosity (Pa·s).

The effective stress tensor of a viscous suspension with a matrix designated *m* and fibres *f* is described by Eq. 9.15; the contribution of the fibres to the stress tensor can be calculated using Eq. 9.16 (Ghanbari and Karihaloo, 2009).

$$\sigma_{ij} = (1-\phi) \cdot (\sigma_{ij})_m + (\sigma_{ij})_f \qquad [9.15]$$

with:

$$\sigma_{ij} = \frac{N}{2} \cdot (F_i \cdot r_j + F_j \cdot r_i) \qquad [9.16]$$

where σ_{ij} = stress tensor of viscous suspension, $(\sigma_{ij})_m$ = stress tensor matrix, $(\sigma_{ij})_f$ = stress tensor fibres and r_i = component of the position vector of the centroid in deformed configuration.

The fibres are considered slender rigid bodies, whose body translation and rotation are restricted as a result of the resistance of a viscous self-compacting concrete. The viscosity of the fibre suspension (Eq. 9.14) is related to the viscosity of the paste. In different steps, parts of the solid fraction (from the finest to the coarsest solid phase) were added in the calculation and hereby, the relative viscosity increased compared to the previous stage. A detailed description of the procedure to determine the viscosity is given by Phan-Thien and Karihaloo (1994) and Ghanbari and Karihaloo (2009). A good agreement between experimental results and Eq. 9.14 was obtained. The authors recommend to re-examine the validity of Eq. 9.14 for aspect ratios higher than 85 and fibres dosages higher than 2.0 vol.-%.

9.5 Developments in fibre concrete and rheology

This section discusses recent developments in the field of fibre rheology and the simulation of the flow of fibre suspensions. Theoretical and experimental studies often target materials commonly applied by industry and can be rather specific (i.e. concerning material characteristics, the level of fibre concentration or the application of shear-thinning/thickening suspending fluids). However, the rheology of fibre suspensions is relevant for many industries with stress vectors and/or the prediction of the orientation evolution being the desired output. Table 9.6 summarises examples of case studies on the rheology and the flow of fibre suspensions presented during the Twelfth International Conference on Rheology (Advani and Ait-Kadi, 1997).

9.5.1 Implementing fibres in flow simulations

Fibres were implemented in a numerical approach of the distinct element method (DEM) by Mechtcherine and Shyshko (2007) (cf. Chapter 11). Since not all

Table 9.6 Theoretical and experimental studies on fibre flow and rheology

Case study	Fibre suspension	Reference
Planar flow/converging flow	Polymer melt: concentrated suspension	Thomasset *et al.*, 1997
Extensional flow	Polymer melt: concentrated suspension	Creasy and Advani, 1997
T-shape branching channel	Water-syrup solution/ vinyl fibres	Nishimura *et al.*, 1997
Channel with change of diameter	Newtonian/polymer melt	Azaiez *et al.*, 1997
Injection moulding	Thermoplastics with tubes/ discs	Vincent *et al.*, 1997
Squeezing flow	Syrup matrix with nylon fibres	Goshawk *et al.*, 1997

particles could be modelled, the fluid had to have model characteristics, which depend on the choice of rigid particles (size, shape and number). Rigid particles were considered to be suspended in a fluid of defined characteristics. By clumping (connecting) smaller spheres long elongated fibres could be modelled. The particle flow code 'ITASCA' was applied to execute the calculations. Positions and motions are transferred to forces and updated at each time step. With the definition of the characteristics of the contact elements, simulations (two- and three-dimensional) could be compared with specific flow situations such as the slump flow or flow through a V-funnel. The calibrated model was then applied to predict more complex flow processes such as mould filling or sprayed concrete. The quantitative analysis of the process still has to be executed and requires further verification.

9.5.2 Verifying flow simulations with a non-destructive measuring technique

The measurement technique, impedance spectroscopy (IS), has been applied to determine the orientation and the distribution of the fibres in cementitious materials (Woo *et al.*, 2005; Ferrara *et al.*, 2010). With this method, conductive steel fibres were subjected to currents (direct or alternate current) at different frequencies. The resistance of the matrix at different locations and defined conditions is a measure of the dispersion of the fibres. By executing measurements in different directions the orientation of the fibres can be quantified. The outputs of computational fluid dynamics (CFD) simulations with the software package POLYFLOW were the orientation and the distribution of fibres (Ferrara *et al.*, 2010). The results of flow simulations and IS measurements were compared and showed encouraging agreement.

9.5.3 Predicting fibre orientation

Martinie and Roussel (2010) carried out flow simulations on the shear flow between two parallel plates. Model fluids with only fibres were implemented. Fibres with various initial orientations were considered to represent the macroscopic orientation process. Jeffrey's equation (Eq. 9.11) was used to deduce the orientation evolution at different time steps. The orientation range was 0 to π and the yield stress of the concrete 50 Pa (self-compacting concrete) and 800 Pa (traditional vibrated concrete), respectively. The plastic viscosity was 80 Pa.s in each case. The fibres oriented quickly along the walls in the concrete with the high yield stress. Orientation mainly occurred within a thin layer of paste sliding along the walls and differences in the orientation numbers were small (Fig. 9.11: left). In contrast, the flow of the concrete with a low yield stress caused a more preferred orientation (Fig. 9.11: right). The different flow patterns of both concrete types are reflected by the differences in the average orientation number: 0.50 at a high and 0.71 at a low yield stress of concrete.

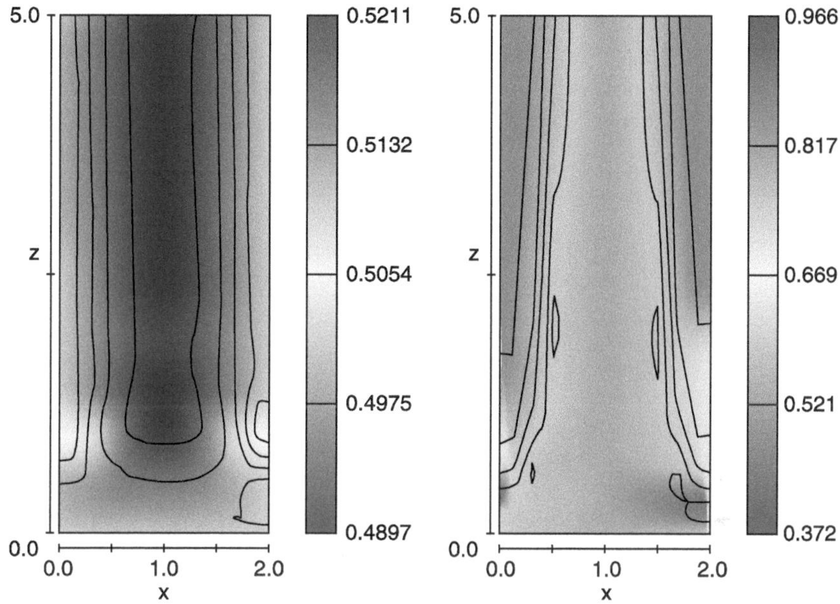

9.11 Orientation factor relative to the flow direction z; yield stress of concrete, left: 800 Pa, right: 50 Pa (Martinie and Roussel, 2010).

9.6 Conclusions

The addition of fibres can improve the performance of cementitious materials in the hardened state with regard to, for example, the tensile strength and crack width control. With an optimized mixture composition and controlled rheology and quality, fibres can become more effective.

Due to the slender shape of the fibres, aspects such as surface area, packing density and fibre flexibility have to be considered for the mix design. The critical fibre content depends on the mixture composition and marks the maximum fibre dosage at which, for example, self-compacting properties, a high workability or a homogenous fibre distribution can be obtained. Fibres decrease the packing density, increase the surface area and/or contribute to the formation of a network and as a result rheological parameters like the yield value and the plastic viscosity are expected to increase. A more than proportional increase of these parameters can be prevented with an adequate mix design.

Significant progress has been made during the past years on the field of flow simulations and the rheology of fibre suspensions. The main difficulties are related to the non-Newtonian behaviour of fibre suspensions (i.e. shear-thinning due to fibre orientation and local flow-induced structures) and the difficulty in predicting and measuring the different contributions of, for example, hydrodynamic effects and mechanical interaction. Fibre-reinforced cementitious materials are highly concentrated suspensions and it is not yet possible to model and to include all particles in simulations. However, with the progress in computer technology, measuring techniques and growing insight, fibres could be considered a common rather than an additive component of cementitious materials.

9.7 References

Advani SG and Ait-Kadi A (1997), Editorial, Rheology and Flow of Fibre Suspensions, Symposium, 12th International Congress on Rheology, Quebec, Canada, Elsevier Science.

Ausias G, Fan XJ and Tanner RI (2006), 'Direct simulation for concentrated fibre suspensions in transient and steady state shear flows', *Journal of Non-Newtonian Fluid Mechanics,* 135, 46–57.

Azaiez J, Guénette R and Ait-Kadi A (1997), 'Investigation of the abrupt concentration flow of fiber suspensions in polymeric fluids', *Journal of Non-Newtonian Fluid Mechanics*, 73, 289–316.

Banfill PFG (2003), 'The rheology of fresh cement and concrete – a review', Proceedings 11th International Congress on the Chemistry of Cement, Grieve G and Owens G (eds.), Vol. 1, Durban, South Africa, pp. 50–62.

Banfill PFG, Starrs G, Derruau G, McCarter WJ and Chrisp TM (2006), 'Rheology of low carbon fibre content reinforced cement mortar', *Cement and Concrete Composites*, 28, 773–780.

Barnes HA, Hutton JF and Walters K (1989), *An Introduction To Rheology*, Amsterdam, Elsevier Science.

Barnett SJ, Lataste JF, Parry T, Millard SG and Soutsos MN (2010), 'Assessment of fibre orientation in ultra high performance fibre reinforced concrete and its effect on flexural strength', *Materials and Structures*, 43, 1009–1023.

Batchelor GK (1977), 'The effect of Brownian motion on the bulk stress in a suspension of spherical particles', *Journal of Fluid Mechanics*, 83, 97–117.

Beaupré D (1994), *Rheology of high performance shotcrete*, Doctoral thesis, University of British Columbia, Vancouver, Canada.

Clarke B (1967), 'Rheology of coarse settling suspensions', *Transactions of the Institution of Chemical Engineers*, 45, 251–256.

Creasy TS and Advani SG (1997), 'A model long-discontinous-fiber filled thermoplastic melt in extensional flow', *Journal of Non-Newtonian Fluid Mechanics*, 73, 261–278.

De Larrard F (1999), *Concrete Mixture Proportioning: A Scientific Approach,* London, E&FN Spon.

Einstein A (1906), 'Eine neue Bestimmung der Moleküldimension' [in German], *Annalen der Physik (Weinheim, Germany)*, 19, 289–306.

Einstein A (1911), 'Berichtigung zu meiner Arbeit: Eine neue Bestimmung der Moleküldimension' [in German], *Annalen der Physik (Weinheim, Germany)*, 34, 591–592.

Fan X-J, Phan-Tien N and Zheng R (1998), 'A direct simulations of fibre suspensions', *Journal of Non-Newtonian Fluid Mechanics*, 74, 113–135.

Ferrara L, Park YD and Shah SP (2007), 'A method for mix-design of fiber-reinforced self-compacting concrete', *Cement and Concrete Research*, 37, 957–971.

Ferrara L, Tregger N and Shah SP (2010), 'Flow-induced fiber orientation in SCSFRC: Monitoring and prediction', in *Design, Production and Placement of Self-Consolidating*

Concrete, Khayat KH and Feys D (eds.), Dordrecht, RILEM Bookseries 1, pp. 417–428.

Ghanbari A and Karihaloo BL (2009), 'Prediction of the plastic viscosity of self-compacting steel fibre reinforced concrete', *Cement and Concrete Research*, 39, 1209–1216.

Giesekus H (1983), 'Disperse systems: dependence of rheological properties on the type of flow with implications for food rheology', in *Physical Properties of Foods*, Jowitt *et al.* (eds.), London, Applied Science, Chapter 13.

Goshawk JA, Navez VP and Jones RS (1997), 'Squeezing flow of continuous fibre-reinforced composites', *Journal of Non-Newtonian Fluid Mechanics*, 73, 327–342.

Groth P and Nemegeer D (1999), 'The use of steel fibres in self-compacting concrete', in First International Symposium on SCC, Stockholm, Skarendahl Å and Petersson Å (eds.), Cachan, RILEM Publications PRO 7, pp. 497–508.

Grünewald S (2004), *Performance-based design of self-compacting fibre reinforced concrete*, PhD thesis, Delft University of Technology, Department of Structural and Building Engineering, Delft University Press.

Grünewald S, Ferrara L and Dehn F (2010), 'Structural design with flowable concrete – a fib-recommendation for tailor-made concrete', in *Design, Production and Placement of Self-Consolidating Concrete*, Khayat KH and Feys D (eds.), Dordrecht, RILEM Bookseries 1, pp. 13–24.

Holschemacher K and Dehn F (2002), 'Faserbeton ein innovativer Baustoff auf dem Weg in die Zukunft' [in German], in *Faserbeton*, König G, Holschemacher K and Dehn F (eds.), Berlin, Bauwerk Verlag, pp. 1–18.

Jeffrey GB (1922), 'The motion of ellipsoidal particles immersed in a viscous fluid', *Proceedings of the Royal Society of London, Series A*, 102, 161–179.

Johnston CD (1996), 'Proportioning, mixing and placement of fibre-reinforced cements and concretes', in *Production Methods and Workability of Concrete*, Bartos PJM, Marrs DL and Cleland DJ (eds.), London, E&FN Spon, pp. 155–179.

Keshtar M, Heuzey MC and Carreau PJ (2009), 'Rheological behaviour of fiber-filled model suspensions: effect of fiber flexibility', *Journal of Rheology*, 53, 631.

Khayat KH and Roussel Y (1999), 'Testing and performance of fiber-reinforced self-consolidating concrete', First International Symposium on SCC, Stockholm, Skarendahl Å and Petersson Å (eds.), Cachan, RILEM Publications PRO 7, pp. 509–521.

Kitano T, Kataoka T and Shirota T (1981), 'An empirical equation of the relative viscosity of polymer melts filled with various inorganic fillers', *Rheologica Acta*, 20, 207–209.

Krage G and Wallevik OH (2007), 'Rheology of synthetic-fiber reinforced SCC', 5th International RILEM Symposium on Self-Compacting Concrete, Ghent/Belgium, Vol. 1, pp. 347–352.

Krieger IM and Dougherty TJ (1959), 'A mechanism for non-Newtonian flow in suspensions of rigid spheres', *Journal of Rheology*, 3, 137–152.

Kristjánsson TI, Wallevik OH, Krage G and Nielsson I (2009), 'Mix design for fiber reinforced SCC', Third International RILEM Symposium on Rheology of Cement Suspensions such as Fresh Concrete, Reykjavik, RILEM Publications, pp. 322–327.

Kuder KG, Ozyurt N, Mu EB and Shah SP (2007), 'Rheology of fiber-reinforced cementitious materials', *Cement and Concrete Research*, 37, 191–199.

Laranjeira F (2010), *Design-oriented constitutive model for steel fiber reinforced concrete*, Doctoral thesis, Universitat Politécnica de Catalunya, Barcelona, Spain.

Laranjeira F, Grünewald S, Walraven J, Blom C, Molins C and Aguado A (2011), 'Characterization of the orientation profile of steel fiber reinforced concrete', *Materials and Structures*, 44, 1093–1111.

Laskar AI and Talukdar S (2008), 'Rheology of steel fiber reinforced concrete', *Asian Journal of Civil Engineering*, 9, 167–177.

Li V (2002), 'Large volume, high-performance applications of fibers in civil engineering', *Journal of Applied Polymer Science*, 83, 660–686.

Markovic I, Grünewald S, Walraven JC and Van Mier JGM (2002), 'Characterization of bond between steel fibres and concrete – conventional fibre reinforced versus self-compacting fibre reinforced concrete', Bond in Concrete – from Research to Standards, Third International Symposium, Budapest, Publishing Company of Budapest University of Technology, pp. 520–528.

Martinie L and Roussel N (2010), 'Fiber-reinforced cementitious materials: from intrinsic properties to fiber alignment', in *Design, Production and Placement of Self-Consolidating Concrete*, Khayat KH and Feys D (eds), Dordrecht, RILEM Bookseries 1, pp. 407–415.

Martinie L, Rossi P and Roussel N (2009), 'Rheology of fiber reinforced cementitious materials: classification and prediction', SCC 2008: 3rd North American Conference on the Design and Use of Self-Consolidating Concrete, pp. 465–469.

Martinie L, Rossi P and Roussel N (2010), 'Rheology of fiber reinforced cementitious materials: classification and prediction', *Cement and Concrete Research*, 40, 226–234.

Mechtcherine V and Shyshko S (2007), 'Simulating the behaviour of fresh concrete using distinct element method', Fifth International RILEM Symposium on Self-Compacting Concrete, Ghent, Belgium, Vol. 1, pp. 467–472.

Nishimura T, Yasuda K and Nakamura K (1997), 'Orientation behaviour of fibers in suspension flow through a branching channel', *Journal of Non-Newtonian Fluid Mechanics*, 73, 279–288.

Pabst W, Gregorová E and Christoph C (2006), 'Particle shape and suspension rheology of short-fiber systems', *Journal of the European Ceramic Society*, 26, 149–160.

Petrie CJS (1999), 'The rheology of fibre suspensions', *Journal of Non-Newtonian Fluid Mechanics*, 87, 369–402.

Phan-Tien N and Karihaloo BI (1994), 'Materials with negative Poisson's ratio: a qualitative microstructural model', *Journal of Applied Mechanics*, 61, 1001–1004.

Philipse AP (1996), 'The random contact equation and its implication for (colloidal) rods in packings, suspensions, and anisotropic powders', *Langmuir*, 12, 1127–1133.

Resplendino J (2011), 'Introduction: What is a UHPFRC?', in *Designing and Building with UHPFRC,* Toutlemonde F and Resplendino J (eds.), London, John Wiley and Sons, pp. 3–14.

Rossi P and Harrouche N (1990), 'Mix design and mechanical behaviour of some steel-fibre-reinforced concretes used in reinforced concrete structures', *Materials and Structures*, 23, 256–266.

Sedran T (1999), *Rheologie et rheometrie des betons. Applications aux betons autonivelants* [in French], PhD thesis, LCPC, Nantes.

Sepehr M, Carreau PJ, Moan M and Ausias G (2004), 'Rheological properties of short fiber model suspensions', *Journal of Rheology*, 48, 1023–1048.

Swamy RN (1975), 'Fibre reinforcement of cement and concrete', *Materials and Structures*, 8, 235–254.

Swamy RN and Mangat PS (1974), 'Influence of fibre-aggregate interaction on some properties of steel fibre reinforced concrete', *Materials and Structures*, 7, 307–314.

Tattersall GH (1991), *Workability and Quality Control of Concrete*, London, E&FN Spon.

Tattersall GH and Banfill PFG (1983), *The Rheology of Fresh Concrete,* London, Pitman.

Thomasset J, Grmela M and Carreau PJ (1997), 'Microstructure and rheology of polymer melts reinforced by long glass fibres: direct simulations', *Journal of Non-Newtonian Fluid Mechanics*, 73, 195–203.

Vandewalle L (1993), 'Fibre reinforced concrete – special types of concrete and applications', Katholieke Universiteit Leuven, Department Burgerlijke Bouwkunde, pp. 77–98.

Vincent M, Devilers E and Agassant J-F (1997), 'Fibre orientation calculation in injection moulding of reinforced thermoplastics', *Journal of Non-Newtonian Fluid Mechanics*, 73, 317–326.

Woo LY, Wansom S, Ozyurt N, Mu B, Shah SP and Mason TO (2005), 'Characterizing fiber dispersion in cement composites using AC-Impedance Spectroscopy', *Cement and Concrete Composites*, 27, 627–636.

Yamamoto S and Matsuoka T (1993), 'A method for dynamic simulation of rigid and flexible fibers in a flow field', *Journal of Chemical Physics*, 98, 644–650.

Yamamoto S and Matsuoka T (1995a), 'Dynamic simulation of flow-induced fiber fracture', *Polymer Engineering and Science*, 35, 1022–1030.

Yamamoto S and Matsuoka, T (1995b), 'Dynamic simulation of fiber suspensions in shear-flow', *Journal of Chemical Physics*, 102, 2254–2260.

Yamamoto S and Matsuoka T (1996), 'Dynamic simulation of microstructure and rheology of fiber suspensions', *Polymer Engineering and Science*, 36, 2396–2403.

Yamanoi M and Maia JM (2010), 'Analysis of the rheological properties of fibre suspensions in a Newtonian fluid by direct fibre simulation. Part 1: Rigid fibre suspensions', *Journal of Non-Newtonian Fluid Mechanics*, 165, 1055–1063.

Yamanoi, M, Maia, JM and Tae-Soo, K (2010), 'Analysis of the rheological properties of fibre suspensions in a Newtonian fluid by direct fibre simulation. Part 2: Flexible fibre suspensions', *Journal of Non-Newtonian Fluid Mechanics*, 165, 1064–1071.

Yu AB and Zou RP (1998), Prediction of the porosity of particle mixtures, *Kona*, 16, 68–81.

Yu AB, Standish N and McLean A (1993), 'Porosity calculation of binary mixtures of non-spherical particles', *Journal of the American Ceramic Society*, 76, 2813–2816.

Zirnsak MA, Hur DV and Boger DV (1994), 'Normal stresses in fiber suspensions', *Journal of Non-Newtonian Fluid Mechanics*, 54, 153–193.

Zou RP and Yu AB (1996), 'Evaluation of the packing characteristics of mono-sized non-spherical particles', *Powder Technology*, 88, 71–79.

9.8 Appendix: notations and symbols

D	rate of strain tensor
E	E-modulus (MPa)
L_f	fibre length (mm)
N	number of fibres (-)
N_{sf}	number of steel fibres (-)
V_f	fibre volume (Vol.%)
W	vorticity tensor
d	grain diameter (mm)
d_f	fibre diameter (mm)
d_p	equivalent packing diameter (mm)

d_v	volume diameter (mm)
f	fibre flexibility (mm^4)
g	parameter correlated with yield value (Nm)
h	parameter correlated with viscosity (Nm·s)
k_f	wall-effect fibres (-)
n	number of fibres per unit volume (-)
p	unit vector
r	aspect ratio: L_f/d_f (-)
r_i	component of the position vector of the centroid in deformed configuration
v_p	perturbed volume in a container (Vol.%)
ϕ	volume fraction of solids/fibres (Vol.%)
ϕ_c	solid volume (loose packing) (Vol.%)
ϕ_{cf}	critical fibre volume (Vol.%)
ϕ_f	percentage of fibres of the granular skeleton (Vol.%)
ϕ_m	solid volume (dense packing) (Vol.%)
Φ	phase volume of solids (Vol.%)
Φ_f	volume fraction of fibres (Vol.%)
Φ_{fc}	random loose packing fibres (-)
Φ_{fm}	random dense packing fibres (-)
Φ_{ms}	dense packing fraction sand (-)
Φ_s	volume fraction of sand (Vol.%)
α	unperturbed packing density (-)
$\bar{\alpha}$	mean packing density (affected by the container size) (-)
α_c	packing density parameter, loose packing (-)
$\alpha_{\vec{d}}$	average orientation number (-)
α_m	packing density parameter, dense packing (-)
$\dot{\gamma}$	shear rate (1/s)
$\varepsilon_{0d,cyl}$	dense (experimental) packing density of fibres (-)
η	viscosity (Pa·s)
(η)	intrinsic viscosity (-)
η_e	plastic viscosity of mix with fibres (Pa·s)
η_s	viscosity of suspending fluid (Pa·s)
λ	parameter dependent on r
μ	plastic viscosity Bingham model (Pa·s)
σ	standard deviation
σ_{ij}	stress tensor of viscous suspension
$(\sigma_{ij})_m$	stress tensor matrix
$(\sigma_{ij})_f$	stress tensor fibres
t	shear stress (Pa)
τ_0	yield value (Pa)
ψ	sphericity (-)

Part III

Casting and the rheological behaviour of concrete

10
Modelling the flow of self-compacting concrete

L. NYHOLM THRANE, Danish Technological Institute, Denmark

Abstract: This chapter gives an overview of the use of numerical tools to simulate the flow of fresh concrete, in particular fluid concrete such as self-compacting concrete (SCC). The techniques that have been applied are very different and may be divided into three main families: homogeneous fluid simulations, modelling of discrete particle flow (distinct element method), and numerical techniques to model particles suspended in a fluid. The chapter describes the general approach of each technique, presents recent examples of their application and briefly discusses future trends in numerical simulations.

Key words: numerical simulations, rheology, CFD, distinct element method, suspension flow, SCC, casting.

10.1 Introduction

In many industries where the manufacturing process involves the casting of a fluid or a suspension, it is essential to have tools predicting the flow and the subsequent hardening process for mould design and process planning. Trial and error may be applied to optimize the casting process when a large number of small samples are produced. However, when casting concrete, especially in situ, trial and error is rarely an option and full control of the casting process is particularly important. The structural size and in situ production often leaves only one attempt to obtain a satisfactory form filling and thereby the required structural quality. Recent developments in concrete technology have paved the way for a rigorous and scientifically based approach to concrete casting. These developments include the development of self-compacting concrete (SCC) and numerical flow modelling tools.

Although rapidly gaining popularity in the precasting industry, SCC as ready mixed concrete has not yet been able to threaten the dominant position of conventional concrete as the preferred construction material. In most of Europe, for instance, SCC accounts for only approximately 1–3% of the total ready mixed concrete production according to statistics from the European Ready Mixed Concrete Organization (ERMCO, 2009). The limited success of SCC for ready mixed concrete may be attributed to a general perception of SCC as a special applications material and to a lack of understanding of SCC in its fresh state. The fluid-like behaviour of SCC distinguishes it significantly from conventional slump concrete. However, even in the range of SCC, the rheological properties may vary significantly and a proper match between the rheological properties and casting technique is essential to obtain the quality and performance required for the specific structure.

The rheological properties of SCC are a key characteristic in order to control and manipulate SCC to the desired effects. Compared with empirical test values, the rheological properties provide an independent physical characterization of the flow properties. However, the use of rheology is far from standard in the concrete industry. The concrete industry has a long tradition of using standard test methods to measure the flow properties and, for SCC, many different methods have been proposed such as the V-funnel, the L-box and the most commonly used, the slump flow test. There are problems associated with these, especially as the parameters measured are empirical, subject to operator error, and do not provide a fundamental link to the concrete behaviour during form filling. Furthermore, it is often difficult to observe the flow during form filling which may lead to misinterpretations of the final result and lack of valid knowledge acquirement in the long run, which does not motivate the further use of SCC. Due to these challenges contractors may be reluctant to use SCC in spite of its inherent advantages over conventional concrete for many applications.

To benefit from the full potential of fluid concretes such as SCC, tools are needed to predict the flow and performance. For instance, numerical simulations may be used to:

- Analyse form filling for defects such as lack of form filling, segregation and blocking in order to specify the optimal rheology and casting technique for a given application. This may then be used in rheological mix design models to optimise the mix composition.
- Analyse fibre orientation and distribution.
- Analyse and optimize mixing procedures, e.g. to study the effectiveness of a given mixer design.
- Analyse empirical test methods to understand the relation between empirical-based parameters and the rheological parameters.
- Analyse the flow in rheometers, e.g. to validate the assumptions normally applied to calculate rheological parameters.

Only simpler cases do not require a computer; for example, analytical expressions may be derived for pipe flow (as in pumping) and steady state shear flow.

The use of numerical simulations will bring a better understanding to SCC processing and help to bridge the gap between material scientists and SCC practitioners, i.e. bring rheology into practical applications (Fig. 10.1). Numerical simulations may be thought of as a scientific supplement to empirical-based specifications of flow properties, casting technique and mix composition for any type of application.

This chapter gives an overview of numerical techniques applied to simulate concrete flow and shows examples of their use. At first, some of the challenges involved in casting fluid concretes are illustrated with examples from different castings. This is followed by a short section to present the basic fluid mechanics and the role of rheology in the governing equations. At the end of the chapter, some future trends in this field are proposed. As the topic of numerical simulations

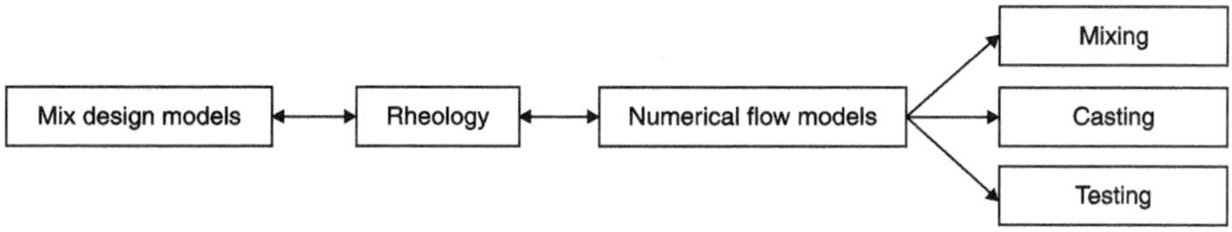

10.1 A scientific-based approach to concrete flow.

is very wide and involves, for example, extensive mathematics, references are given in each section for more detailed reading.

10.1.1 Challenges of fluid concrete castings: some examples

Some examples to illustrate the challenges of casting fluid concretes like SCC are described below. In the first example, SCC was used to cast the basement walls of a new apartment building. The contractor had used SCC for some time, also for vertical applications, and had generally found the results to be positive in terms of quality, productivity and working environment. The walls were 4 m high and included different openings for doors, ventilation, etc. By using SCC, the contractor did not need to prepare channels for the poker vibrator to access the space under the recesses. For most of the castings the concrete was cast directly from the conveyor belt or discharged from a crane bucket. To avoid too high drop heights, the inlet was placed above the highest positioned opening, though high drop heights could not be avoided completely. In general, the outcome was positive, with complete form filling and nice surfaces (Fig. 10.2).

However, for some of the walls, problems occurred and complete repair was required. The problems were lack of form filling below the recesses and openings, honeycombing, poor surface quality and poor encapsulation of the reinforcement (Fig. 10.3). In one case, the yield stress was too high. Job site control had not been applied as part of standard procedure, but the slump flow was measured during casting. The slump flow was very low, approximately 450–500 mm, corresponding to a relatively high yield stress and below the normal SCC range.

Simulations of the casting could have been used to specify the maximum and minimum yield stress allowed, which would serve as a target for job site control. In other words, to find a concrete which would be just sufficiently fluid to fill the formwork bearing in mind that additional and thus unnecessary fluidity will always have a cost, for example, in terms of increased amount of superplasticizer, increased porosity causing loss of mechanical resistance and durability, and increased risk of segregation (Roussel *et al.*, 2007a).

Examples of segregation are shown in Fig. 10.4. To the left is an example of a high fluidity SCC used for a horizontal application. Clear segregation is observed during casting where a front of paste is building up. At the bottom, segregation is

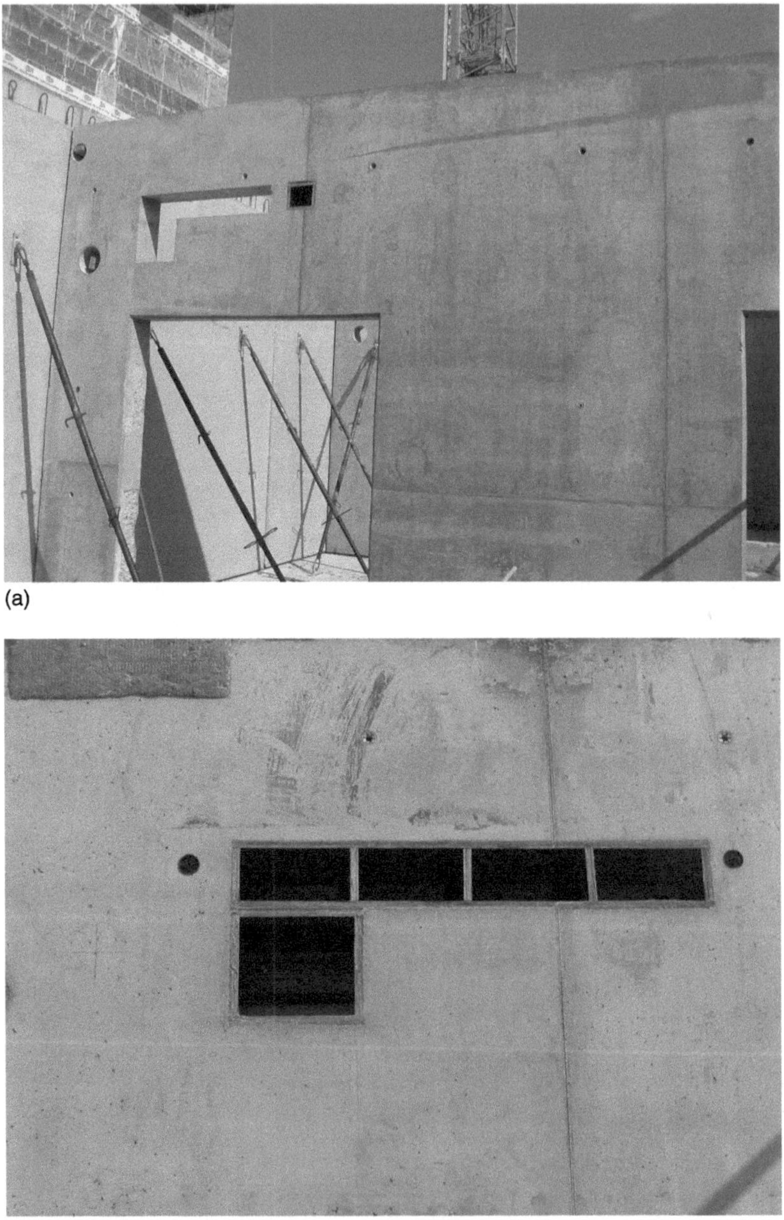

(a)

(b)

10.2 (a) Example of SCC wall casting; (b) close up near the recess where complete form filling was achieved.

(a)

(b)

10.3 Example of a SCC wall casting with severe defects mainly due to a high yield stress: (a) SCC was poured from a crane bucket connected to a hose; (b) view of the concrete during the filling process;

(c)

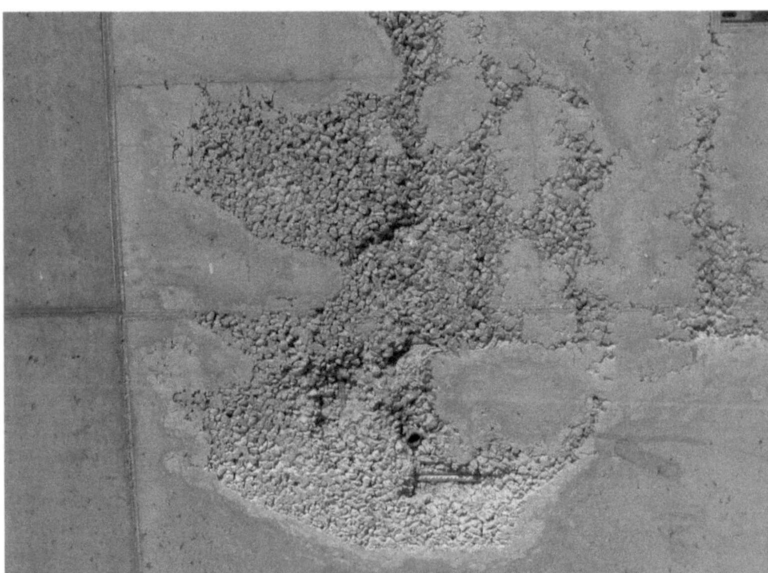

(d)

10.3 Continued (c) view of the wall after demoulding; (d) severe honeycombing was observed;

(e)

(f)

10.3 Continued (e) poor form filling under the recesses; (f) in some places the reinforcement was not encapsulated.

(a)

(b)

10.4 (a) Segregation due to a very fluid, low yield stress SCC. (b) Segregation in a vertical casting due to the combined effect of gravity and flow induced segregation.

observed in a vertical casting with a 10 cm top paste layer. The concrete did not suffer from instability before the casting, indicating the importance of dynamic segregation or flow-induced segregation (Thrane *et al*., 2007).

Thus, form filling ability, segregation resistance, formation of arches between reinforcement bars (blocking) and surface finish are some of the important physical phenomena to be addressed in simulations. All these aspects condition the future hardened properties of the concrete.

10.1.2 Fluid mechanics and material behaviour law

This section presents the most general equations of fluid mechanics, which describe the flow of ideal fluids subjected to laminar flow (fluid flow in parallel layers with no disruption between the layers). Water, polymeric liquids and oils may be thought of as ideal fluids (Bird *et al.*, 1960; Acheson, 1990). Suspensions, on the other hand, depend on the properties and interactions between the different phases on a local scale, i.e. fluid–particle and particle–particle interactions (Barnes *et al.*, 1989; Wallevik, 2003). When the particles remain homogeneously distributed, suspensions are comparable to ideal fluids on a macroscopic scale (Malvern, 1969; Lai *et al.*, 1993).

The three main equations in fluid mechanics are derived from conservation of mass, momentum and energy (Bird *et al.*, 1987). For a region of space occupied by the fluid, the fluid is characterized by a velocity field, density field, pressure field and temperature field. The temperature field is derived from the energy equation, which may be discarded if temperature variations are negligible (isothermal conditions). Also, if constant density is assumed (the so-called incompressibility constraint) the governing equations are given by:

$$\frac{\partial v_i}{\partial x_i} = 0 \qquad [10.1]$$

where v_i is velocity. After Cauchy, the equation of motion for a non-Newtonian fluid is given by:

$$\rho\left(\frac{\partial v_t}{\partial t} + v_i\frac{\partial v_t}{\partial x_j}\right) = \frac{\partial \sigma_{ij}}{\partial x_j} + \rho F_i \qquad [10.2]$$

where σij is the stress tensor, Fi the volume forces per mass unit, ρ is the density and t is the time (Malvern, 1969; Lai *et al.*, 1993). The stress tensor is given by:

$$\sigma_{ij} = -P\delta_{ij} + \tau_{ij} \qquad [10.3]$$

where P is the isotropic pressure and τ_{ij} the deviatoric stress tensor. The term δ_{ij} is the unit dyadic (Malvern, 1969; Lai *et al.*, 1993). For many fluids in nature such as air and water, the relation between stress and strain rate is linear with the constant of proportionally, the viscosity μ. Such fluids are called Newtonian. For Newtonian fluids, the governing equations are often designated as the Navier–Stokes equations (Bird *et al.*, 1987). If the viscosity is not constant, the fluid is referred to as a non-Newtonian fluid. These fluids are called non-Newtonian since the relation between stress and strain rate is non-linear. The generalized equation for the viscosity incorporates the idea of a shear rate dependent viscosity, the apparent viscosity η:

$$\tau_{ij} = \eta(\dot{\gamma})\dot{e}_{ij} \qquad [10.4]$$

where is $\dot{\gamma}$ the scalar magnitude of the strain rate tensor, which in shearing flows is called the shear rate (Bird *et al.*, 1987).

Different apparent viscosity functions have been proposed to describe the non-Newtonian rheological properties. One of them is the two-parameter Bingham model which is often applied to cement based materials and to describe viscoplastic behaviour. The apparent viscosity is given by:

$$\begin{cases} \eta = \infty & (\tau \leq \tau_0) \\ \eta = \eta_{pl} + \dfrac{\tau_0}{\dot{\gamma}} & (\tau \geq \tau_0) \end{cases} \qquad [10.5]$$

where τ_0 is the yield stress and η_{pl} the plastic viscosity (Hohenemser and Prager, 1932; Oldroyd, 1947; White, 1977).

The next three sections present the three overall computational concepts that have been applied to simulate the flow of cement based materials, the homogeneous fluid approach, the discrete flow approach (distinct element method) and the suspension flow approach (Roussel *et al.*, 2007a).

10.2 Homogeneous fluid approach (computational fluid dynamics)

The term computational fluid dynamics (CFD) is often used to cover numerical methods and algorithms to solve the governing equations of the ideal fluids presented in Section 10.1.2. For an introduction to CFD see, for example, Wendt (1992) and Anderson (1995).

One of the first papers of CFD was published in 1967 by Hess and Smith (1967), and today CFD is applied in many different industries for mould filling analyses, for example, in metal forming to avoid gas entrapment and in thermoplastic injection to predict flow patterns and avoid mould cavities (Tie *et al.*, 2006; Kermanpur *et al.*, 2007). If homogeneity is assumed, concrete may be treated in a similar way as a one-phase fluid and simulated within the framework of CFD. The scale of observation is of great importance to choose whether or not such a homogenous approach is legitimate. The order of magnitude of a formwork's smallest characteristic size is around 0.1 m, while the order of magnitude of the size of the coarsest particles is around 0.01 m. This means that if, as a first approximation, the presence of the rebars is neglected, the flow in a typical formwork can be considered as the flow of a homogenous fluid (Fig. 10.5).

Different approaches exist for discretization of the non-linear partial equations in Section 10.1.2, namely the finite difference, finite element and finite volume technique (Wendt, 1992; Anderson, 1995). The overall objective is to reduce the

10.5 The homogenous or ideal fluid approach to simulation of concrete flow. Scale of observation (left), local scale (middle) and homogeneous fluid assumption (right).

continuum problem (infinite number of degrees of freedom) to a discrete problem (finite number of degrees of freedom) described by a system of algebraic equations. The procedure begins with the division of the continuum region of interest into a number of elements, the so-called meshing procedure. The density of the mesh is important and is always a balance between accuracy and computational time. Typically, the solution will converge toward a constant solution when increasing the mesh density. Further increasing the mesh density will be at the cost of increased computational time.

The mesh may be fixed, a so-called Eulerian description in which the free surface flow may be captured using the so-called using volume of fluid technique, which is widely used in CFD (Malvern, 1969; Hirt and Nichols, 1981; Donea *et al.*, 2004). The advantage of this approach is its ability to capture larger distortions and complex free surface flow, but generally at the expense of precise interface definition and resolution of flow details (Donea *et al.*, 2004). The mesh may also be deforming, a so-called Lagrangian description, which is mainly used in structural mechanics. Its weakness is its inability to follow large distortions of the computational domain without recourse to frequent remeshing operations (Donea *et al.*, 2004). Attempts to combine the two methods have also been proposed (Lewis and Taylor, 1997; Souli and Zolesio, 2001). The most appropriate technique relies on the material and deformation characteristics.

To start calculations, material properties (viscosity and density) and initial/boundary conditions must be defined. Also, a proper iterative solution procedure and time-stepping method must be selected (Wendt, 1992; Anderson, 1995). One of the numerical difficulties is the treatment of the yield stress. The apparent viscosity as defined in Eq. 10.5 approaches infinity when the shear rate approaches zero. O'Donovan and Tanner (1984) proposed a bi-viscosity approach where a constant apparent viscosity, η_{max}, is applied below some critical shear rate. A higher value of η_{max} gives a better imitation of the Bingham fluid. In principle the flow will never stop; however, it may be assumed that flow has stopped (yield stress reached) when the flow has slowed down significantly. Another, more direct approach, is an elasto-viscoplastic model applied by Roussel (2006) where the material behaves as an elastic solid up to the point of the yield stress.

10.2.1 Examples of homogeneous fluid approach

Mori and Tanigawa (1990) were the first to apply a homogeneous approach for simulation of concrete flow using their so-called viscoplastic finite element method (VFEM) and the viscoplastic divided element method (VDEM), for example, to study flow in test methods. Test methods have since been the subject to many researchers including Christensen (1991), Thrane *et al.* (2004, 2005), Uebachs and Brameshuber (2005), Roussel and Coussot (2005), van Waarde *et al.* (2007), Tregger *et al.* (2008), Gram (2009), Cremonesi *et al.* (2010) and Bras (2010). For instance, Fig. 10.6 shows L-box simulations compared with experimental observations (Thrane *et al.*, 2004). The finite element based CFD code Fidap® was used to simulate both slump flow and the L-box. A three-dimensional (3D) Bingham model and yield criterion was used to describe the tested fluid behaviour and a no slip condition was applied to the boundaries. Combined simulations and experiments indicated that the flow evolution in both methods can be simulated applying the same Bingham parameters. To obtain the results it was important to implement a proper moving boundary condition to simulate lifting of the cone in the slump flow test and the opening gate in the L-box.

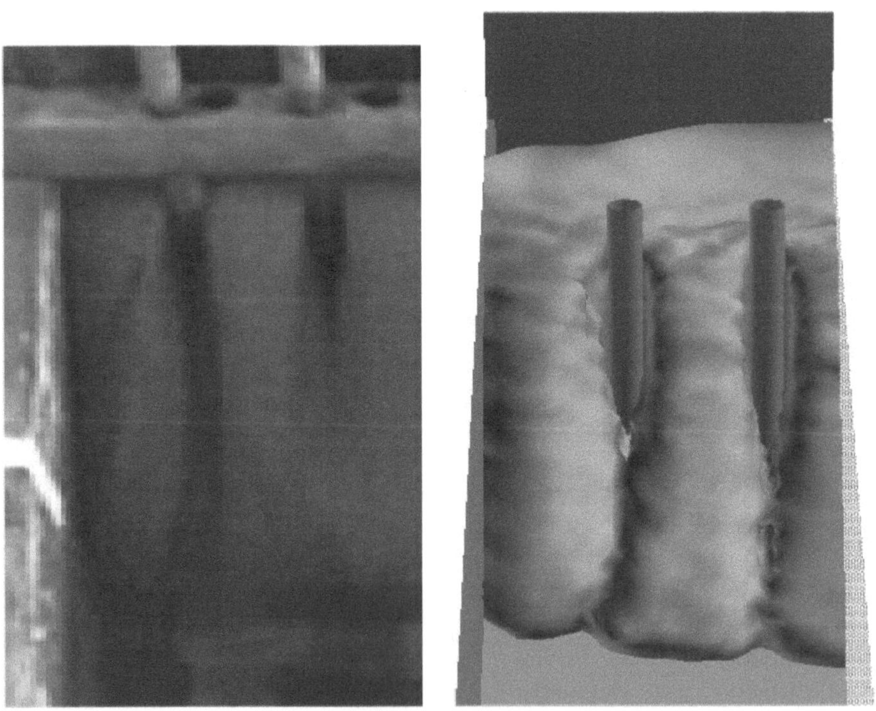

(a) (b)

10.6 Examples of the free surface shape in the L-box. Experimental observations (a,c) and simulations (b,d) (Thrane *et al.*, 2004).

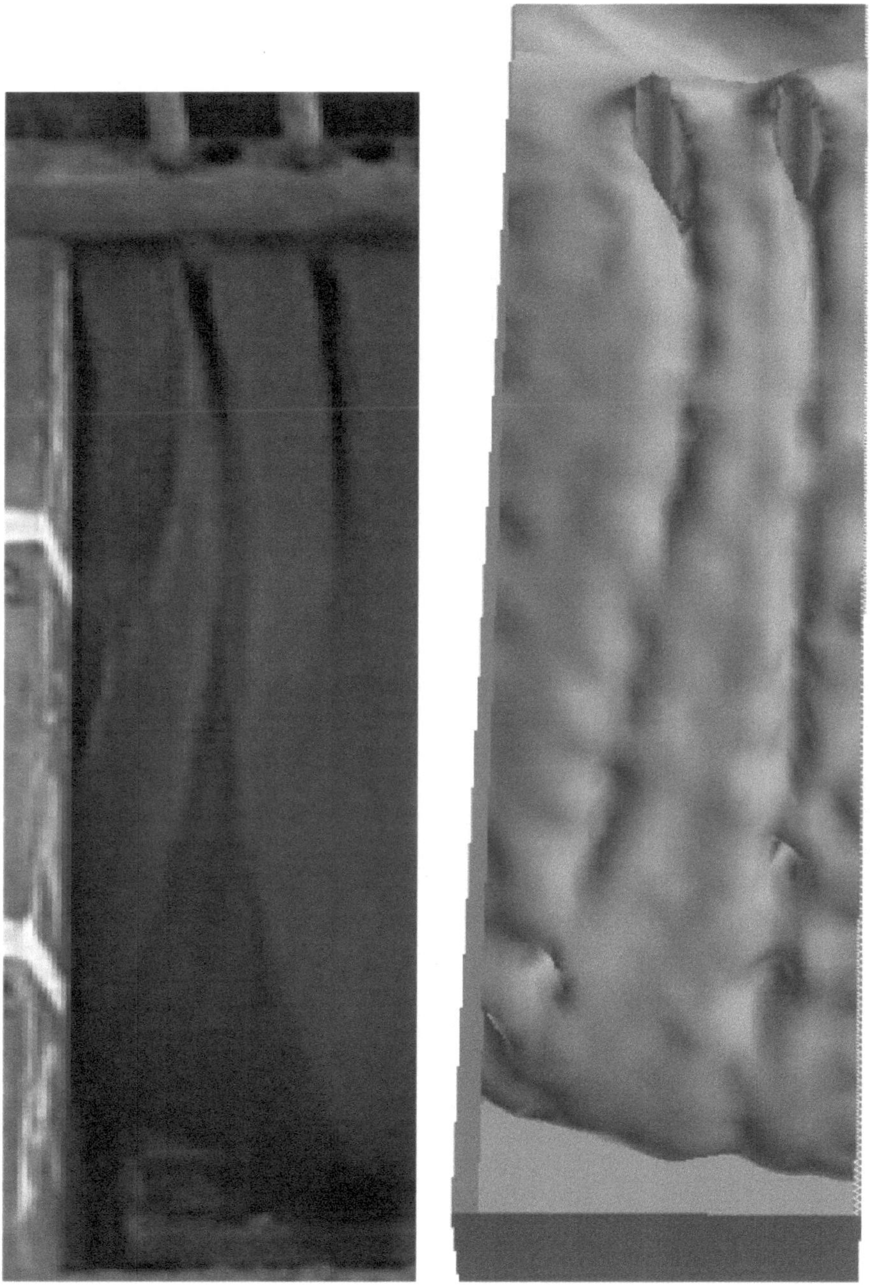

(c) (d)

10.6 Continued

In a similar way, Roussel (2006) used the finite volume based CFD code Flow3D® to study 3D stoppage flow in different slump test methods. The simulations were carried out for yield stresses between 25 and 5500 Pa, and good agreement between numerical and experimental results was obtained for the mini cone test and for the ASTM slump test. The obtained numerical results were compared with two analytical solutions valid in two asymptotic cases: when there is a very small slump (purely extensional flow) and when there is a large spread (purely shearing flow) (Roussel and Coussot, 2005). In general, numerical simulations of some of the most common test methods have verified that the final shape is dependent only on yield stress (and density) except for those cases where the free surface may be sensitive to kinetic energy, for example, when testing low viscosity fluids in the L-box test. However, flow evolution is also sensitive to the moving boundary conditions, which are important to consider in particular when the test results are used to evaluate the plastic viscosity e.g. in the 4C-Rheometer (Thrane *et al.*, 2010).

CFD has also been used to study flow in rheometers to evaluate the geometry and assumptions made about them. Hu *et al.* (1996) used finite element methods to study the BTRHEOM rheometer. The authors were able to simulate slippage or not at the walls on the rheometer container, as well as determine the uncertainty of the results. Wallevik (2003, 2008) applied his own and freely available software VVPF to study steady state and transient flow conditions (velocity and shear stress profiles) in selected rheometers such as the ConTec BML Viscometer 3, ConTec Viscometer 4 and the BTRHEOM. The simulations have been used to compare rheometers with regard to, for example, variations in shear rate and particle migration.

More and more examples of casting simulations are appearing such as Ovarlez and Roussel (2006), Roussel *et al.* (2007a), Thrane *el al.* (2007), van Waarde *et al.* (2007), Bakker (2009), Gram (2009) and Patzak and Bittnar (2009). For instance, for various values of Bingham parameters, Roussel *et al.* (2007b) simulated the filling process of a pre-cambered composite beam, which helped to determine the

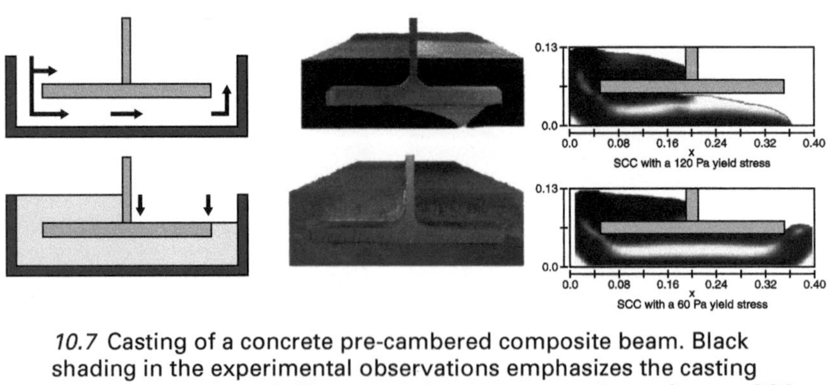

10.7 Casting of a concrete pre-cambered composite beam. Black shading in the experimental observations emphasizes the casting defect (upper picture). Numerical simulations are shown for two SCC with yield stresses equal to 120 Pa (upper figure) and 60 Pa (lower figure). Units are metres (Roussel *et al.*, 2007b).

critical maximum yield stress to avoid cavities. Figure 10.7 shows a schematic view of the casting technique, observations from trial castings and numerical simulations using different yield stress values. Cavities could be avoided at yield stress equal to 60 Pa.

Thrane *et al.* (2007) studied full-scale wall castings when the simulated flow patterns provided a qualitative means of understanding the flow-induced segregation and the development of surface air voids observed in the experiments. It appeared that segregation was initiated when the concrete was forced sharply upwards against gravity, while more horizontally oriented, layered flow patterns caused less segregation. It was speculated that flow-induced segregation could be traced back to the combined effect of shear-induced particle migration and gravity-induced segregation due to differences in density. With respect to surface quality it was concluded that shearing at the form surface contributed positively to reduce the number of air voids. Ovarlez and Roussel (2006) also studied the flow pattern in vertical castings to estimate the extent of the zone where SCC is at rest in the formwork in order to take thixotropy into account for formwork pressure predictions.

Studies have also been carried out on mixing. For instance, Mokéddem *et al.* (2009) carried out simulations of a so-called Viscoprobe inside the mixer. The Viscoprobe is a spherical probe mounted on one of the blades, which continuously measures the resistance during mixing. Potentially, better understanding of the probe response may lead to better interpretations of the evolution of the flow properties and a more robust quality system, for example, for adjustment during mixing before discharging the concrete.

Most of the applications have focused on those with no or only little reinforcement. When dense reinforcement systems are applied, the viscous flow resistance is increased and it may no longer be a valid assumption to neglect the reinforcement. However, detailed modelling of the reinforcement will increase the

Table 10.1 Short overview of the CFD approach for simulation of concrete flow

Main motivations for using CFD	Challenges of CFD	Software (examples)
Rheological parameters, which can be measured, are direct input parameters to the simulation. A number of rheological behaviour laws can be applied. CFD gives an insight into flow patterns that are difficult, expensive or impossible to study using experimental techniques. CFD is the fastest way to simulate flow in particular in 3D over larger scales of time and length.	Solid particles are not included, i.e. it is not possible to simulate heterogeneous phenomena such as segregation and blocking.	Flow3D© Fluent© Fidap© Polyflow© OpenFoam (free) VVPF (free) OOFEM (free)

mesh density significantly and make computations extremely time consuming. Recently Vasilic *et al.* (2011) proposed a promising new approach to overcome this obstacle in which the reinforcement layout is assumed to be similar to a porous medium. A smaller section of reinforcement bars were simulated and the porous medium approach showed good correlation with both simulations based on direct approach and experimental observations using a model material. A short overview of the CFD approach for simulations of concrete flow is shown in table 10.1

10.3 Distinct element method (DEM)

When the scale of observation no longer permits neglecting the variations in velocity between the particles and the fluid, the situation is said to be in the discrete regime (Roussel *et al.* 2007a). Where the ideal fluid concept captures macroscopic flow and flow patterns, the overall objective of the discrete particle flow approach is to capture the motion of individual particles. It may be used to study, for example, blocking, segregation and the flow of concrete near boundary walls where the scale of particles is close to the scale of the flow. In general, discrete particle flow simulations are based on interaction laws defined between the particles (Fig 10.8). When contacts occur between particles, a local constitutive law determines the contact forces and consequently the resulting motions of the particles involved in the contact. The various discrete simulation methods in the mechanics literature can be found in Kishino (2001).

The distinct element method (DEM) was initially developed to better understand the behaviour of dry granular materials, powder mechanics and rock mechanics, for instance, segregation in dynamic applications such as chutes, hoppers and belt conveyers (see Cleary, 1998; Vu-Quoc *et al.*, 2000; Fortin *et al.*, 2005; Kruse and Lemmon, 2005; Balevicius *et al.*, 2011). The DEM method evolved from early molecular dynamic algorithms and was first introduced by Cundall (1971) for the analysis of rock mechanics problems and then applied to soil problems. DEM consists of modelling the granular medium as a collection of particles whose trajectories are studied separately. Each calculation cycle includes two steps: the determination of the contact forces between solid particles and the calculation of the motion of each particle determined by the application of Newton's second law. Cundall and Strack (1979) and Walton (1982) developed the first simple contact

10.8 Discrete particle flow approach for simulation of concrete flow. Interaction laws are defined between the particles. Scale of observation (left), local scale (middle) and discrete particle flow assumption (right).

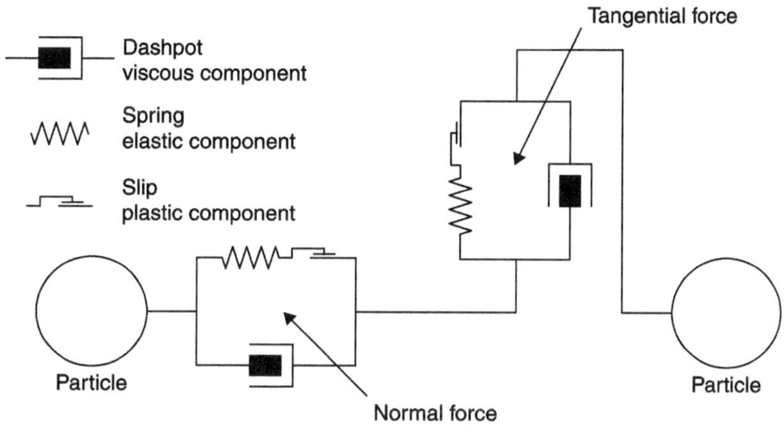

10.9 A standard contact law between two particles in DEM. According to the separating distance and relative velocity of the particles, normal and tangential interaction forces between particles can be calculated (Roussel *et al.*, 2007a).

algorithms between two-dimensional circular (2D) disks. Much research has focused on the development and improvement of contact laws and particle shape flexibility as well as extension to three dimensions (Lin and Ng, 1995; Langston *et al.*, 2004; Zhao *et al.*, 2006). For most applications, each contact point between particles is replaced by a set of normal and shear springs, normal and shear dashpots, normal and shear no-tension joints and shear sliders (Fig. 10.9).

Contrary to granular media, concrete is typically considered to be a two-phase suspension, for example, aggregates (particles) and paste (liquid). To handle two phases, different approaches have been proposed. The two phases may be separated or coupled in one particle (Quoc and Uomoto, 2008). In the first approach, two types of particles will then be present: aggregate particles and paste particles. The contact parameters are different for the two types of particles. For instance, using the code PFC3D©, this approach has been used to simulate slump flow, L-box, V-funnel and J-ring (Chu *et al.*, 1996; Noor and Uomoto, 1999; Petersson and Hakami, 2001; Petersson, 2003). The method seemed applicable for both 2D and 3D simulations. In the second approach, each particle consists of a hard inner core with a surrounding paste layer. This approach can reduce the number of particles to be modelled and allows for larger volumes of SCC to be simulated. As above, the contact parameters for the aggregate and the mortar phase are different. For instance, this approach was adapted by Gram (2009) to simulate the slump flow test and J-Ring using PFC3D© (Fig. 10.10). However, the results indicate that blocking rarely jam, even for cases when blocking is likely to happen. The cut-off line between the mortar phase and the aggregates was 2 mm and six different particle sizes were chosen to represent the true particle size distribution. Having calibrated the contact parameters, the

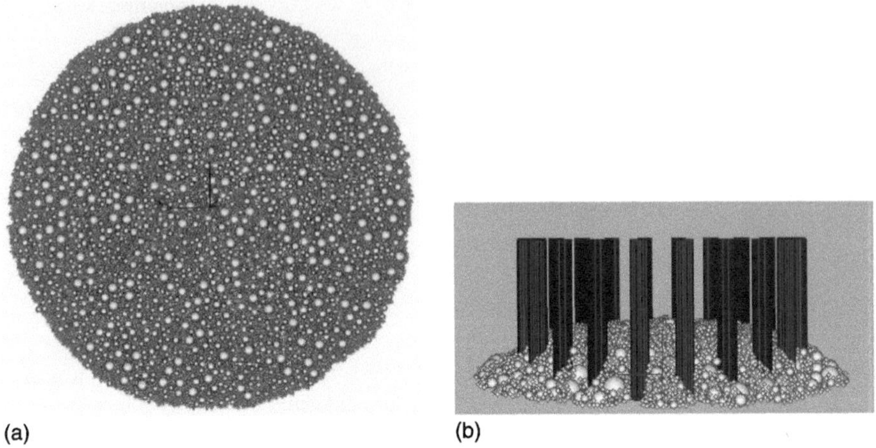

10.10 (a) DEM simulation of slump flow test (Gram and Silfwerbrand, 2011); (b) DEM simulation of the J-Ring test (Gram and Silfwerbrand, 2011).

simulation was compared with experimental observations and good agreement was found. The cut-off value could be increased to decrease the size of the model and number of computations.

DEM modelling to simulate fresh concrete flow has developed considerably over the past years and results shows its potential to simulate the macroscopic flow and trajectories of individual aggregate; see also Schwabe (2005), Mechtcherine and Shyshko (2006, 2007) and Krenzer and Schwabe (2009). Schwabe and Kuch (2005) used PFC3D© to model and analyse the blending of the grain ingredients within a concrete mixer. The different grain fractions are filled successively into the mixer and the analysis of the mixture quality is realized by a virtual extraction of a probe (sampling) within the mixing box. A validation of the obtained numerical results was achieved by using an experimental laboratory in which mixture quality and material flow were measured. Experimental analyses with different construction designs in planetary mixers at real scale confirmed the simulations.

Mechtcherine and Shyshko (2007) used PFC3D© to include fibres in the model. Fibres were created by attaching aggregate particles together to form the prober fibre geometries. A 2D mould filling of a beam was carried out and simulations indicated the possibility of simulating fibre orientation.

One of the main challenges of DEM is the selection of contact parameters. It is not possible to measure these contact parameters, which may be as many as 13 as in the slump flow test mentioned above (Fig. 10.10). Therefore, a real experiment is carried out to which the model is compared. Fixing the DEM parameters is an iterative procedure, which may be both difficult, time consuming and subjected to error (Krenzer and Schwabe, 2009). Furthermore, the contact parameters are particle size dependent so small changes in DEM model requires a completely

Table 10.2 Short overview of the DEM approach for simulation of concrete flow

Motivations for using DEM	Challenges of DEM	Software
Particle trajectories can be modelled, e.g. to study segregation, blocking and mixing. Useful when the scale of observation is of the same order as the size of the coarsest particles. Detailed modelling of the free surface.	Physical meaning of the system, for instance, how to define and set the contact law parameters between solid particles and mortar particles. Calibration needed. Computational costs in particular in 3D and for large number of particles.	PFC3D© EDEM©

new parametric study. Thus, a correlation between Bingham parameters and the contact parameters are not easy to find and the contact parameters are somewhat imaginary constants (Zhuguo, 2007). Finally, the computational capacity may set limits on the size of the flow domain and the number of particles it is possible to simulate. A short overview of the DEM approach for simulations of concrete flow is shown in table 10.2.

10.4 Suspension flow

The last of the three family categories may be referred to as suspension flow. Compared to the CFD and DEM methods presented in the previous sections, these methods aim to capture the suspension nature of concrete (particles in a matrix phase) (Fig 10.11). Much research is now going in this direction to study heterogeneities such as particle migration, blocking and fibre distribution and orientation. The first model was probably proposed by Mori and Tanigawa (1992) using their so-called viscoplastic suspension element method (VSEM). The mortar and coarse aggregate was selected as the fluid phase and particles, respectively. Kurokawa *et al.* (1996) compared 2D simulations to the measured flow during an L-flow and slump test for concretes with yield stresses between 50 and 300 Pa. They found the numerical approach

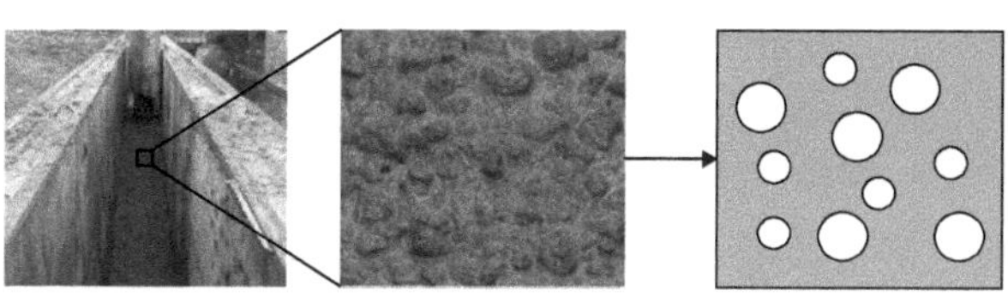

10.11 Suspension flow approach for simulation of concrete flow. Scale of observation (left), local scale (middle) and suspension flow assumption (right).

to overestimate the initial rate and underestimate the final flow. Detection of heterogeneous flow phenomena such as segregation was not evident from these studies. More recently, examples of suspension flow modelling may be found in Dufour and Pijaudier-Cabot (2005), Martys (2005), Modigell *et al.* (2007), Martinie and Roussel (2010), Spangenberg *et al.* (2010) Skocek *et al.* (2011) and Svec *et al.* (2011).

Dufour and Pijaudier-Cabot (2005) applied the so-called FEMLIP method developed by Moresi *et al.* (2003) to simulate SCC flow in the slump flow test (Fig. 10.12). It is possible to use the model for homogeneous flow (no particles) and to introduce different Lagrangian particles to simulate the particle behaviour. Any rheological model can be applied to the matrix phase.

Martys (2005) applied so-called dissipative particle dynamics (DPD), which is based on molecular dynamics with mesoscopic particles. For instance, it was used to study the flow between rebars of two different monosized aggregate suspensions. Blocking was observed for the aggregates sized half of the free space whereas the aggregate sized a fifth of the free space seemed to flow freely (Martys and Ferraris, 2002). DPD has also been used to study the influence of particle size and particle shape on the rheological behaviour, for example, to show that replacing crushed aggregate with spherical shapes can decrease the viscosity by a factor of approximately two (Martys, 2005)

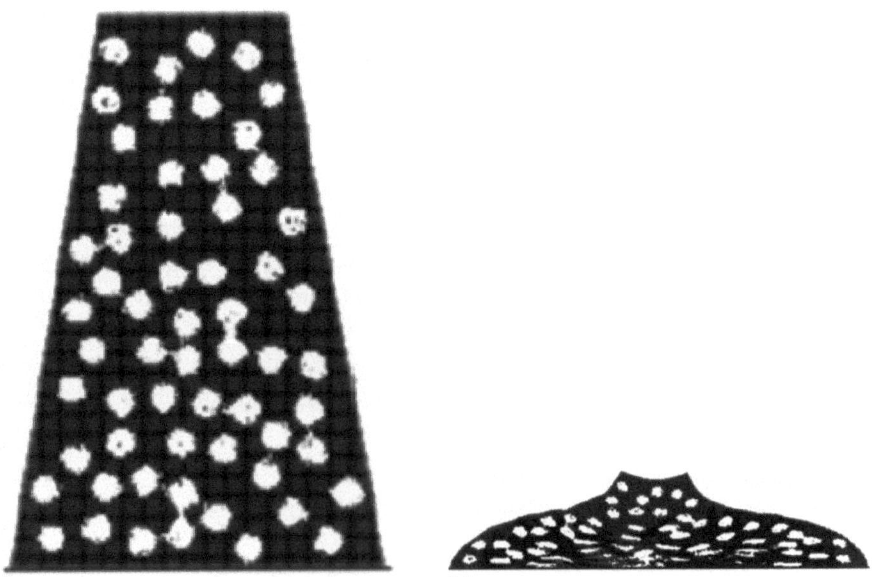

(a) (b)

10.12 (a) Initial condition for numerical simulation of the slump flow test using FEMLIP (Dufour and Pijaudier-Cabot, 2005); (b) the FEMLIP simulation of the slump flow test after some time (Dufour and Pijaudier-Cabot, 2005).

Modigell *et al.* (2007) suggested a two-phase model for simulation of SCC flow. Here, a so-called pseudo-continuum approach was applied consisting of two phases: the continuous liquid matrix and the dispersed solid phase, modelled as a pseudo-liquid. The method was used to compare simulated flow in L-shaped geometry between homogeneous fluid approach using FLUENT© and the two-phase simulation using PETERA©. Some differences between the flow propagation were observed, indicating potential phase separation in the two-phase model.

Spangenberg *et al.* (2010) developed a model to simulate segregation of the coarsest particles in the casting process. The numerical model to simulate the particle flow was developed within the framework of CFD also using Flow3D©. The position of the particles was calculated explicitly with a one-way momentum coupling between the continuous phase and the particles (i.e. the particles 'feel' the continuous phase but not vice versa). Also, no interaction between the particles was assumed. Based on variations in aggregate volume fractions, the model can be used to produce maps of properties in the hardened state of a structural element.

Martinie and Roussel (2010) developed a model to simulate the evolution of a finite group of slender fibres within the framework of CFD using Flow3D©. The method was applied to a yield stress fluid with yield stress of 300 Pa and plastic viscosity of 50 Pa·s flowing between two parallel walls and in a U-shaped channel. The results show that orientation occurs very quickly with very high orientation factors close to the walls. In the almost unsheared zone in the central and inner part of the element, fibres stay isotropic.

Skocek *et al.* (2011) and Svec *et al.* (2011) developed models based on the lattice Boltzmann method (LBM) to simulate rigid bodies such as aggregate and fibres in yield stress fluids. The model has been used to simulate shear flow and flow in a slump flow test and L-box. The results indicate that it is possible to

Table 10.3 Short overview of suspension flow modelling of concrete flow

Motivations for using suspension method	Challenges of suspension method	Software
These methods capture the nature of concrete (particles submerged in a matrix phase). The rheological properties of the matrix phase are direct input to the simulations. Different rheological models can be applied to the matrix phase. Modelling of heterogeneities such as segregation, blocking and fibre orientation is possible. Reasonable computational time.	Development of robust numerical algorithms for each different method. Modelling of fully coupled flow for a large number of particles and polydisperse particle size distributions. To include more than two phases, e.g. advance to include air as an additional phase.	None

capture the effect of particle volume fraction on the effective viscosity and potentially simulate dynamic segregation of aggregates and fibre orientation even for quite large concentrations. Furthermore, it seems that computations are relatively fast using this technique. A short overview of suspension flow modeling of concrete flow is shown in table 10.3.

10.5 Future trends

Numerical simulation of cement-based materials has been a fast-growing research field over the last ten years and approximately every second month a new paper is published. Different approaches have been applied each having their advantages and disadvantages. The use of these tools has brought further understanding into the flow and behaviour of cement-based materials in casting, mixing and testing. Computational fluid dynamics (CFD) assuming a homogeneous fluid behaviour is probably the most developed method in terms of its transfer from the research field into practical industrial applications where a few examples are already available. However, even with today's computer processors, the calculation time at an industrial scale is still a limiting factor but the first attempts to handle, for example, complex reinforcement layouts have been proposed and will be further developed and verified in the future. As a new person in this field, CFD may be a good starting point for which both commercial and free types of software are available.

DEM has also improved significantly and can bring answers to the local behaviour of particle flow, although important issues such as calibration of model parameters and computation cost is still a major issue in particular for multi-range particle sizes. Commercial software is also available in this field. One future possible improvement would be to combine CFD and DEM to benefit from the potential of each method and avoid their drawbacks.

Finally, the last years have seen the development of suspension flow techniques, which aim to capture both the macroscopic flow behaviour and the interaction between the fluid and rigid bodies such as aggregates and fibres. The different models proposed show promising results on aspects such as dynamic segregation, rheological modelling and fibre orientation. Much research is expected to go in this direction and it will be very interesting to follow the future developments. For instance, the adoption of the lattice Boltzmann method to concrete flow is one way to potentially capture the coupled effect of large volumes of fibres and aggregate at relatively low computational costs.

In general, some of the aspects of concrete flow that may be subject to more research in the future are:

- Boundary conditions: very often simulations apply a no-slip boundary condition. However, formwork materials, coatings, castings techniques and mix compositions can be very different and more advanced interface models may be needed. Detailed experimental studies could bring further insight to the flow characteristics of the interface between the concrete and the formwork.

- Air voids: Air in concrete, which may represent up to 6 vol%, is known to affect the flow and rheological properties of concrete. Advanced numerical models may potentially be able to account for the effect of air voids, their size and distribution.
- Surface finish: Surface finish is an important topic and more research is needed to look into the combined effect of flow patterns, formwork materials and mix compositions.

As already mentioned the field of numerical flow modelling is promising and growing rapidly. Therefore, as a final comment, the need for experimental validation of the models is becoming more and more important. This will be a crucial step in order for the models to advance from the research level to practical and trustworthy tools to be used in, for example, casting process planning.

10.6 Sources of further information and advice

For further reading please refer to the references given throughout the chapter. In addition, a state-of-the-art report to be published in 2011 as part of the RILEM TC–222 SCF initiative, 'Simulation of fresh concrete', will contain more details on the different techniques and a section on potential errors in numerical simulations, i.e. errors which may lead to misinterpretations of the flow behaviour.

10.7 References

Acheson D J (1990), *Elementary Fluid Dynamics*, Oxford Applied Mathematics and Computing Science Series, Clarendon Press, Oxford.

Anderson J D (1995), *Computational Fluid Dynamics: The Basics with Applications*, McGraw-Hill, New York and London.

Bakker P (2009), *Grout backfill process of a nuclear waste disposal gallery – a study into the most suitable filling strategy*, Master thesis, Delft University of Technology, The Netherlands.

Balevicius R, Kacianauskas R, Mroz Z and Sielamowicz I (2011), 'Analysis and DEM simulation of granular material flow patterns in hopper models of different shapes', *Adv Powder Technol*, 22, 226–235.

Barnes H A, Hutton J F and Walters K (1989), *An Introduction to Rheology*, Elsevier Science, Amsterdam.

Bird R B, Stewart W E and Lightfoot E N (1960), *Transport Phenomena*, John Wiley and Sons, New York.

Bird R B, Armstrong R C and Hassager O (1987), *Dynamics of Polymeric Liquids. Volume 1 Fluid Dynamics*, John Wiley and Sons, New York.

Bras A (2010), *Grout optimization for masonry consolidation*, PhD thesis, Universidade Nova de Lisaboa, Portugal.

Christensen G (1991), *Modelling the flow of fresh concrete: The slump test*, PhD thesis, Princeton University.

Chu H, Machida A and Suzuki N (1996), 'Experimental investigation and DEM simulation of filling capacity of fresh concrete', *Trans Jap Concr Inst*, 16, 9–14.

Cleary P (1998), 'Discrete element modelling of industrial granular flow applications', *TASK Q*, 2, 385–416.

Cremonesi M, Ferrara L, Frangia A and Perego U (2010), 'Simulation of the flow of fresh cement suspensions by a Lagrangian Finite Element approach', *J Non-New Fluid Mech*, 165, 1555–1563.

Cundall P A (1971), 'A computer model for simulating progressive large scale movements in blocky rock system', *Proc Symp Int Soc Rock Mech*, 2, 8, 129.

Cundall P A and Strack O D L (1979), 'A discrete numerical model for granular assemblies', *Geotechnique*, 29, 47–64.

Donea J, Huerta A, Ponthot J Ph and Rodriguez-Ferran A (2004), 'Arbitrary Lagrangian–Eulerian methods', in *Encyclopedia of Computational Mechanics*, Stein E, de Borst R and Hughes T J R (eds.), John Wiley and Sons, Chichester.

Dufour F and Pijaudier-Cabot G (2005), 'Numerical modeling of concrete flow: homogeneous approach', *Int J Numer Anal Methods Geomech*, 29, 395–416.

ERMCO (2009), 'European ready-mixed concrete industry statistics, 2009', European Ready Mixed Concrete Organization, http://www.ermco.eu [Accessed December 2010].

Espanol P (2005), 'Dissipative particle dynamics', in *Handbook of Materials Modeling*, Yip S (ed.), Springer, Dordrecht, pp. 2503–2512.

Fortin J, Millet O and Saxce G (2005), 'Numerical simulation of granular materials by an improved discrete element method', *Int J Numer Methods Eng*, 62, 639–663.

Gram A (2009), *Numerical modelling of self-compacting concrete flow, discrete and continuous approach*, Licentiate thesis, Royal Institute of Technology, Stockholm.

Gram A and Silfwerbrand J (2011), 'Numerical simulation of fresh SCC flow: applications', *Mater and Struct*, 44, 805–813.

Hess J L and Smith A M O (1967), 'Calculation of potential flow about arbitrary bodies', *Proc Aero Sci*, 8, 1–138.

Hirt C and Nichols B (1981), 'Volume of fluid (vof) method for the dynamics of free boundaries', *J Comput Phys*, 39, 201–225.

Hohenemser K and Prager W (1932), 'Fundamental equations and definitions concerning the mechanics of isotropic continua', *J Rheol*, 3, 16.

Hu C, Larrard F de, Sedran T, Boulay C, Bosc F and Deflorenne F (1996), 'Validation of BTRHEOM, the new rheometer for soft-to-fluid concrete', *Mater Struct*, 29, 620–631.

Kermanpur A, Mahmoudi Sh and Hajipour A (2007), 'Three-dimensional numerical simulation of metal flow and solidification in the multi-cavity casting moulds of automotive components', *Int J ISSI*, 4, 8–15.

Kishino Y (ed.) (2001), *Powders and Grains 2001: Proceedings of 4th International Conference on Micromechanics of Granular Media*, Sendai, Japan, A.A. Balkema.

Krenzer K and Schwabe J H (2009), 'Calibration of parameters for particle simulation of building materials, using stochastic optimization procedures', *Proceedings 3rd International RILEM Symposium of Rheology of Cement Suspensions such as Fresh Concrete*, Reykjavik, pp. 135–142.

Kruse D and Lemmon R (2005), 'Using the discrete element method as an everyday design tool', *Bulk Solids Handl*, 25, 358–367.

Kurokawa Y, Tanigawa Y, Mori H and Nishinosono Y (1996), 'Analytical study on effect of volume fraction of coarse aggregate on binghams constants of fresh concrete', *Trans Jap Concr Inst*, 18, 37–44.

Lai W M, Rubin D and Krempl E (1993), *Introduction to Continuum Mechanics*, 3rd

edition, Pergammon Press, Oxford.
Langston P, Al-Awamleh M, Fraige F and Asmar B (2004), 'Distinct element modelling of non-spherical frictionless particle flow', *Chem Eng Sci*, 59, 425–435.
Lewis R and Taylor S N (1997), 'A mixed Lagrangian–Eulerian approach to modelling fluid flow during mould filling', *Int J Numer Methods Fluids*, 25, 931–952.
Lin X and Ng T (1995), 'Contact detection algorithms for three dimensional ellipsoids in discrete element modelling', *Int J Numer Analy Methods Geomech*, 19, 653–660.
Malvern L E (1969), *Introduction to the Mechanics of Continuous Medium*, Prentice-Hall, New Jersey.
Martinie L and Roussel N (2010), 'Fiber-reinforced cementitious materials: From intrinsic isotropic behavior to fiber alignment', Proceedings of SCC2010, Montreal, Canada, pp. 407–415.
Martys N S (2005), 'Study of a dissipative particle dynamics based approach for modeling suspensions', *J Rheol*, 49, 401–424.
Martys N and Ferraris C F (2002), 'Simulation of SCC flow', *Proceedings of 1st North American Conference of the Design and Use of SCC*, Chicago, pp. 27–30.
Mechtcherine V and Shyshko S (2006), 'Modelling the behaviour of fresh and hardened concrete using the distinct element method', *2nd International Symposium on Advances in Concrete through Science and Engineering*, Quebec City, Canada.
Mechtcherine V and Shyshko S (2007), 'Simulating the behaviour of fresh concrete using distinct element method', *Proceedings of 5th International RILEM Symposium on SCC*, Ghent, pp. 467–472.
Modigell M, Vasilic K, Brameshuber W and Uebachs S (2007), 'Modelling and simulation of the flow behaviour of self-compacting concrete', *Proceedings of 5th International RILEM Symposium on SCC*, Ghent, pp. 387–392.
Mokéddem S, Roquet N, Cazacliu B and Lanos C (2009), 'Perturbation patterns and velocity modelisation of a fresh concrete around a spherical probe Viscoprobe® immersed in concrete mixers', *Proceedings 3rd International RILEM Symposium of Rheology of Cement Suspensions such as Fresh Concrete*, Reykjavik, pp. 155–163.
Moresi L, Dufour F and Mühlhaus HB (2003), 'A Lagrangian integration point finite element method for large deformation modeling of viscoelastic geomaterials', *J Comput Phys*, 184, 476–497.
Mori H and Tanigawa Y (1992), 'Simulation methods for fluidity of fresh concrete', Memoirs of the School of Engineering, Nagoya University, 44, 71–134.
Noor M A and Uomoto T (1999), 'Three-dimensional discrete element simulation of rheology tests of self-compacting concrete', *Proceedings of 1st International RILEM Symposium on SCC*, Stockholm, pp. 35–46.
O'Donovan E J and Tanner R I (1984), 'Numerical study of the Bingham squeeze film problem', *J Non-Newtonian Fluid Mech*, 15, 75–83.
Oldroyd J G (1947), 'A rational formulation of the equations of plastic flow for a Bingham solid', *Proc Cambridge Philos Soc*, 43, 100–105.
Ovarlez G and Roussel N (2006), 'A physical model for the prediction of lateral stress exerted by self-compacting concrete on formwork', *Mater Struct*, 39, 269–279.
Patzák B and Bittnar Z (2009), 'Modeling of fresh concrete flow', *Comput Struct*, 87, 962–969.
Petersson Ö (2003), 'Simulation of self-compacting concrete – laboratory experiments and numerical modeling of testing methods, J-ring and L-box tests', *Proceedings of 3rd*

International RILEM Symposium on SCC, Reykjavik, pp. 202–207.

Petersson Ö and Hakami H (2001), 'Simulation of SCC-laboratory experiments and numerical modeling of slump flow and L-box tests', *Proceedings of 2nd International RILEM Symposium on* SCC, Tokyo, pp. 79–88.

Quoc P H D and Uomoto T (2008), 'Review on simulation of fresh concrete using distinct element method', *3rd ACF International Conference ACF/VCA,* pp. 585–591.

Roussel N (2006), 'Correlation between yield stress and slump: Comparison between numerical simulations and concrete rheometers results, *Mater Struct*, 39, 501–509.

Roussel N and Coussot P (2005), 'Fifty-cent rheometer for yield stress measurements: from slump to spreading flow', *J Rheol*, 49, 707–718.

Roussel N, Geiker M R, Dufour F, Thrane L N and Szabo P (2007a), 'Computational modeling of concrete flow: general overview', *Cem Concr Res*, 37, 1298–1307.

Roussel N, Staquet S, D'Aloia Schwarzentruber L, Le Roy R and Toutlemonde F (2007b), 'SCC casting prediction for the realization of prototype VHPC-precambered composite beams', *Mater Struct*, 40, 877–887.

Schwabe J H and Kuch H (2005), 'Development and control of concrete mix processing procedures', *Proceedings 18th BIBM International Congress BFBN*, pp. 108–109.

Skocek J, Svec O, Spangenberg J, Stang H, Geiker M R, *et al*., (2011), 'Modeling of flow of flow of particles in non-Newtonian fluid using lattice Boltzmann method', Proceedings of 13th International Congree on Chemistry of Cement (ICCC), Madrid, 2011.

Souli M and Zolesio J (2001), 'Arbitrary Lagrangian–Eulerian and free surface methods in fluid mechanics', *J Comput Methods Appl Mech Eng*, 191, 451–466.

Spangenberg J, Roussel N, Hattel J H, Thorborg J, Geiker M R, *et al*., (2010), 'Prediction of the impact of flow induced inhomogeneities in self compacting concrete (SCC)', Proceedings of SCC2010, Montreal, Canada, pp. 209–215.

Svec O, Skocek J, Stang H, Olesen J F and Poulsen P N (2011), 'Flow simulation of fiber reinforced self compacting concrete using lattice Boltzmann method', Proceedings of 13th International Congree on Chemistry of Cement (ICCC), Madrid, 2011.

Thrane L N (2007), *Form filling with SCC*, PhD thesis, Department of Civil Engineering, Technical University of Denmark.

Thrane L N, Szabo P, Geiker M, Glavind M and Stang H (2004), 'Simulation of the test method 'l-box' for self-compacting concrete', *Annu Trans Nordic Rheol Soc*, 12, 47–54.

Thrane L N, Pzabo P, Geiker M, Stang H and Pade C (2005), 'Simulation and verification of flow in test methods', *Proceedings 2nd North American Conference on the Design and Use of SCC and the 4th International RILEM Symposium on SCC*, Chicago, pp. 551–556.

Thrane L N, Stang H and Geiker M (2007), 'Flow induced segregation in full scale castings with SCC', *Proceedings of 5th International RILEM Symposium on SCC*, Ghent, pp. 449–454.

Thrane L N, Pade C and Nielsen C V (2010), 'Determination of rheology of self-consolidating concrete using the 4C-rheometer and how to make use of the results', *J ASTM Int*, 7, 10pp.

Tie G, Dequn L and Huamin Z (2006), 'Three-dimensional finite element method for the filling simulation of injection molding', *Eng Comput*, 21, 289–295.

Tregger N, Ferrara L, Shah S P (2008), 'Identifying cement rheological properties from the mini-slump-flow test', *ACI Mater J*, 105, 558–566.

Uebachs S and Brameshuber W (2005), 'Numerical Simulation of the flow behaviour of self-compacting concretes using fluid mechanical methods', *Proceedings 2nd North American Conference on the Design and Use of SCC and the 4th International RILEM Symposium on SCC*, Chicago, pp. 597–605.

Van Waarde F, Koenders E A B, Nijeholt L and Walraven J C (2007), 'Theoretical and practical investigations on SCC formwork', *Proceedings of 5th International RILEM Symposium on SCC*, Ghent, pp. 417–423.

Vasilic K, Meng B, Kühne H C and Roussel N (2011), 'Flow of fresh concrete through steel bars: A porous medium analogy', *Cem Concr Res*, 41, 496–503.

Vu-Quoc L, Zhang X and Walton O (2000), 'A 3-D discrete-element method for dry granular flows of ellipsoidal particles', *Comput Methods Appl Mech Eng*, 187, 483–528.

Wallevik J E (2003), *Fresh concrete, mortar and cement paste with various types of lignosulfonates*, PhD thesis, Department of Structural Engineering, Norwegian University of Science and Technology.

Wallevik J E (2008), 'Minimizing end-effects in the coaxial cylinders viscometer: Viscoplastic flow inside the ConTec BML Viscometer 3', *J Non-Newtonian Fluid Mech*, 155, 116–123.

Walton O (1982), 'Explicit particle dynamics model for granular materials', *Proceedings 4th International Conference on Numerical Methods in Geometry*, Edmonton, Canada, pp. 1261–1268.

Wendt J F (1992), *Computational fluid Dynamics: An Introduction*, Springer, Berlin.

White J L (1977), 'A plastic-viscoelastic constitutive equation to represent the rheological behavior of concentrated suspensions of small particles in polymers melts', *J Non-Newtonian Fluid Mech*, 45, 177–190.

Zhao D, Nezami E, Hashash Y and Ghaboussi J (2006), 'Three-dimensional discrete element simulation for granular materials', *Eng Comput*, 23, 749–770.

Zhuguo L (2007), 'State of workability design technology for fresh concrete in Japan', *Cem Concr Res*, 37, 1308–1320.

11
Thixotropy: from measurement to casting of concrete

N. ROUSSEL, Université Paris Est, IFSTTAR, France

Abstract: Interactions among cement particles can lead to the formation of microstructures, which may reversibly evolve with the stress applied to the system. This chapter first discusses the origin of this thixotropic behaviour in the case of cementitious materials and then describes thixotropy of concrete in industrial practice. Finally rheological models able to predict the thixotropic behaviour of cement pastes and concretes are presented and the consequences of this behaviour on casting processes are considered.

Key words: thixotropy, yield stress, structuration rate, casting.

11.1 Introduction

Cement-based suspensions such as cement pastes, mortars and concretes are granular pastes (cf. Chapter 2). Most rheological properties find their origins at the scale of the colloidal suspension of fine particles dispersed in water located between the coarsest grains. Interactions among these particles due to either inter-particle forces or the nucleation of hydration products may lead to the formation of microstructures in the suspension at rest or during slow flows. Depending on how the structure responds to an applied shear stress, one can observe different types of macroscopic flow behaviour such as yield stress behaviour or thixotropy (Mewis and Spaull, 1976; Bird and Yarusso, 1982; Coussot and Ancey, 1999).

Shear thinning bodies are said to be thixotropic if (i) after a long rest when a shear stress or strain rate is applied suddenly and then held constant, the apparent viscosity is a diminishing function of the time of flow and (ii) the body recovers its initial state following a long enough interval after cessation of the flow (Roussel, 2005). This time-dependant decrease in the viscosity may be explained by a reversible change of the suspension microstructure during shear. In the absence of shear, the damaged structure rebuilds. Thus yield stress or shear-thinning behaviour might be considered as a particular case of thixotropy where both structural breakdowns under shear and structure rebuilding take place simultaneously.

From an industrial point of view and in the case of concrete, thixotropy is of interest for the following non-exhaustive list of applications:

- Self-compacting concrete (SCC) pressure formwork: during placing, the fresh SCC behaves as a fluid but, if cast slowly enough or if at rest, it builds up an internal structure and has the ability to withstand the load from concrete cast above it without increasing the lateral stress against the formwork (cf. Chapter 13 for more details on this specific topic).

- Distinct-layers casting: during placing, a layer of SCC has a short time to rest and build up a structure before a second layer of concrete is cast above it. If its apparent yield stress increases above a critical value, then the two layers do not mix and, as vibration is prohibited in the case of SCC, a weak mechanical interface appears in the final structural element. Loss of strength of more than 40% has been reported (Roussel and Cussigh, 2008).
- Stability of SCC: during placing, the constitutive cement paste is destructured because of mixing and casting. This allows for easy placement of the material. However, as soon as casting is over and before setting, gravity may induce sedimentation of the coarsest particles. A thixotropic cement paste will build up a structure once at rest. Its apparent yield stress will increase and will be sufficient to prevent the particles from settling (Roussel, 2006a).

This chapter is organised as follows: we first discuss the origin of the thixotropic behaviour. In a second part, we describe the thixotropy of concrete in industrial practice. We then describe the rheological models able to predict the thixotropic behaviour of cement pastes. Finally, we describe a thixotropy model for concrete and give some examples of applications.

11.2 Origin of thixotropic behaviour

In the case of many cementitious materials, a reversible evolution of the material rheological behaviour is often noted during the dormant period of the hydration reaction: the apparent yield stress (or static yield stress) continuously evolves (Otsubo *et al.*, 1980; Banfill and Saunders, 1981; Jarny *et al.*, 2005; Coussot and Roussel, 2006; Roussel, 2006a, 2006b). As this evolution can be erased by a strong shearing of the paste and the material can be brought back to a reference state, it is often described as thixotropy. The origin of this evolution is, however, very unclear. The word 'structuration' is often used to describe the consequences of this thixotropic behaviour as this word is not associated to any physical phenomena (Coussot, 2005). It only implies that a structure is being built or destroyed within the material all long its flow history. As these evolutions are macroscopically reversible, most explanations available in the literature focus on microscopic reversible physical phenomena, such as reversible colloidal flocculation and deflocculation. The phenomena at the origin of thixotropy in other material in nature or industry can, however, be very varied. Of course reversible colloidal flocculation may explain thixotropy in many suspensions, but reversible entanglement of polymer chains or reversible orientation of Brownian slender particles are also potential explanations for macroscopic thixotropic behaviour (Barnes, 1997).

As thixotropy is a reversible macroscopic phenomenon, it is therefore often associated to reversible physico-chemical phenomena such as flocculation and deflocculation under the shear of colloidal suspensions. It has, however, to be kept in mind that thixotropy is defined at a purely macroscopic level: it only implies that the macroscopic evolutions of the behaviour of the material are reversible (Barnes, 1997; Coussot, 2005). In the case of cement pastes, this means that, as long as shear is sufficient to bring the material back to a reference state, one does not have to care whether the evolution of the material behaviour is due to colloidal flocculation or to calcium hydrosilicate (CSH) bridges between particles. Even if, at a microscopic scale, it is a non-reversible chemical reaction that creates bonds between particles, these may be weak enough to be broken by shear whereas new bonds may spontaneously appear again at rest (Jiang *et al.*, 1995; Nonat *et al.*, 1997). Their formation is hence not incompatible with a macroscopic reversible evolution. From a practical point of view, hydration may therefore have reversible macroscopic consequences as long as the available mixing power is sufficient to break the CSH links between cement particles. Hydration is, however, at the origin of workability loss as soon as the available mixing power becomes insufficient to break these inter-particle connections (Jiang *et al.*, 1995; Nonat *et al.*, 1997; Roussel *et al.*, 2010), if we neglect here any non-compatibility phenomena between HRWRA and cement (Flatt and Houst, 2001).

Although it is still under debate, recent results tend to show that the role of CSH nucleation in the thixotropic behaviour is stronger than initially described in (Tattersall and Banfill, 1983). The experimental facts supporting the role of CSH in the thixotropy of cementitious materials are:

- Increased temperature increases the structural build-up rate. If hydration is at the origin of thixotropy, increasing the temperature will increase the nucleation rate of CSH.
- Addition of fine particles such as silica fume increases the structural build-up rate. If hydration is at the origin of thixotropy, adding particles known for their nucleation properties will also increase the nucleation rate of CSH.
- Lowering the water to cement ratio increases the structural build-up rate. If hydration is at the origin of thixotropy, reducing the average distance between cement particles and increasing the average number of neighbouring particles will indeed increase the strength of the network formed by cement particles linked by nucleated CSH bridges.
- Addition of superplasticizer decreases the structural build-up rate. If hydration is at the origin of thixotropy, covering the particles with adsorbed polymer molecules will naturally decrease the nucleation rate of CSH.
- Addition of accelerators strongly increases the structural build-up rate. If hydration is at the origin of thixotropy, it is obvious that increasing the chemical activity in the system will increase the nucleation rate of CSH between the cement grains.

Micro-scale analysis along with oscillation rheometry is currently used to bring more definitive answers to the question of the origin of thixotropy of cementitious materials.

Finally, it has to be kept in mind that another way to create or enhance thixotropy in cementitious systems could lie in the use of polymers that adsorb on the cement grains and get entangled in the interstitial solution. The resulting macroscopic thixotropy could then find its origin in the entanglement/disentanglement processes.

No matter the origin of the thixotropic behaviour, the particle interaction forces determine for each particle a potential energy well (i.e. there is an equilibrium position for each particle for which the potential energy is minimum) (Coussot, 2005). As long as the energy ΔE given to the system is lower than a given value, the particle does not leave this well. When the applied stress or strain stops, the particle comes back to its initial position (elastic solid behaviour). However, if the energy given to the system is higher than a given value, the particle is then able to leave this potential energy well and flow occurs (yield stress behaviour). In the case of systems displaying thixotropic behaviour, the depth of the potential energy well increases at rest with time. The energy $\Delta E'$ needed for the particle to leave the well increases. This will correspond to an increase in the apparent yield stress. However, if all particles leave their respective wells, these wells come back to their initial depth and the system is fully rejuvenated.

11.3 Thixotropy in practice

As the thixotropic behaviour is reversible, one needs to have an agreed reference state around which the variations of the rheological behaviour can be studied. The most frequent reference state in practice is the 'most destructured state'. This is reached at the time point in the material flow history when it is submitted to the highest shear rate. In the laboratory, in the case of cement pastes, this corresponds most often to what is called the pre-shear phase often used before any rheological measurements. In industrial practice and in the case of concrete, it has to be noted that, apart from the case when concrete is pumped, the most destructured state is reached during mixing (Roussel, 2006b). This shear rate can, however, vary significantly depending on the mixer type, mixer charge and rotational velocity, and also strongly affect the structure state reached by the material and thus its rheological properties. A mix proportioning suitable for a given mixer will not be suitable for another mixer in terms of filling ability. Moreover, as the available power of the mixing tools decreases from the making of the concrete (plant mixer) to the building site (mixing truck), the structure of the material will increase and its fluidity will decrease.

Finally, considering the shear rates that can be applied by most concrete rheometers (cf. Chapter 4), none of these devices is able to completely break the structure of the material after a resting period and bring it back to its destructured state in the mixer. The maximum shear rate they can apply to the material is indeed always lower than the shear rate during mixing (Roussel, 2006b). This means that the rheological behaviour that is measured immediately after mixing will never be measured again if the sample stays in the rheometer. As a consequence, the most suitable reference state for a given concrete is the structuration state immediately after mixing. This is the most destructured state in the concrete flow history. However, it should be kept in mind that, because of the limitations in rotation speed of the rheometers, this reference state and an apparent irreversible evolution of the material (which is not due to the hydration process but to the rheometer's limitations) will be measured. Of course, if the sample is re-mixed in the initial mixer between each measurement, this artefact will not appear.

In practice, a concrete is called thixotropic if it seems to build up a structure rather quickly at rest and it becomes apparently more and more fluid while flowing during typically several tens of seconds. It has to be noted that, to be rigorously correct, all concretes are thixotropic. Indeed, a pure yield stress behaviour can be considered as a thixotropic behaviour with very low structuration rate (very long structuration characteristic time) and very fast de-structuration rate (very short de-structuration characteristic time), so that there is no apparent increase of the yield stress at rest and steady state is reached almost instantaneously. However, in the opinion of the present author, it can be admitted that, in practice, a 'thixotropic concrete' is a concrete displaying a rather short structuration characteristic time (typically several minutes) and a de-structuration characteristic time of several tens of seconds in the 1–10 s^{-1} shear rate range.

It has finally to be noted that, between the two aspects of thixotropy (structuration at rest and de-structuration under shear), the understanding and measuring of the first is far more important in terms of potential applications. In the three points of interest listed in the introduction (pressure formwork, distinct-layers casting, and stability), concrete is not flowing. It is at rest and what really matters is the increase of the apparent yield stress or the apparent yield stress of the cement paste in the case of stability. That is why recent approaches to quantify thixotropic behaviour have focused only on the structural build-up phenomenon.

11.4 Rheological models for cement paste

In order to describe the rheological behaviour of cementitious materials in steady state, several non-Newtonian models have been proposed and tested with success and a large number of experimental and theoretical studies of the rheological characteristics of fresh cementitious materials have been published. However, the available data thixotropic behaviour are not as numerous.

The steady state viscosity of cement pastes depends on strain rate. Shear-thinning or shear-thickening behaviours may be observed. Bingham or modified Bingham, Hershel Bulkley, Ellis, Casson or Eyring models are more or less suitable to describe the steady state behaviour of fresh cements pastes (Atzeni *et al.*, 1985; Yahia and Khayat, 2001).

Cheng and Evans (1965) suggested a general mathematical form of the equation of state of a thixotropic material. Equation 11.1 links the shear stress to the shear rate:

$$\tau = \eta(\lambda, \dot{\gamma})\dot{\gamma} \qquad [11.1]$$

An evolution equation is added where λ is a structural parameter related to the structuration level inside the material.

$$\frac{d\lambda}{dt} = f(\lambda, \dot{\gamma}) \qquad [11.2]$$

Papo (1988) derived from these ideas and his own experimental data a rather complete constitutive equation. The steady state model adopted in Papo (1988) is the Casson model.

$$\tau = \tau_{00} + \eta_\infty \dot{\gamma} + 2(\tau_{00}\eta_\infty \dot{\gamma})^{\frac{1}{2}} + \left[(\tau_{01} - \tau_{00}) + 2(\eta_\infty \dot{\gamma})^{\frac{1}{2}}\left(\tau_{01}^{\frac{1}{2}} - \tau_{00}^{\frac{1}{2}}\right)\right]\exp(-k_b t) \qquad [11.3]$$

where η_∞ is the viscosity at infinite shear rate, k_b is the constant rate of destructuration, and τ_{00} and τ_{01} are the steady state yield stress and the initial yield stress, respectively. The de-structuration rate is assumed not to depend on the strain rate. The structural rebuilding process is neglected compared to the de-structuration phenomenon. This model involves four parameters, which is only one more than a Hershel Bulkley model, for instance, which only predicts steady state flow. However, this model does not predict structuration at rest. τ_{01} can not indeed be predicted in terms of resting time and has to be fitted.

The interest of the recent approach developed by Wallevik (2003) is that it is derived from a rather complete physical description of the flocculation and dispersion of the cement grains. Using these mechanisms, it is demonstrated that the steady state and transient behaviour of fresh cement pastes can be mathematically described. Instead of using an evolution equation of a structure parameter, fading memory integrals are used to take the flow history into account. However, the five parameters of this model have to be fitted using numerical simulations through a rather complex process.

The model used by Roussel (2005) and developed by Coussot *et al.* (2002) is based on the same assumptions as those of Cheng and Evans (1965). The basic added proposal consists of considering that the rate of change of λ is equal to the sum of a 'natural' material restructuration term and a rate of de-structuration due to flow, which is proportional to the rate of shear and the flocculation state. The model of Coussot *et al.* (2002) uses these simple basic ideas to predict shear stress. In this phenomenological model, the apparent viscosity is an increasing function of the structure parameter λ or 'degree of jamming'.

$$\tau = \mu_0(1 + \lambda^n)\dot{\gamma} \qquad [11.4]$$

$$\frac{\partial \lambda}{\partial t} = \frac{1}{T} - \alpha\lambda\dot{\gamma} \qquad [11.5]$$

where μ_0 is the viscosity at infinite shear rate when λ tends towards zero, n is a constant positive parameter, $1/T$ is the structuration term and the second term in

Eq. 11.5 can be associated with the de-structuration rate. One may note that there is no explicit yield stress in this model. However, when the shear rate tends towards zero, the apparent viscosity increases as in any other shear-thinning model. This model has several advantages and drawbacks:

- It predicts structuration at rest. This should allow the prediction of τ_{max} in terms of the resting time. The model is, however, not able to predict quantitatively the structuration state for resting times longer than one minute.
- It describes the de-structuration phenomenon under an applied shear rate (constant or variable in time).
- It describes steady state as an equilibrium state between de-structuration and structuration phenomena.
- The number of parameters is low (only one more parameter than most models only describing steady state behaviour). This makes this model suitable for a practical use.
- It is easy to implement in numerical simulations. As no yield stress is involved, no mathematical discontinuity appears in the calculations.

This model predictions were compared to experimental measurements obtained with a coaxial viscometer (HAAKE ViscoTester ® VT550) (Roussel, 2005).

11.5 Rheological models for concrete

In the case of a model suitable to describe flows of concrete *and* useful from a practical point of view, advantages other than scientific exactitude have to be considered. First the model has to be simple. Second, the number of parameters has to stay low and these parameters have to be easily measured.

The model proposed by Roussel (2006b) is based on the principles described in the previous section and writes:

$$\tau = (1+\lambda)\tau_0 + \mu_p\dot{\gamma} \qquad [11.6]$$

$$\frac{\partial\lambda}{\partial t} = \frac{1}{T} - \alpha\lambda\dot{\gamma} \qquad [11.7]$$

where λ is the flocculation state of the material. This state depends on the flow history. Just after mixing, if the mixing phase is considered as the phase when the applied shear rate is maximal, λ is equal to zero. This means that the thixotropic apparent yield stress due to flocculation λ and $_{\tau 0}$ is also equal to zero. Through the successive steps in the casting process (rest phase, re-mixing phase, pumping phase, etc), λ will evolve from its initial zero value to a positive value according to the evolution Eq. 11.7 and an apparent yield stress greater than the initial yield stress will appear. At rest, when shear rate equals zero, the evolution of the apparent yield stress is $\tau_0(t) = (1 + \lambda)\tau_0 = \tau_0 (1 + t/T) = \tau_0 + A_{thix}t$ where $A_{thix} = \tau_0/T$ is the structuration rate in Pa/s.

11.5.1 Formwork pressure

As described in Chapter 12, SCC flows readily under its own weight and achieves good consolidation without any mechanical vibration. During casting, given the high fluidity of this type of concrete, it can be expected that a hydrostatic pressure will be reached in the formwork. Formworks are prudently designed by taking into account this high pressure. Such an approach, however, increases the cost of the formwork and limits the maximum allowable placement height, which was advertised as an advantage of SCC. In cases around the world, the pressure was, when monitored, reported as being very high; however, in other cases, the opposite results were reported. It was concluded that the thixotropic behaviour of the SCC had to play a role (cf. Chapter 13). During placing, the material behaves indeed as a fluid but, if cast slowly enough or if at rest, it builds up an internal structure and has the ability to withstand the load from concrete cast above it without increasing the lateral stress against the formwork.

Using the model presented above, it was demonstrated by Ovarlez and Roussel (2006) that the relative formwork pressure (or relative lateral stress as the system is not hydrostatic) during casting of a wall of thickness e, height H at a casting rate of R could be predicted using the following relationship:

$$\frac{P}{\rho g H} = 1 - \frac{H A_{thix}}{\rho g e R} \qquad [11.8]$$

where R (m/s) is the casting rate, H (m) is the formwork height, e is the formwork width and ρ is the concrete density. This relation was compared to experimental measurements by Ovarlez and Roussel (2006), Tchamba *et al.* (2008) and Perrot *et al.* (2009). It should be kept in mind that highly thixotropic SCC is the most suitable for this application.

11.5.2 Distinct-layer casting

During placing, a layer of SCC has a short time to rest and build up a structure before a second layer of concrete is cast above it. If its apparent yield stress increases above a critical value, then the two layers do not mix at all and, as vibration is prohibited in the case of SCC, a weak interface appears in the hardened concrete. Loss of strength of more than 40% was reported by Roussel and Cussigh (2008). This phenomenon was studied using the concrete thixotropy model described above. It was shown that the time Δt_c after which the two layers do not remix is of the order of:

$$\Delta t_c = \frac{\mu_p V}{A_{thix} h} \qquad [11.9]$$

where V is the concrete flowing speed and h is the thickness of the second layer.

If thixotropy is mastered from a mix proportioning point of view, specific concrete displaying suitable structuration rates could be prepared according to the element to be cast. As the distinct-layers casting problem is dominant in concrete slabs, the SCC should be as non-thixotropic as possible (low structuration rate). In the case of walls, where the formwork pressure problem dominates, the SCC should be as thixotropic as possible (high structuration rate). The equations above can then be used to calculate the needed rate of structuration A_{thix} for a given application, which will become a target value for the mix-proportioning engineer.

11.6 References

Atzeni, C., Massidda, L. and Sanna, U. (1985), 'Comparison between rheological models for Portland cement pastes', *Cem Concr Res*, 15, 511–519.

Banfill, P.F.G. and Saunders, D.C. (1981), 'On the viscosimetric examination of cement pastes', *Cem Concr Res*, 11, 363–370.

Barnes, H.A. (1997), 'Thixotropy – a review', *J. Non-Newtonian Fluid Mech*, 70, 1–33.

Bird, M.G.H. and Yarusso, B.J. (1982) 'The rheology and flow of viscoplastic material', *Rev Chem Eng*, 1, 1–70.

Cheng, D.H. and Evans, F. (1965), 'Phenomenological characterization of the rheological behaviour of inelastic reversible thixotropic and anti-thixotropic fluids', *Br J Appl Phys*, 16, 1599.

Coussot, P. (2005), 'Rheometry of pastes, suspensions, and granular materials', John Wiley & Sons, New Jersey.

Coussot, P. and Ancey, C. (1999) 'Rheophysical classification of concentrated suspensions and granular paste', *Phys Rev Lett E*, 59, 4445–4457.

Coussot, P. and Roussel, N. (2006), 'Quantification de la thixotropie des matériaux cimentaires et de ses effets' [in French], *Revue Européenne de Génie Civil*, 10, 45–63.

Coussot, P, Nguyen, Q.D., Huynh, H.T. and Bonn, D. (2002), 'Viscosity bifurcation in thixotropic, yielding fluids', *J Rheol*, 46(3), 573–589.

Flatt, R.J. and Houst, Y.F. (2001), 'A simplified view on chemical effects perturbing the action of superplasticizers', *Cem Concr Res*, 31, 1169–1176.

Jarny, S., Roussel, N., Rodts, S., Le Roy, R. and Coussot, P. (2005), 'Rheological behavior of cement pastes from MRI velocimetry', *Cem Concr Res*, 35, 1873–1881.

Jiang, S.P., Mutin, J.C. and Nonat, A. (1995), 'Studies on mechanisms and physico-chemical parameters at the origin of cement setting. 1. The fundamental processes involved during the cement setting', *Cem Concr Res*, 25, 779–789.

Mewis, J. and Spaull, A.J.B. (1976), 'Rheology of concentrated suspensions', *Adv Colloid Interface Sci*, 6, 173–200.

Nonat, A., Mutin, J.C., Lecoq, X. and Jiang, S.P. (1997), 'Physico-chemical parameters determining hydration and particle interactions during the setting of silicate cements', *Solid State Ionics*, 101–103, 923–930.

Otsubo, Y., Miyai, S. and Umeya, K. (1980), 'Time-dependant flow of cement pastes', *Cem Concr Res*, 10, 631–638.

Ovarlez, G. and Roussel, N. (2006), 'A physical model for the prediction of lateral stress exerted by self-compacting concrete on formwork', *Mater Struct*, 39, 269–279.

Papo, A. (1988), 'The thixotropic behaviour of white Portland cement pastes', *Cem Concr Res*, 18, 595–603.

Perrot A., Amziane, S., Ovarlez, G. and Roussel, N. (2009), 'SCC formwork pressure: Influence of steel rebars', *Cem Concr Res*, 39, 524–528.

Roussel, N. (2005), 'Steady and transient flow behaviour of fresh cement pastes', *Cem Concr Res*, 35, 1656–1664.

Roussel, N. (2006a), 'A theoretical frame to study stability of fresh concrete', *Mater Struct*, 39, 81–91.

Roussel, N. (2006b), 'A thixotropy model for fresh fluid concretes: theory, validation and applications', *Cem Concr Res*, 36, 1797–1806.

Roussel, N. and Cussigh, F. (2008), 'Distinct-layer casting of SCC: the mechanical consequences of thixotropy', *Cem Concr Res*, 38, 624–632.

Roussel, N., Lemaître, A., Flatt, R.J. and Coussot, P. (2010), 'Steady state flow of cement suspensions: A micromechanical state of the art', *Cem Concr Res*, 40, 77–84.

Tattersall, G.H. and Banfill, P.G.F. (1983), *The Rheology of Fresh Concrete*, Pitman, London.

Tchamba, J.C., Amziane, S., Ovarlez, G. and Roussel, N. (2008), 'Lateral stress exerted by fresh cement paste on formwork: laboratory experiments', *Cem Concr Res*, 38, 459–466.

Wallevik, J.E. (2003), *Rheology of particle suspensions, fresh concrete, mortar and cement paste with various types of lignosulfonates*, PhD thesis, Norwegian University of Science and Technology.

Yahia, A. and Khayat, K.H. (2001), 'Analytical models for estimating yield stress of high performance pseudo-plastic grout', *Cem Concr Res*, 31, 731–738.

12
Understanding formwork pressure generated by fresh concrete

P. BILLBERG, Swedish Cement and Concrete Research Institute, Sweden

Abstract: Correct prediction of the pressure on formwork from the fresh concrete to avoid under- or overestimating its value is vital in minimizing the cost of the formwork. Existing models for formwork pressure do not include the behaviour of the fresh concrete at rest, which is related to its consistency in terms of shear strength expressed as the slump value. With the introduction of self-compacting concrete (SCC), research efforts shifted to this material property at rest, i.e. the structural build-up. This chapter discusses existing models for conventional concrete, key parameters influencing formwork pressure and new models based on the structural behaviour of concrete at rest. Laboratory methods (more scientific) and industrial methods (more empirical) used to measure the static yield stress are presented.

Key words: formwork pressure, structural build-up, static yield stress, laboratory methods, industrial methods, prediction models.

12.1 Introduction

As a construction material, concrete is used on a global scale and is overall the most widely used construction material on the planet. Aside from its obvious hardened properties, an important key feature of concrete is that while the material is still fresh, it can be made to adopt nearly any shape. Assuming that the right type of concrete for the application is used, the only limit to what final shape can be achieved is lack of imagination.

Whether the shape of the hardened concrete is determined due to required mechanical properties, durability issues or architectural demands, the formwork into which the concrete is cast must meet certain basic requirements. The most obvious requirement is that it withstands the loads from the freshly placed concrete in order to avoid collapse. This would completely jeopardize the construction and can be very dangerous for the workers on the construction site. It is also extremely important for the formwork to not deform uncontrollably under the concrete load since the positive property of concrete is that it will follow the shape of the mould. Thus, a deformed formwork will lead to a concrete shape not meeting the required measure tolerance. Investment in construction projects is substantial for both infrastructure and residential construction. The share of the investment for a concrete construction represented by the formwork system can be as high as 60% (ACI, 1989). From this perspective also, the prediction of concrete pressure on formwork is very important in that optimized formwork can lead to considerably lower cost and high quality concrete structures. The obvious importance of correctly predicting formwork pressure is also evident in the large number of international reports and articles published on this topic, beginning as early as the 1900s (Shunk, 1909; Rodin, 1952; Clear and Harrison, 1985; Gardner, 1985;).

When discussing formwork pressure from the perspective of the concrete material, it is worthwhile to examine the difference between conventional vibrated concrete (CVC) and self-compacting concrete (SCC), as existing codes and recommendations of how to calculate formwork pressure (with one exception discussed in Section 12.5.5) relate only to CVC. In many countries, the existing formwork design codes were developed decades ago and do not incorporate such fluid concretes like SCC; to be more precise, the codes incorporate fluid concrete, but the recommendations are to design the formwork for hydrostatic pressure.

SCC development began in Japan in the mid-1980s and use of this concrete is today spreading around the world and continuously increasing. This is especially the case in the precast industry, for which SCC is nothing less than a truly revolutionary material. However, the question of formwork pressure is by far the most important in cast-in-place applications and uncertainty in predicting formwork pressure when using SCC is one of the obstacles to increased use of this superb concrete. Reported measurements have shown, however, that pressures considerably lower than hydrostatic pressure can be observed when using SCC and this fact has been the driving force behind the intense research undertaken in this field during the last ten years. In general, the design method for CVC has changed over the last few decades due to the development of new extremely effective superplasticizers, often based on a polycarboxylate ether (PCE). Instead of regulating the consistency by adjusting water content in the mixture, as was done previously, it is more common today to do this by adjusting the superplasticizer dosage for the mixture. This leads to considerably different concrete mixtures, especially in terms of particle concentration and thus structural behaviour at rest. Therefore, the older formwork design codes most likely overestimate pressures for today's CVCs and need to be updated. A few examples of existing codes for calculation of formwork pressure are presented in Section 12.2.4.

In this chapter, we focus more on SCC rheological properties and the influence they have on formwork pressure than on CVC and the older formwork design codes. In theory, however, the basic mechanisms are the same for CVC and SCC, so that from a phenomenological point of view, discussing SCC and leaving out CVC is no real restriction. Finally, we focus exclusively on concrete lateral pressure on vertical formwork.

12.2 Factors affecting formwork pressure

In reality, there are a large number of factors influencing formwork pressure. Theoretically, any imaginable factor involved in the casting process could influence the pressure but, in practice, most existing formwork design models only include a limited number of factors. Throughout the last century a few review papers were published that related back to the previous decades. In their reviews, Rodin (1952) and Gardner (1985) listed a number of factors, as did also the authors of CIRIA Report 108 (Clear and Harrison, 1985). All these lists overlap to some extent, but the factors can be split into three groups: concrete properties, casting technique and formwork characteristics. The way these factors influence formwork pressure is discussed and briefly explained below.

12.2.1 Concrete properties

The literature identifies the following concrete properties and the constituent materials affecting these properties as being related to formwork pressure:

- concrete density
- concrete consistency (shear strength)
- chemical and mineral admixtures
- concrete temperature
- aggregate shape.

Concrete density influences the lateral formwork pressure when the concrete is in a fluid state and thus lacking shear strength. This is the case for CVC directly after casting and during vibration or when concrete is re-vibrated. This obviously applies to SCC as well, since initially after casting SCC is very fluid.

When concrete is left at rest in the formwork it develops shear strength which increases with time, or as discussed nowadays, the material static yield stress, τ_S (Pa), grows. This time-dependent structural behaviour at rest is in fact a key property for both CVC and SCC in relation to lateral formwork pressure; it is discussed in more detail and in relation to formwork pressure in the next section. The liquid-like SCC is capable of filling the formwork without compaction work (vibration), partly because of its characteristic low internal structure. CVC, on the other hand, has relatively higher shear strength despite the fact that when it is vibrated, the structure breaks down and it also acts as a liquid. Therefore, anything that affects the shear strength also affects formwork pressure. Consequently, lowering the temperature will slow down the structural build-up of the material as the chemical and physical mechanisms involved are delayed.

General discussion of the role of admixtures focuses mainly on their influence on the time-dependent change in shear strength. Thus, any admixture that increases retardation leads to higher pressure while any accelerating admixture lowers the pressure due to the respective influence on structure development.

Aggregate shape relates to the friction between particles influencing the shear resistance of concrete. The influence of coarse aggregate on pressure in SCC was investigated by Assaad (2004) and found to be significant. An increase in the coarse-to-total aggregate ratio reduced the initial pressure, as well as a decrease of pressure with time. This is due to interlocking of aggregate particles causing an arching phenomenon. Maximum size was also investigated and its influence was found to relate to packing, in that favourable packing generates lower pressure due to less mobility of the settled SCC at rest (Assaad, 2004).

12.2.2 Casting technique

It is easy to understand from the earlier discussion of the time-dependent growth of the material's structure at rest that the concrete casting technique is important for formwork pressure. The following casting-related actions are discussed in the literature:

- vibration and re-vibration
- casting rate
- pumping concrete from the top or from the base
- impact from concrete discharge, in air or under water.

When concrete is vibrated, it becomes fluidized since the structure in the material is broken down. Concrete is locally liquefied and the pressure will be hydrostatic at the geometrical extent of the vibration work. Figure 12.1 illustrates well the influence of vibration on formwork pressure in the vibrated zone.

Casting rate is, together with construction height and concrete static yield stress, the most important parameter and is included in most existing models for

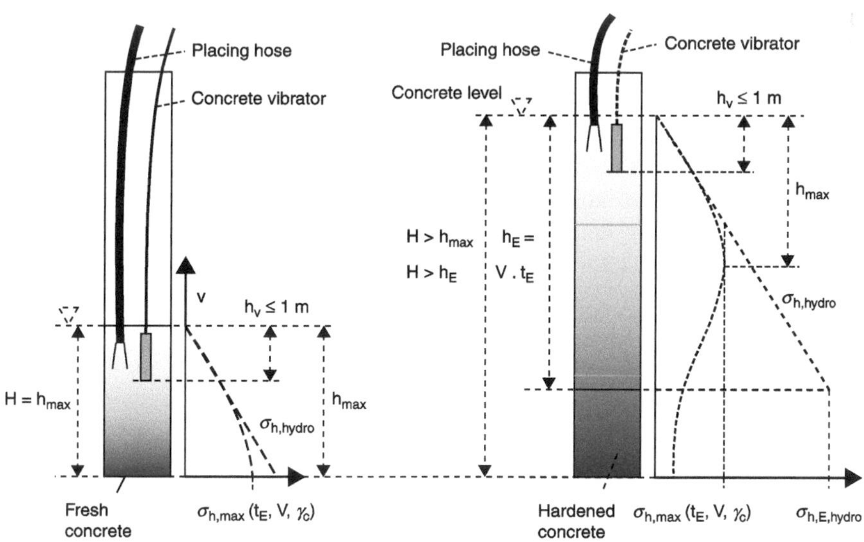

12.1 Illustration of the influence of vibration on formwork pressure (Proske and Graubner, 2010).

formwork pressure. Since this property is inversely related to concrete resting time, an increased casting rate reduces this time, resulting in higher pressures due to lower levels of concrete static yield stress (shear strength).

In cases where concrete is pumped from the base of the formwork, the design value for the pressure should be hydrostatic pressure plus the pump pressure. The reason is, of course, that the concrete is in motion during the entire casting process without any possibility for concrete shear strength to develop. The discharge height of the concrete (pouring height) influences impact forces that locally liquefy the previously cast concrete layer below. Finally, casting under water enables to account for the buoyant weight density, that is to say the density of concrete minus the density of water.

12.2.3 Formwork characteristics

The formwork itself and its characteristics have an influence on lateral pressure. Some of these factors are:

- formwork geometry
- reinforcement
- formwork permeability or water tightness
- roughness of the sheathing material
- stiffness of the formwork
- formwork slope.

Formwork geometry is important, not only in terms of the obvious influence of construction height, but also in terms of structure width in that the effect of shear forces at the wall due to friction between concrete and wall decreases as formwork width increases. The friction between concrete and formwork is discussed further in Section 12.3.2 and in some of the models for formwork pressure using SCC in Section 12.5.

Discussion of friction consequently leads to an examination of the impact of denser reinforcement configuration on lateral pressure. The increase in vertical surfaces due to the presence of rebars contributes to these supporting friction forces and reduces the lateral stresses against the wall. This phenomenon has been investigated by Perrot *et al.* (2009), who reported that the presence of steel rebars can significantly reduce formwork pressure.

The permeability and roughness of the formwork material also influence the friction between the formwork surface and the concrete. Permeability decreases the pore water pressure due to leakage of water through the formwork surface while increased roughness has an obvious impact on elevated friction.

In general, stiffer formwork results in higher pressures. This can be explained by the fact that pressure reduction due to yielding of the formwork allows arching to occur, as suggested by Rodin (1952). If one side of the formwork is sloped (i.e. non-parallel formwork sides), this implies that the plane area varies with height and thus the vertical casting rate is not constant even though the cast volume per time unit is constant. Recalculation is thus needed.

12.2.4 Factors included in existing codes for formwork pressure using conventional vibrated concrete (CVC)

It is obvious that, in practice, all the factors known to affect formwork pressure cannot be included in codes for designing formwork. The formulas would then be both complicated and based on too many parameters to be characterized for each project. Instead, many countries that have developed national codes for formwork design have chosen specific parameters and included large enough safety margins. An examination of existing formwork codes for Sweden (published 1968), Germany (published 1980), the UK (published 1985), Japan (published 2002) and the USA (published 2005) enables us to list those parameters included (Table 12.1). Note that included parameters are only highlighted if they appear in the codes as variables. It should be stressed that casting rate and concrete temperature are included in all the selected codes as variables, while the influence of aggregate and reinforcement is not included. If the parameters included in a planned casting extend beyond the range of limits established for each of the respective codes, all of them state that hydrostatic pressure is to be used.

Table 12.1 Parameters included in some existing formwork design codes

Parameter	Country				
	Sweden[1]	UK[2]	Germany[3]	USA[4]	Japan[5]
Temperature	X	X	X	X	X
Casting rate	X	X	X	X	X
Concrete density		X	X	X	
Formwork geometry		X	X	X	X
Effect of retarder		X	X	X	
Effect of binder type		X		X	
Concrete consistency	X		X		
Consistency range*	S ≤ 150		F ≤ 500	S ≤ 175	S ≤ 100

Source: McCarthy *et al.* (2007)
Notes:
[1] Backsell and Bydemar (1968)
[2] Clear and Harrison (1985)
[3] DIN (1980)
[4] Hurd (2005)
[5] Ueda (2002)

12.3 Relation of concrete structural behaviour at rest to lateral pressure

It is reasonable to begin by elaborating upon and sorting out what is meant rheologically by structure in terms of concrete. We then deal with how the material parameter theoretically relates to lateral formwork pressure.

12.3.1 Reversible and irreversible structures

As discussed in previous chapters, fresh concrete is very often characterized by the Bingham model (Eq. 12.1)

$$\tau = \tau_0 + \mu_{pl} \cdot \dot{\gamma} \quad [12.1]$$

where τ is the shear stress in Pa, τ_0 the yield stress in Pa, μ_{pl} the plastic viscosity in Pa·s, and $\dot{\gamma}$ the rate of shear in s^{-1}. The yield stress according to the Bingham model is referred to subsequently as the dynamic yield stress, since this material parameter is only correctly characterized when the material is at steady state in the range of the evaluated shear rate interval. The dynamic yield stress thus correlates very well to, for example, the slump flow of SCC (as shown in Chapter 4). In other words, the dynamic yield stress determines how far a material can flow before the stresses in the material become lower than the dynamic yield stress, such as when a material goes from a state of motion to a state of rest.

The shear strength, however, relates to what stress has to be exceeded in the material for it to go from a state of rest to a state of motion. This stress is referred to subsequently as the static yield stress (τ_{S} in Pa). The structural build-up of the concrete at rest is by definition the time-dependent growth of the static yield stress and has been shown to be linear in a reasonable time range of the order of an hour at rest (Billberg, 2006; Ovarlez and Roussel, 2006; cf. Chapter 11). In the literature, the structural build-up is given many designations by different authors, such as $\dot{\tau}_s$ by Billberg (2006), A_{thix} by Ovarlez and Roussel (2006) and τ_{0rest} (t) by Omran (2009). Despite these different designations, the meaning is exactly the same for all the examples shown, that is to say it refers to the linear increase of the static yield stress (in Pa) with time (t) at rest. The time unit is most often seconds.

During the last decade, in which discussion of formwork pressure when using SCC has generated considerable interest, the material property most often elaborated upon has been thixotropy. However, thixotropy is by definition a time-dependent, isothermal and reversible process (Mewis, 1979) and if we are discussing thixotropy alone, the three-dimensional structure building up in the concrete at rest could hypothetically be fully broken down and the state of structure before rest could be regained. In such a case, the structural change of concrete influencing formwork pressure would indeed be due to thixotropy only. However, at the moment of intermixing between cement and water, another time-dependant process begins, namely the chemical hydration process of the cement.

Depending on the mix design in question, primarily the combination of cement type and type of chemical admixtures (superplasticizers) used, this process can be either negligible or quite pronounced. The chemical hydration process is in practice an irreversible process that mainly affects the dynamic yield stress; this phenomenon is, in everyday language, referred to as slump loss.

Therefore, in focusing on what is relevant for formwork pressure, we should consider not only the reversible thixotropic part of the structure, but also the irreversible slump loss part, i.e. the sum of these parts which constitutes the growth of static yield stress with time. We will see later on, in Section 12.5, that the material property used in different approaches to model and predict formwork pressure when using SCC is indeed the structural build-up.

The material structure is evidently able, to some extent, to carry the vertical load of the concrete above without resulting in horizontal stress against formwork equal to the vertical stress. This ability increases with time at rest as the structure accentuates. All types of concrete build up a structure at rest, but the degree of this structural build-up is naturally greatly dependent upon the actual mix design. Therefore, there is no principal difference between CVC and SCC in this respect. As mentioned earlier, directly after casting, there is still an insignificantly low level of structure in the fresh concrete, especially in the case of SCC and in the case of conventional concrete during vibration, where the structure in practice is more or less absent. Therefore, initially, the formwork pressure will be equal to hydrostatic, but depending on the structural build-up property of the material, the height of the structure and the casting rate (i.e. the time at rest), the pressure can be far lower than hydrostatic. The physico-chemical mechanism behind this structural build-up is discussed in detail in Chapter 10 and therefore not elaborated upon further here.

12.3.2 The silo theory

But how do the concrete properties interact with the formwork to influence formwork pressure? It will be shown here that this is a matter of the friction between the resting concrete material and the formwork panel.

The silo theory is actually based on the theory presented in 1885 by H. A. Janssen (Janssen, 1885), which originally dealt with the pressure in silos due to storage of granular materials. In fact, this theory is the basis for several models for formwork pressure when using SCC (Vanhove *et al.*, 2004; Ovarlez and Roussel, 2006; Beitzel, 2009; Proske and Graubner, 2010). This model relates the horizontal stress to the vertical stress using the Janssen parameter, K:

$$P_{horizontal} = K \cdot P_{vertical} \qquad [12.2]$$

The way this model is used in terms of concrete lateral pressure was first discussed by Ovarlez and Roussel (2006). They used the Janssen model based on the assumptions that this model is also valid for a confined elastic material with frictional contacts or yield stress at the walls and that the maximum stress in the material reaches the material yield stress, according to the Tresca plasticity criterion.

The elastic theory, considering small deformations and an isotropic material, is used with the boundary conditions that: (1) the stress at the walls is equal to τ_0 and that (2) the walls are infinitely rigid. In this approach, it is shown that the Janssen parameter, K, is related to the Poisson ratio, v_p. If the confined material is incompressible, the Poisson ratio is equal to 0.5 and consequently the Janssen parameter is equal to 1. In this case, the horizontal stress is equal to the vertical stress. But concrete in general is indeed compressible since there is always a certain volume of air in concrete. The so-called 'background' volume, without air-entraining agent, is normally estimated to be around 2%. When air is deliberately entrained, the aim is to reach roughly 5% air content. Thus, to an extent depending on the air content, concrete is compressible and the Janssen parameter can be lower than 1. The values of K are computed by Ovarlez and Roussel using the Mori–Tanaka homogenization scheme and become equal to 0.931 if the air content is 5%.

The yield stress of the concrete at rest in the form is not constant but is found to increase linearly with time, and since the initial yield stress is small in relation to its increase with time, it can be neglected. Ovarlez and Roussel introduce their expression of the time-dependent yield stress as A_{thix} (Pa/s). They further show that the lateral pressure at a certain depth, H (m), is equal to the hydrostatic pressure reduced by the vertical stress at the walls, which is between 0 and the concrete yield stress, τ_0. They also assume that the vertical deformation due to the concrete's own weight (settlement) is sufficient to increase this shear stress to be equal to τ_0. Considering that the casting rate, R (m/h), is constant, the concrete height can be expressed as $H = R \cdot t$, where t is the time (h). Numerical simulation of the disturbed zone of the top layer of concrete during pouring shows it to be of the order of the thickness of the formwork, e (m). Thus, at the bottom of the concrete resting zone, the resting time is $(H-e)/R$ and, at the top, it is equal to zero. By integrating the depth-dependent yield stress, incorporating the Janssen parameter and finally considering that the thickness (e for walls or d as the diameter for columns) is neglected, as it is much smaller than H, the ratio of lateral stress to the hydrostatic, σ, is calculated using the following expression:

$$\sigma = \frac{P_{horizontal}}{P_{vertical}} = K\left(1 - \frac{HA_{thix}}{\rho geR}\right) \qquad [12.3]$$

Those readers wishing to follow the steps taken by Ovarlez and Roussel (2006) to deduce this equation in more detail are strongly advised to read the original reference.

12.4 Characterization of fresh concrete structure

The parameters influencing formwork pressure are discussed in Section 12.2, where apart from the density of concrete, the development of static yield stress with time in fresh concrete resting in the formwork is the only material property discussed relative to formwork pressure. How to characterize and measure this specific concrete property is more thoroughly discussed below in terms of both laboratory and industrial approaches.

12.4.1 Laboratory approaches to the measurement of static yield stress

In his doctoral project, Billberg (2006) evaluated several different configurations using a stress-controlled rheometer (Physica MCR 300, Fig. 12.2) equipped with concentric cylinder geometry. In a preliminary series of tests, the structural build-up of four different cement pastes was tested at water to cement (W/C0 ratios ranging from 0.36 to 0.42. The rheometer was configured to increase the stress by 3.33 Pa/s up to a maximum of 300 Pa or until the structure broke below that stress level. The criterion for considering the structure as broken down was when the shear rate (i.e. the response when using the stress-controlled feature) reached 0.5 s^{-1}. When the structure broke down, the rheometer was stopped to let the paste rest for ten minutes until the next stress growth. This was continued until a total of 40 minutes at rest was reached.

Results for all four pastes are illustrated in Fig. 12.3 and show clearly that the W/C ratio, or more correctly, the particle concentration, influences the structural build-up significantly. The structural build-up ranges from 2.5 Pa/min for the paste with W/C 0.42 up to 5.1 Pa/min for the paste with W/C 0.34. Note also the almost perfect linear response in terms of static yield stress with time at rest for all pastes.

12.2 The Physica MCR 300 rheometer (left) and the serrated bob and cup (right).

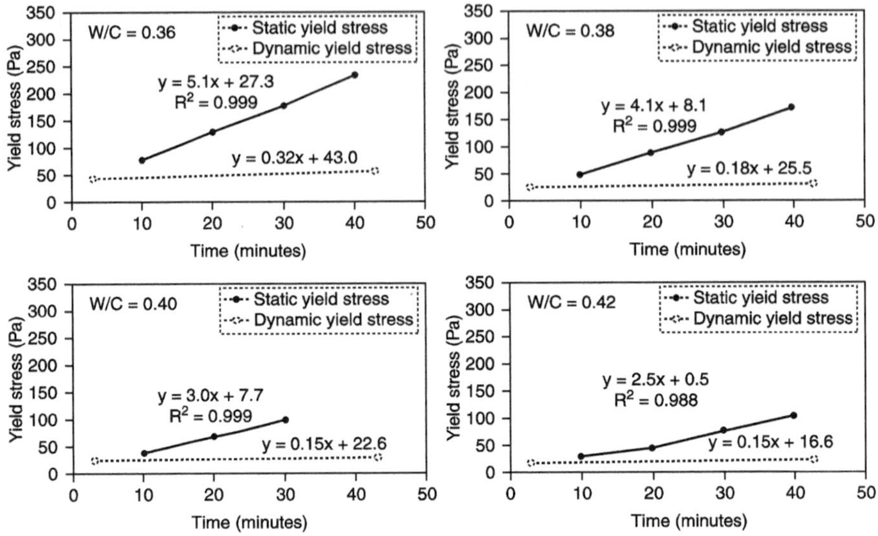

12.3 Development with time after mixing of the dynamic yield stress, τ_0 (dotted lines) and the static yield stress, τ_S (filled symbols) for four cement pastes with W/C ratios ranging from 0.36 to 0.42 (Billberg, 2006).

According to the discussion in Section 12.3.1, there are reversible thixotropic phenomena occurring in the cement paste at rest, but also an irreversible change in dynamic yield stress, as shown in Fig. 12.3. By calculating the reversible structure, that is to say the area under the static yield stress curve reduced by the area under the dynamic yield stress curve and then plotting these values vs. the corresponding W/C ratio, the perfectly exponential relation shown in Fig. 12.4 is obtained.

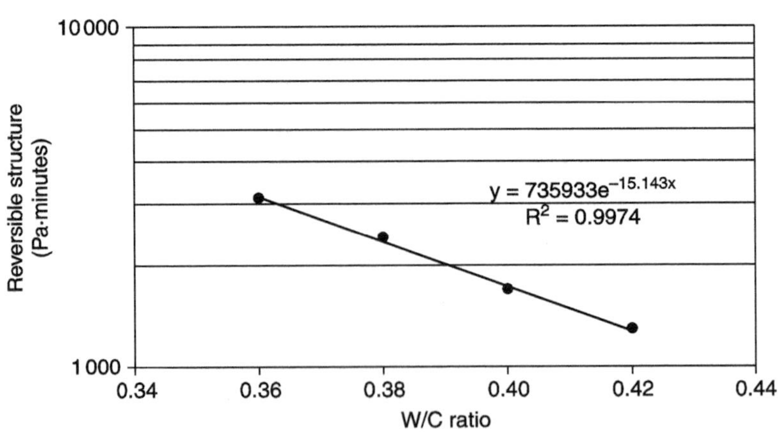

12.4 Relationship between reversible structures vs. W/C ratio for the four tested cement pastes shown in Fig. 12.3.

Amziane *et al.* (2008) developed another very simple yet still scientifically sound and elegant method to measure the structural build-up of cementitious materials. In this case, examples are taken from tests on cement paste. The test set-up shown in Fig. 12.5 (left) consists of a beaker, a plate and two scales. However, only one of the scales is required for structural build-up measurements (Amziane *et al.*, 2008). The plate (3 mm thick, 35 mm wide and 210 mm long) was covered with sand paper of 200 μm in roughness. The hypothesis used is that the settling of the suspension will generate a stress on the plate. Therefore, the basic and necessary phenomenon for this method is that the cement paste is not completely stable without any movement. Naturally, nothing would be detectable for the plate if the material into which it is immersed did not move at all. The authors state that a critical settlement is in the order of 0.0005 to fully mobilize the shear stress. The forces acting on the plate are shown in Fig. 12.5 (right).

Considering a homogeneous material layer settling with time, the following relation between measured plate mass changes and shear stress of the material was deduced (Eq. 12.4):

$$\tau(t) = \frac{g}{2}\left[\frac{\Delta M(t) + \alpha(T)\cdot t}{l\cdot H} - e\cdot\rho\right] \qquad [12.4]$$

where $M(t)$ is the apparent mass measured by the scale, $\alpha(T)\cdot t$ is a correction term for the evaporation of the liquid, H is the height of the immersed part of the plate, l is the plate width and e is the plate thickness, and τ is the shear stress acting on the plate–paste interface.

A cement paste with W/C equal to 0.35 was prepared and, simultaneously with measurements of stress growth with the immersed plate using Eq. 12.4, the static yield stress was measured with a vane rheometer. The results are shown in Fig. 12.6 and it is clear that both methods measure exactly the same structural build-up. The similarity of the described approach to the discussion of the silo

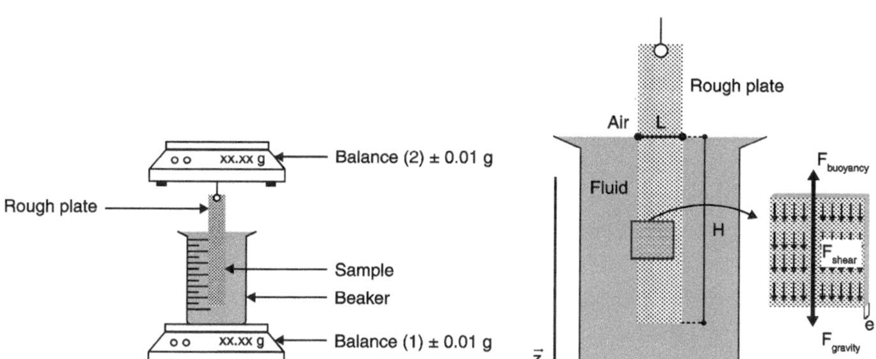

12.5 Test setup (left) and static equilibrium of the plate in fluid (right) (Amziane *et al.*, 2008).

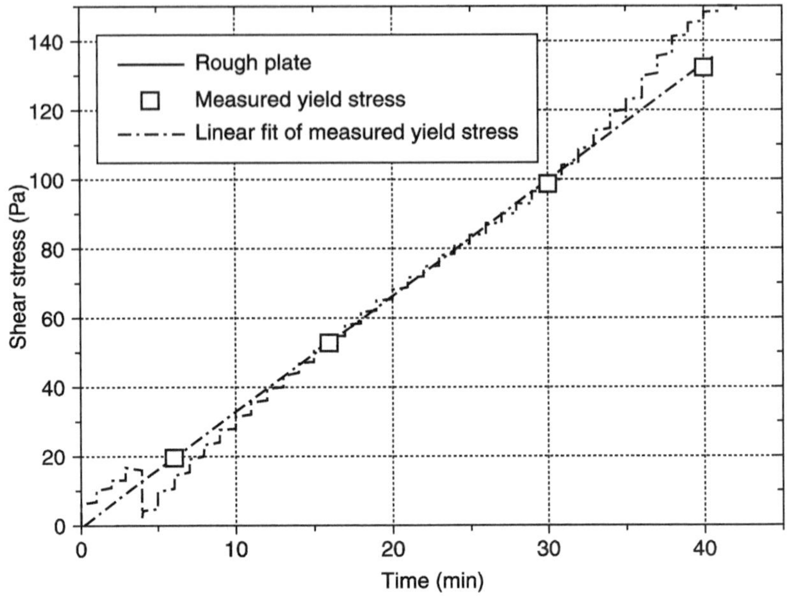

12.6 Results from measurement of cement paste structural build-up using the plate device and a vane rheometer (Amziane *et al.*, 2008).

theory (Section 12.3.2) should be noted in that stress in the material reaches the level of the static yield stress due to friction between concrete and the form wall panel. Ferraris *et al.* (2007) used the same type of approach as that developed by Amziane *et al.* (2008) in that they measured the forces on a paste-immersed plate. However in their case, instead of using a static plate, they pulled the plate slowly (0.05 mm/s) out of the cement paste while measuring the weight (force) using a scale.

Measurements of structural build-up of SCC itself have been reported several times, and different types of concrete viscometers have been used. Ovarlez and Roussel (2006) used a parallel plate BTRHEOM rheometer to measure the structural build-up while Koehler (2004) and Omran (2009) used the ICAR rheometer and modified Tattersall Mk-III rheometer, respectively. The latter two use so-called wide-gap geometries and four-blade vanes as discrete inner cylinders, concentric with the outer cylinders. Billberg (2006) used a ConTec 4 viscometer with concentric cylinders geometry (Fig. 12.7, left). For more detailed information on these rheometers, see the report by ACI Committee 238 (2008).

Regardless of the type of rheometer or geometry used, the essential requirement for accurate measurement is to control the device in order that it rotates relatively slowly while at the same time sampling torque readings frequently enough to secure that the peak stress is captured. Thus, the stress in the material should be slowly increased, but not so slowly that restructuring of the material occurs during the measurement itself (Billberg, 2006).

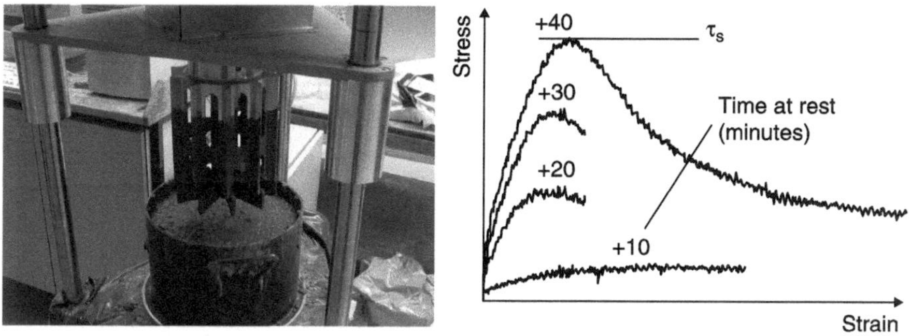

12.7 Close-up of the ConTec 4 viscometer cylinder system (left) and a typical response from the SCC at rest in terms of stress vs. strain (right) (Billberg, 2006).

Billberg (2006) controlled the outer cylinder of the ConTec 4 viscometer to rotate very slowly at 0.736·× 10^{-3} rps, which corresponds to 2.65 revolutions per hour. With the inner cylinder radius equal to 100.0 mm and the gap equal to 44.1 mm, the strain rate in the gap at this rotational speed is 0.015 rad/s. In Fig. 12.7, the cylinder system of the ConTec 4 is shown (left), as well as a typical response in terms of stress vs. strain (right). After the inner cylinder was immersed into the concrete, it was left to rest for ten minutes before rotation was started. By carefully following the increase of stress, rotation was cancelled manually as soon as the peak stress was reached. This was done to break down as little of the structure as possible. In between each test, the concrete rested ten minutes with the last measurement performed after a total of 40 minutes at rest. It should be noted that most peak stresses were reached after 5–6 seconds, which corresponds to a strain of approximately 0.1 rad (Billberg, 2006).

As examples of results using the described methodology, two series from Billberg (2006) are shown in Fig. 12.8. In both series, limestone filler was used, with series Mix 1–3 incorporating a median particle size of 40 microns while

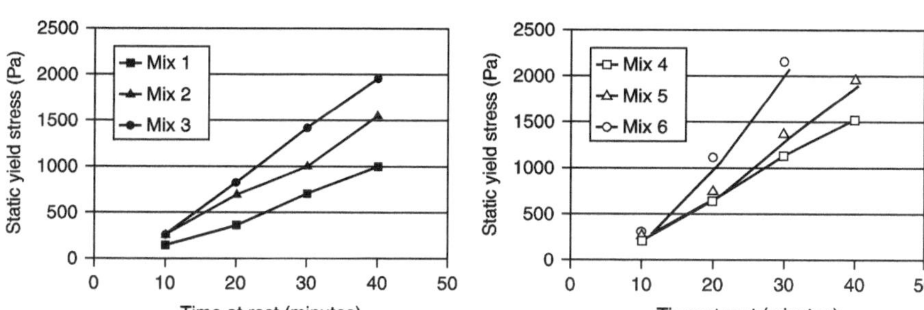

12.8 Structural build-up with time for SCC mixtures 1–3 (left) and for mixtures 4–6 (right).

series Mix 4–6 used a median particle size of 25 microns. All mixtures had slump flow values of 700±40 mm directly after mixing. In each series, the volume concentration of particles in the micro-mortar phase of the concrete was varied: 51.2% for Mixes 1 and 4, 52.3% for Mixes 2 and 5 and 53.3% for Mixes 3 and 6. A PCE-based superplasticizer was used in both series. The structural build-up at rest for the two series is shown in Fig. 12.8. Two observations from the results are obvious:

- Increased particle concentration (reduced water to powder ratios) increases the rate of structural build-up.
- The same goes for the increased fineness of the filler material.

Increased particle concentration and increased particle fineness both reduce inter-particle distances. The results are fully in line with the hypothesis stating that the narrower the particle system becomes, the easier it is for flocculation or coagulation to form a three-dimensional structure (cf. Chapter 11).

12.4.2 Industrial approaches to the measurement of static yield stress

Among the possible faster methods suitable for measuring the static yield stress of a resting concrete is that based on the use of the vane tool. This method originates from geotechnical science in which it was initially developed to measure the shear strength of cohesive soils, such as clay. A few versions of this method, adapted to deliberately capture the structural build-up of SCC, can be found in the literature, of which two are briefly described here.

The first example of the use of a vane tool is shown in Fig. 12.9. As can be seen, the equipment needed basically consists of a four-blade vane tool and a torque meter but yet the ability to characterize the structural build-up at rest of different SCC is still clear (Fig. 12.9, right). Different buckets with concrete were used for the tests and the concrete was left undisturbed in each bucket before each

12.9 Geometry of the 4-blade vane (left), measurement of concrete in a bucket (middle) and results of structural build-up (right) (Roussel and Cussigh, 2008).

respective test. A similar approach was used by Billberg and Österberg (2001), though they measured the structural build-up using the same sample. The device used enabled shifting the level of the four-blade vane to three different vertical positions so that each measure of the static yield stress could be performed on an almost undisturbed part of the sample. The shifting was done directly after a measurement, enabling ten minutes of rest at the new level.

Another example of a vane tool, the portable vane test (PV) developed by Khayat *et al.* (2008) and Khayat and Omran (2010) is shown in Fig. 12.10 (left). Here four vanes of different sizes are used in order to measure torques that vary little, even though the structure increases with time for the tested concrete. The largest vane is used for the weakest structure (i.e. for the concrete rested the shortest amount of time) and gradually smaller vanes with increased time at rest. Typical responses using this method are shown in Fig. 12.10 (right).

A last example of an empirical method to measure the structural build-up of SCC at rest is the so-called inclined plane test (IP). This method was originally introduced by Coussot and Boyer (1995) and developed specifically for SCC (Khayat *et al.*, 2008; Khayat and Omran, 2010). The method is simple in that four cylinders are filled with SCC and lifted to let the SCC spread onto the planes, which are roughened to avoid any slippage. The spread SCC is covered, and at different times of rest, the planes are lifted until the concrete starts flowing. The static yield stress can then be calculated using Eq. 12.5.

$$\tau_y = \rho \cdot g \cdot h \cdot \sin\alpha \qquad [12.5]$$

where ρ (g/cm^3) is the density of the concrete, g (m/s^2) is the gravitational acceleration, h (mm) is the characteristic height of the spread sample and α (degrees) is the angle of the plane when concrete flow is initiated. The test setup is shown in Fig. 12.11 together with some typical responses.

One important difference between these industrial approaches is that, for the vane tool methods, there is no theoretical limit as to the upper limit of measured

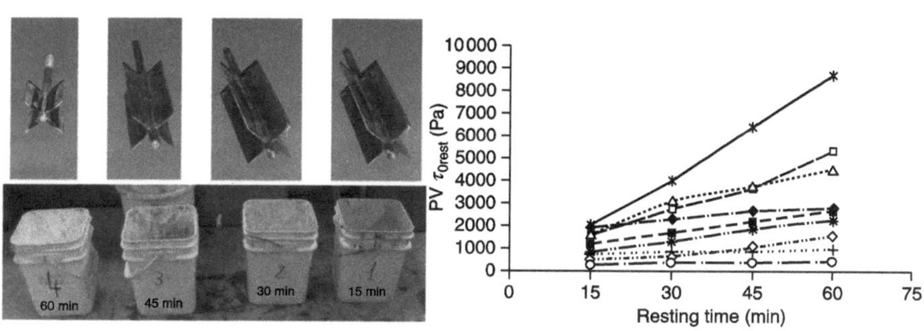

12.10 The four square buckets and the differently sized vanes (left), as well as typical results in terms of structural build-up (right) (Khayat *et al.*, 2008; Khayat and Omran, 2010).

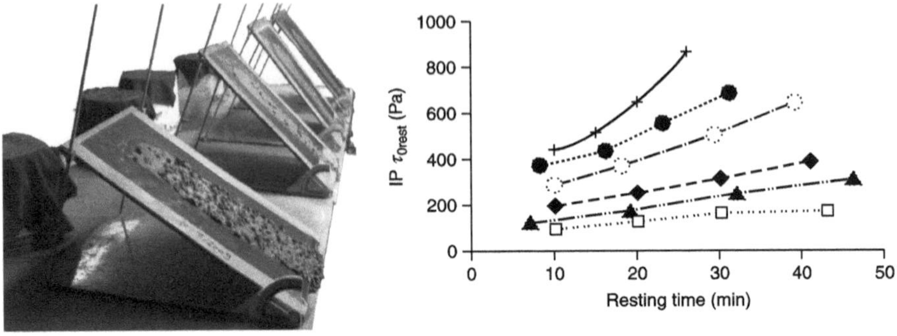

12.11 Inclined planes at different time at rest (left) and typical results in terms of structural build-up using the inclined plane method (right) (Khayat *et al.*, 2008; Khayat and Omran, 2010).

static yield stress. The practical limit is established by the torque meter used. For the inclined plane method, however, the shear stress in the sample increases with the slope of the plane and thus the limit is at 90 degrees (vertical position of the plane). At this extreme slope, the stress depends only on the characteristic height and density, and for normal SCC, neither of those parameters varies considerably between different mixtures at comparable slump-flow measures.

12.5 Modelling of lateral pressure

During the last decade, many attempts have been made to increase the understanding of how to relate the properties of SCC to formwork pressure. Some are conducted as masters or doctoral projects, some as industrial projects, and others as merely theoretical studies. The models selected for presentation here are those published by Assaad *et al.* (2003), Assaad (2004), Tejeda-Dominguez (2005), Ovarlez and Roussel (2006), Proske (2007), Lange *et al.* (2008), Omran (2009), Khayat and Omran (2010), Khayat *et al.* (2010), Proske and Graubner (2010), and Khayat and Omran (2011). For obvious reasons, the models cannot be thoroughly described in minute detail and thus the descriptions of the methods are relatively brief. Readers are strongly advised to consult the original documents referred to in order to obtain more in-depth information regarding the models.

12.5.1 Empirical model based on breakdown area

In his doctoral work, Assaad (2004) developed a rheological approach – which he refers to as the breakdown area, A_b (J/m^3·s) – as the thixotropic characteristic of material to relate to formwork pressure. By using a concrete rheometer, in this case a modified Tattersall Mk-III rheometer, he broke down the concrete structure at different shear rate levels and then combined the initial shear stresses and the

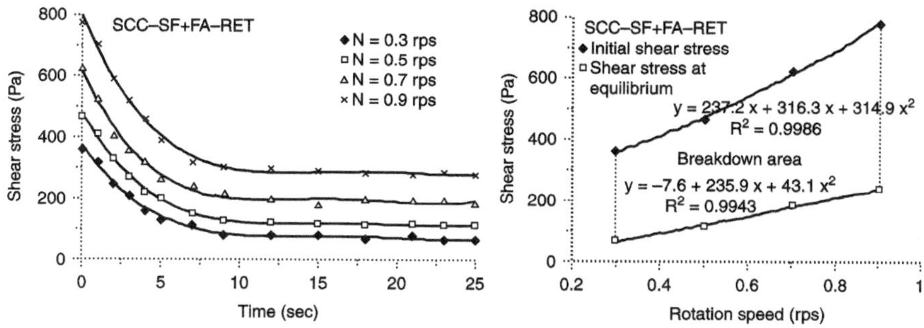

12.12 The breakdown curves at different shear rate levels (left) and the area between the initial shear stresses and the equilibrium stresses representing the breakdown area (right) (Assad *et al.*, 2003).

corresponding equilibrium shear stresses for all shear rate levels, respectively, to calculate the breakdown area. An example of this procedure is shown in Fig. 12.12; the typical breakdown measurements are shown in the left part of the figure and the combined results in the right part. The shear rate levels used were 0.3, 0.5, 0.7 and 0.9 rps. The breakdown area is, the area between the curve representing the combined initial shear stresses and the curve representing the equilibrium stresses (Fig. 12.12, right). Note that, by definition, the lower equilibrium curve is the steady state material response of the material, following the linear Bingham model.

In all, the procedure to establish the breakdown area took some 30 minutes to complete. Because of this, Assaad managed to measure three breakdown areas per mixture representing the times 0–30, 60–90 and 120–150 minutes – designated A_{b1}, A_{b2} and A_{b3}, respectively. The formwork pressure was measured in a cylindrical 2.8 m high plastic column form with an inner diameter of 200 mm and with pressure sensors mounted at heights of 50, 250, 450, 850 and 1250 mm from the base of the column. This extensive investigation focused on the following concrete characteristics:

- use of set-modifying admixtures
- coarse aggregate properties
- type and content of binder
- type and dose of viscosity-modifying admixtures (VMA)
- type of superplasticizer
- water-to-cementitious material ratio (w/cm)
- consistency
- concrete temperature.

A total of 70 SCC mixtures, all having slump flow values of 650±15 mm, a temperature of 20±3°C, densities of 2200±200 kg/m³ and air contents of 7±2%, respectively, were evaluated.

Three parameters were identified as the most important regarding prediction of form pressure:

1. The maximum initial lateral pressure directly at the end of casting, which is the design load for the form
2. The rate of pressure drop with time, which is important for planning subsequent lifts (particularly important for high and voluminous castings)
3. The time before pressure cancellation, which determines the earliest time possible for formwork removal.

The casting rate was set at 10 m/h for all 70 trials. The pressure directly after casting is represented by a parameter K_0 in % of hydrostatic pressure and subsequently, K_{100} and K_{200} represent these values after 100 and 200 minutes, respectively. From all 70 trials, with simultaneous measurements of pressure and thixotropy (breakdown area), the following models (Eq. 12.6–12.8) for estimating lateral pressure at different time intervals were developed:

$$K_0 = -0.047 \cdot A_{b1} + 105.8 \qquad [12.6]$$

$$K_{100} = -0.099 \cdot A_{b2} + 112.2 \qquad [12.7]$$

$$K_{200} = -0.125 \cdot A_{b3} + 116.8 \qquad [12.8]$$

The results from these models are shown in Fig. 12.13 (left). A strong relation was also found between A_{b1}, and respectively A_{b2} and A_{b3}, as shown in Fig. 12.13 (right). Thus, it was concluded that development with time could be predicted only from the first thixotropy measurement after mixing, i.e. from A_{b1}. This project also highlighted that the change in time-dependent structure of SCC is due to an irreversible change in terms of slump loss plus a reversible change due to thixotropy. Naturally, slump loss development alone is not sufficient to explain the development of form pressure.

12.5.2 Empirical model based on characteristic pressure decay

In his Masters work, Tejeda-Dominguez (2005) used a different approach and developed a model based on the characteristic pressure decay of SCC mixtures, i.e. from the maximum pressure at the end of casting to when the pressure eventually drops to zero. This was done using a PVC column (Fig. 12.14) of 920 mm in height and 250 mm in diameter, instrumented with a flush-mounted pressure cell installed at 152 mm from the base. Starting with the expression for hydrostatic horizontal pressure, $P=\gamma h$, where $\gamma = \rho g$ is the concrete unit weight, Tejeda-Dominguez replaced the time (t) and casting-rate (R) dependent concrete height (h) by Rt and finally introduces a function representing the pressure decay, $C(t)$, arriving at the expression:

$$P_h = C(t)\gamma Rt \qquad [12.9]$$

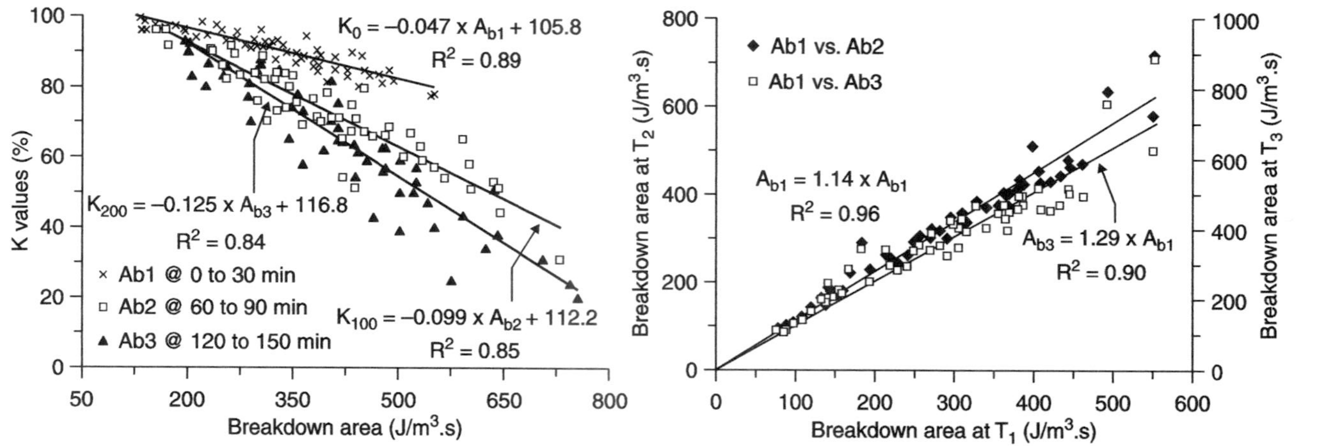

12.13 Relation between breakdown areas at different times corresponding to K-values (left) and the mutual relations between the breakdown areas at different times after mixing (right) (Assad, 2004).

12.14 The PVC column used to characterize the pressure decay of SCC at rest (Tejeda-Dominguez, 2005).

The function $C(t)$ was found to best fit the characteristic measured pressure decay profile of the tested mixtures with a hyperbolic function:

$$C(t) = \frac{C_0}{(at^2 + 1)^{\alpha}} \quad [12.10]$$

where C_0 is the initial pressure and the time-dependent variables a and α are used to fit the function to the pressure decay. The expression for lateral pressure in the case of a continuous casting becomes:

$$P_h = \gamma Rt \frac{C_0}{(at^2 + 1)^{\alpha}} \quad [12.11]$$

An example of how the model applies to a case where a 6 m tall wall is cast at 1.2, 2.4 and 4.8 m/h, respectively, is shown in Fig. 12.15. Here C_0 = 0.95 (initial pressure 95% of hydrostatic pressure) and the time-dependent variables were, for this SCC, found to be $a = 3.16 \cdot 10^{-6}$ and $\alpha = 12$. Note that the pressure in Fig. 12.15 is shown in psi (1 psi = 6.9 kPa).

Results from using this method are reported in a large number of field studies and show that it works satisfactorily (Lange *et al.*, 2008). Even though the method

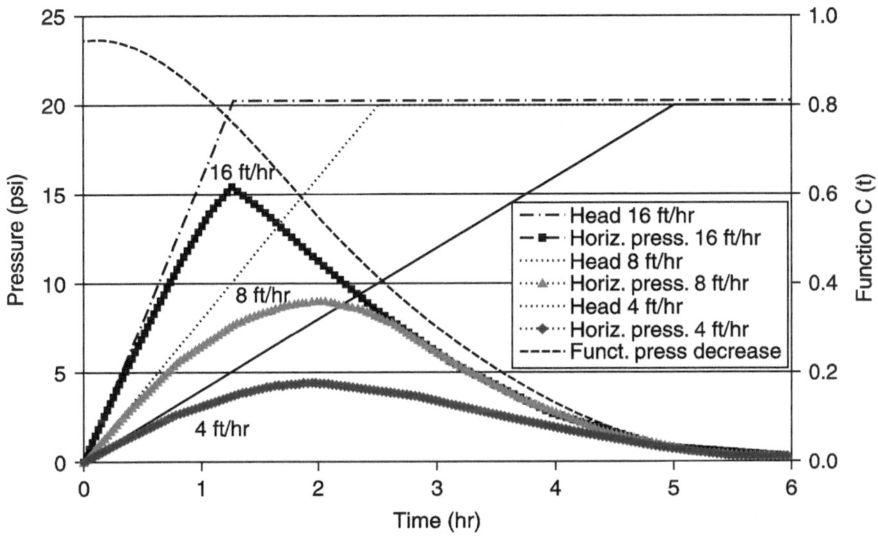

12.15 Pressure profiles for 6 m tall wall cast at 1.2, 2.4 and 4.8 m/h, respectively (Tejeda-Dominguez, 2005).

slightly overestimates the measured pressure, it does so in a limited enough way and is also advantageous in terms of safety margins (Lange *et al.*, 2008).

12.5.3 Analytical model based on structural build-up

The background for this model, which was developed by Ovarlez and Roussel (2006), is described in Section 12.3.2. Here we give a few examples of how this model has been validated.

Ovarlez and Roussel (2006) calculated the theoretical yield stresses necessary to result in lateral pressures measured on a 10 m high, 5.44 m long and 0.20 m wide formwork. Concrete (SCC) was injected from the bottom at 21.4 m/h, which of course resulted in hydrostatic pressure, not only due to the high casting rate, but also due to the fact that the concrete was in constant motion and no part of the concrete was able to build any structure. Pressure sensors were mounted at 0.55, 1.95 and 3.36 m from the base of the formwork. From the initial hydrostatic pressure, the following pressure decay was monitored. As the concrete rested in the formwork, the structural build-up was simultaneously measured using a BTRHEOM rheometer. The static yield stress growth was then compared with the theoretical shear stress at the formwork wall and found to match perfectly (see Fig. 12.16, left). The model was used to compare computations with experimental results by Billberg (2003) and Khayat *et al.* (2005); it was found that the value of A_{thix} was around 0.6 (Pa/s) for the SCC used by Billberg (2003) and 0.2 (Pa/s) for the SCC used by Khayat *et al.* (2005). The results are shown in Fig. 12.16 (right).

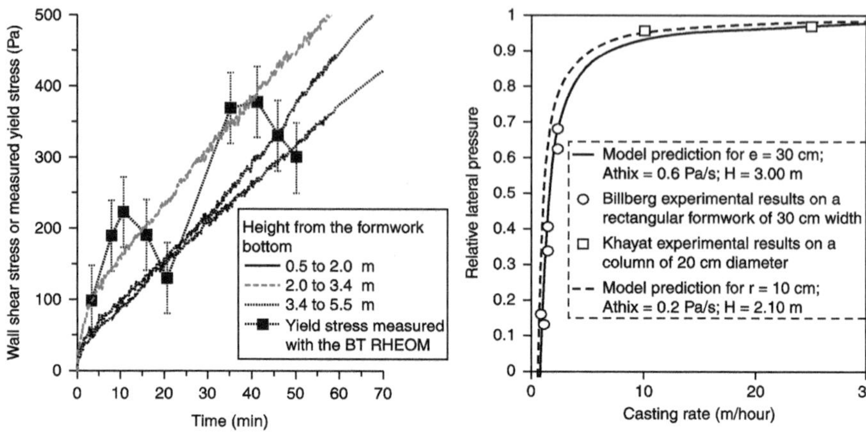

12.16 Comparison between shear stress at the wall needed to obtain the measured pressure and the static yield stress in terms of resting time (left) and comparison between experimental results from the literature and predictions of the model proposed in the case of a rectangular formwork (Khayat *et al.*, 2010) and a column (Khayat *et al.*, 2005).

In another investigation, Tchamba *et al.* (2008) studied the lateral stress on formwork of cement paste and used the model proposed by Ovarlez and Roussel (2006). They used a 1.3 m high column measuring 0.1 m in diameter (Fig. 12.17). The device for measuring the static yield stress of the paste is the same as that described in Section 12.4.1, developed by Amziane *et al.* (2008), but here both smooth and rough plate surfaces were used. The same goes for the inner surface of the column in that it was either smooth or rough. The task was to compare the shear stress at the respective plate and the corresponding column inner surface (smooth–smooth and rough–rough) and also compare it to the static yield stress measured with a vane rheometer. Shear stress at the column inner surface was calculated from the measured loss of vertical load on the bottom surface with time. The smaller the load measured at the bottom, the greater the part of the vertical load that is consequently carried through inner wall frictional forces.

As a first step, it was found that the ability of the plate surfaces to capture the structural build-up was strongly dependent on the roughness of the plates. The shear stress using the rough plate correlated perfectly with the measured static yield stress while the material at the smooth plate obviously could not mobilize the same level of stress. Secondly, and interestingly enough, the same results were obtained using the smooth inner surface of the column. In fact, the shear stresses at the smooth plate and at the smooth inner column surface matched (Fig. 12.18, left). The same perfect match of the shear stresses measured using rough surfaces, the plate, as well as the inner column wall, were also found (Fig. 12.18, right).

As mentioned in Section 12.2.3, it can also be intuitively deduced that reinforcement configuration has an influence on lateral concrete pressure. This is

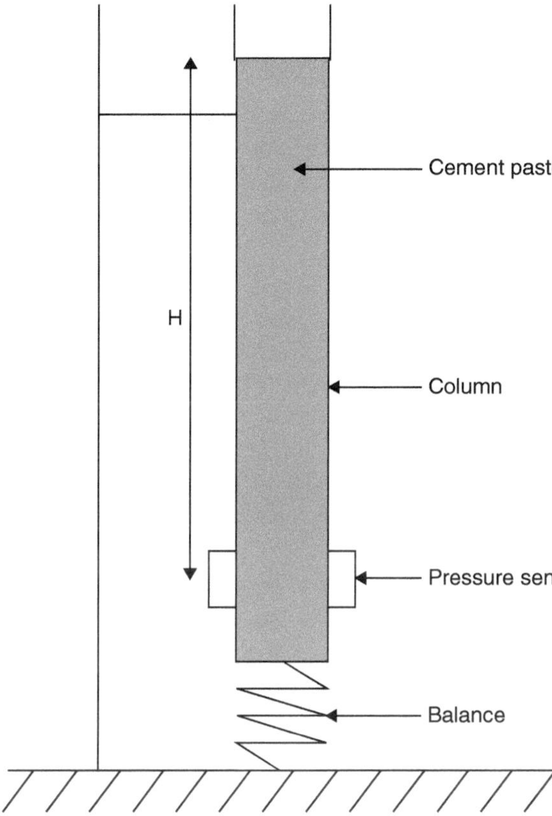

12.17 Setup for the evaluation of yield stress at the inner wall of the column (Tchamba *et al.*, 2008).

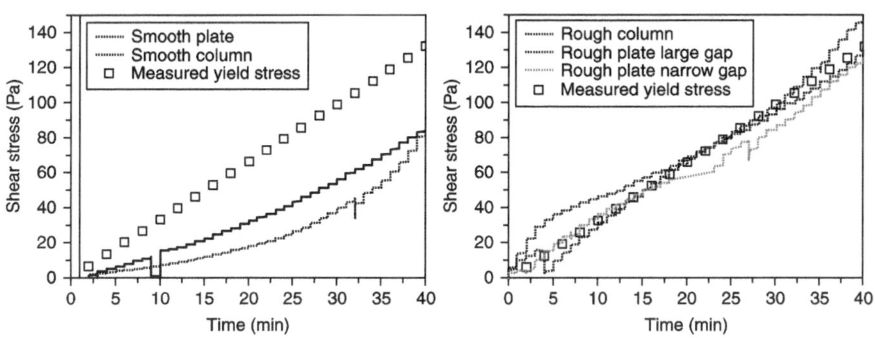

12.18 Measured shear stresses at smooth plate and smooth column compared to static yield stress (left) and equivalent measurements for the rough plate and rough column (right) (Tchamba *et al.*, 2008).

particularly evident when the impact on lateral pressure of material shear stress and friction at the interface is so well established, as described in this section. The influence of reinforcement on the lateral formwork pressure was investigated by Perrot *et al.* (2009), who reported that the presence of steel rebars can indeed significantly reduce formwork pressure. The same theory proposed by Ovarlez and Roussel (discussed above) is used here, also incorporating the surfaces of the rebars. Incorporating the rebars into the calculations can reduce lateral pressure by half (Perrot *et al.*, 2009).

12.5.4 Empirical model based on structural build-up

Within the scope of his doctoral project, Omran (2009) carried out a large number of tests and also developed and evaluated several test methods for both capturing structural behaviour at rest and measuring the lateral pressure of SCC. The two most promising empirical methods for SCC structural build-up are those already described in Section 12.4.2, i.e. the portable vane test (PV) and the inclined plane test (IP).

For measuring formwork pressure, a 0.7 m high PVC pressure column named UofS2, with an inner diameter of 200 mm, was designed to simulate concrete heights of up to 13 m by air overpressure (Fig. 12.19). A sensor for measuring concrete pressure was placed 63.5 mm from the base, flush with the inside of the column, and a second one closer to the top to check the overpressure. The principle was to fill the column with 0.5 m of concrete, seal the column, and then by regulating the overpressure, simulate concrete height and casting rate. The validity of such a limited concrete volume (and height) was verified carefully by comparing results to those obtained with other devices – a similar overpressure column measuring 1.0 m in concrete height and a 3 m free-standing column. A large experimental plan was completed, relating the lateral pressure to many SCC mix design characteristics. One conclusion drawn was, however, that all these mix design parameters can be excluded from possible prediction formulas for formwork pressure and instead be replaced by (included in) the structural build-up property of the SCC at rest.

Besides the structural build-up property of SCC, the following placement characteristics were identified for inclusion into a possible prediction formula for formwork pressure:

- effect of temperature
- casting rate
- waiting period between successive lifts
- minimum formwork dimension.

Many tests were again conducted, and from all the results obtained, derivations of formulas were carried out on five parameters using statistical software (NCSS 2007). Using this software, Omran (2009) was able to establish formulas to

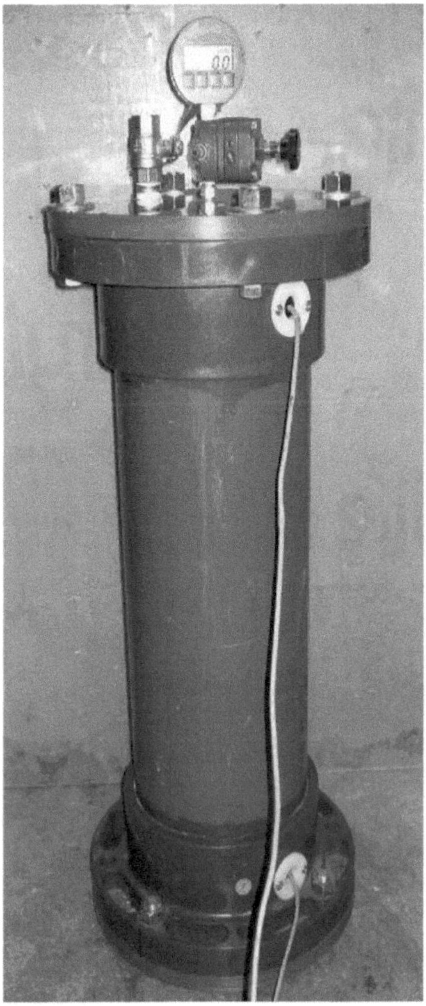

12.19 UofS2 pressure column (Khayat *et al.*, 2008; Khayat and Omran, 2010).

predict maximum formwork pressure, P_{max} (kPa), including the parameters: casting height (*H*), casting rate (*R*), concrete temperature (*T*), minimum lateral dimension of formwork (D_{min}) and structural build-up. The latter parameter, here selected for particular attention, is determined by the PV empirical test method, which is either used at the actual concrete temperature or at 22±2°C, but then established using a correction factor for actual temperature. The effect of a waiting period between successive lifts (*WP*) and maximum size of aggregate (*MSA*) are also included in the formulas as correction factors. The recommended formula for K_0 (% of hydrostatic pressure), resulting in the best correlation between the predicted and measured responses, determined at the same concrete temperature as when casting the concrete, with the highest R^2 value is:

$$K_0(\%) = 98 - 3.82H + 0.63R + 11D_{min} - 0.021PV_{\tau 0rest@15min} \cdot fMSA \cdot fWP \quad [12.12]$$

Equation 12.12 is valid for the following range of included parameters:

- $H = 1\text{–}13$ m
- $R = 2\text{–}30$ m/h
- $T = 12\text{–}30 \pm 2$°C
- $D_{min} = 0.20\text{–}0.35$ m
- $PV_{\tau 0rest@15min} = 0\text{–}2000$ Pa.

A number of field tests were carried out to validate this model at wall castings with wall heights of 3.7 and 4.4 m, respectively, and at the casting of circular columns (height of 3.66 m with a diameter of 0.61 m) (Omran, 2009; Khayat and Omran, 2011). From the results shown in Fig. 12.20 it is evident that the model predicts the pressure satisfactorily. The highlighted symbols represent values where, accidentally, the shear history of the concrete in the buckets of the PV

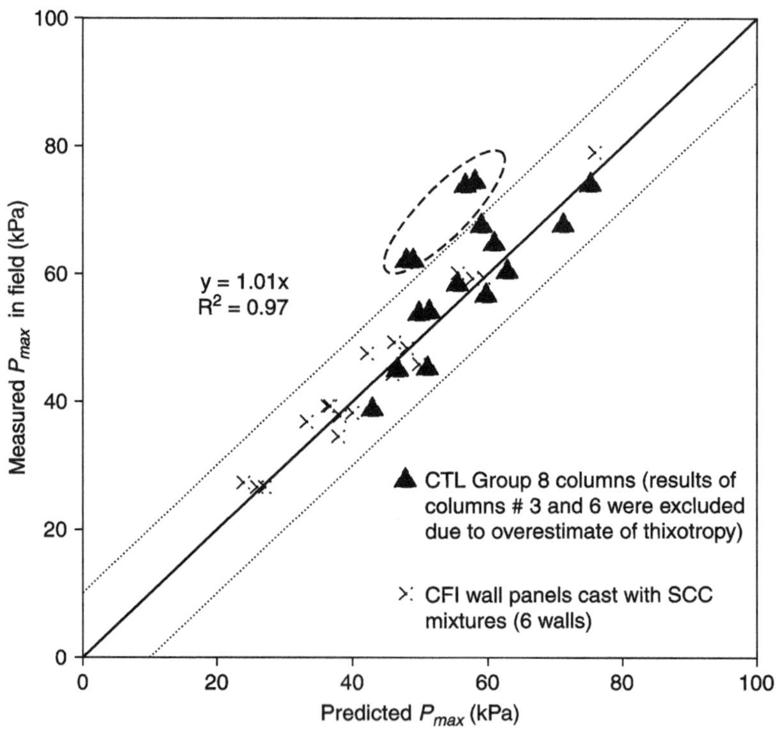

12.20 Measured and predicted lateral pressures at wall and column field castings (Khayat and Omran, 2011).

method became very different relative to the concrete cast in the columns (Omran, 2009; Khayat and Omran, 2011). Thus, the measured pressure was found to be higher than that predicted previously.

12.5.5 Empirical model based on setting time

The model developed by Proske (2007) assumes, as does the one developed by Ovarlez and Roussel (2006), that form pressure is influenced by the friction between the concrete and the wall and the time-dependent behaviour of the concrete at rest. However, Proske (2007) relates this structural behaviour to setting time instead of the development of static yield stress. By considering the Janssen model, and thus the Janssen parameter, K, as the ratio between horizontal and vertical stresses and the shear stress between the concrete and the wall (depending on the friction coefficient, μ), they set up a differential equation for the horizontal pressure, which they solved using the Lagrange method (for details see Proske, 2009).

The time- and load-dependent values for the friction between concrete and wall and the Janssen parameter were measured simultaneously in a special device (Fig. 12.21), where the vertical pressure, P_{vertical} (denoted F(v)), was increased with a hydraulic cylinder and the horizontal pressure, $P_{\text{horizontal}}$ (denoted F(h)), was measured with load cells. Simultaneously, an inserted steel blade was pulled out and the force, F(r), was measured. Thus, the friction coefficient, μ, and Janssen parameter K could be calculated from these measurements.

Several mixtures were tested having different water contents, aggregate, slump flow values and setting times. The calculated result representing an SCC mixture, of μ and K, are shown in Fig. 12.22 where these parameters are plotted against a standardized time, i.e. the actual time relative to the setting time, t/t_E (–). Note that

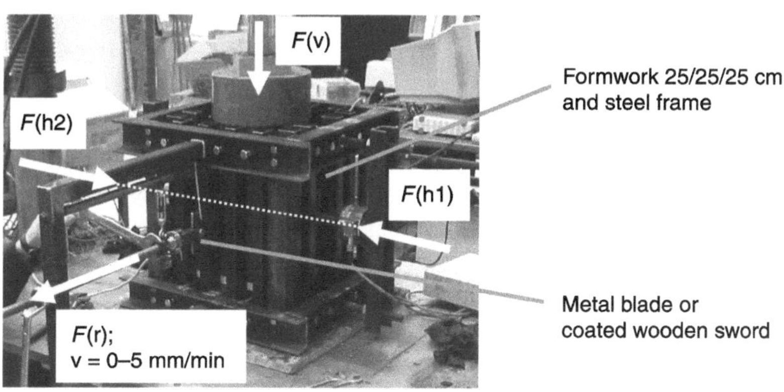

12.21 Device for simultaneous measurement of the horizontal stress and the friction between concrete and a metal or wooden blade (Proske, 2007).

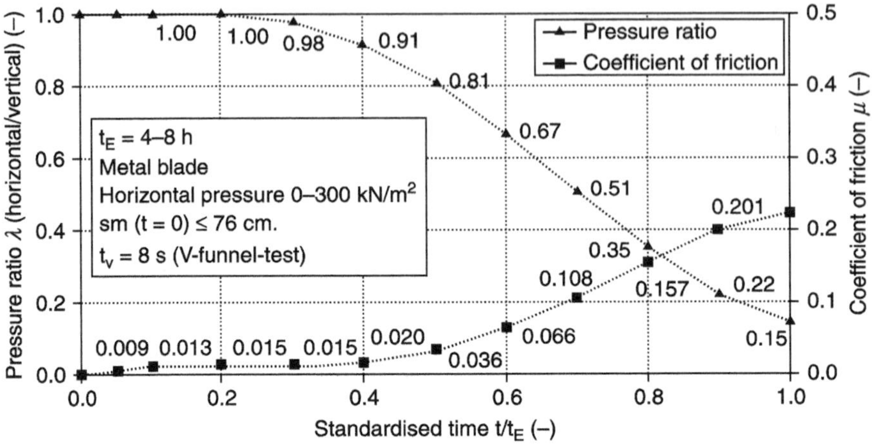

12.22 Time-dependent values $\mu(t)$ and $\lambda(t)$ (Proske, 2007).

the Janssen parameter K is denoted λ in Fig. 12.22. The results of μ and λ were then used to calculate the time-dependent pressure distribution over the height of the formwork.

A German national project aimed at producing new codes for formwork design loads from concrete was completed in 2008 (Graubner *et al.*, 2008). Indeed, the new DIN 18218: 2010–01 (DIN, 2010) standard was published in 2010 as a direct result of this project. A brief presentation of the results and discussions from this project, leading to proposed new formulas for concrete lateral pressure, are given below.

Proske and Graubner (2010) used the device shown in Fig. 12.21 to measure the formwork pressure for a large number of concrete mixtures for SCC with a slump flow of 740±20 mm and V-funnel times of 5±1 s (low viscosity) and 13±2 s (high-viscosity), as well as for concrete of consistency classes F5 and F6 with initial spreads of 600 and 680 mm according to DIN EN 12350–5, respectively. They simulated both stiff and soft formwork by displacement of the formwork panels, as well as the influence of vibration and casting rate (1, 2, 4 and 8 m/h). The results are shown in Fig. 12.23 where the maximum pressure, $\sigma_{h,max}$ (kPa) is plotted vs. the product of final setting time, t_E (h) and casting rate, v (m/h).

Hence, they also relate the structural change of the concrete to the setting time. The lines in Fig. 12.23 represent the normalized pressure, $\bar{\sigma}_{h,E,max}$ (Eq. 12.13)

$$\bar{\sigma}_{h,E,\max} = \frac{\sigma_{h,\max}}{v \cdot t_E \cdot \rho_c \cdot g} = \frac{\sigma_{h,\max}}{v \cdot t_E \cdot \gamma_c} \qquad [12.13]$$

where γ_c is the specific concrete weight in kN/m^3 ($=\rho_c \cdot g$).

A normalized pressure equal to 1.0 represents hydrostatic pressure and for the highly workable concretes, including SCC, the normalized pressure was found to

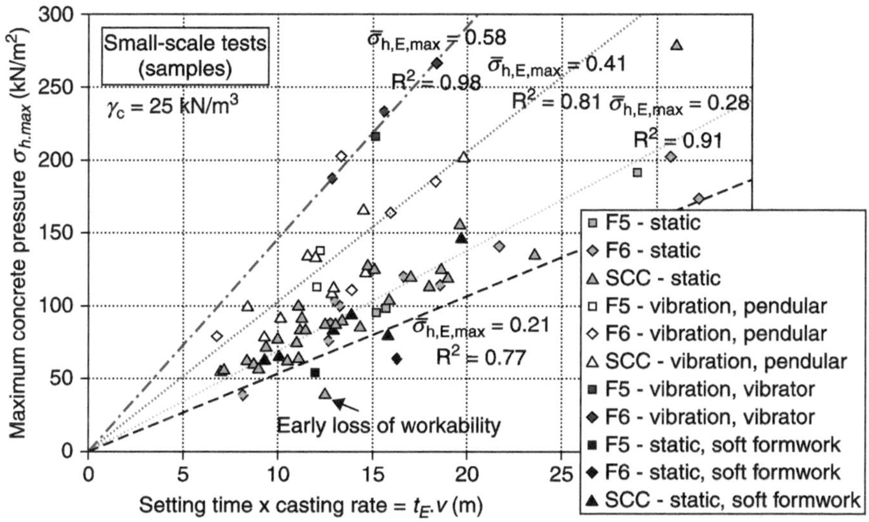

12.23 Formwork pressure vs. $t_E \cdot v$ for different SCC and all F5 and F6 mixtures (Proske and Graubner, 2010).

be 0.28. The impact of vibration increased this value to 0.58 while displacement of the formwork panels, simulating soft formwork, decreased it to 0.21 (Fig. 12.23). It was concluded that, with constant boundary conditions, the normalized pressure is to a large extent independent of the mix design. But since concrete of consistency class F5 and F6 is supposed to be compacted by vibration while SCC is not, the following expressions (Eq. 12.14 and Eq. 12.15) were extracted from Fig. 12.23.

$$\bar{\sigma}_{h,\max} = 0.18 + \frac{1}{h_E} \leq 1 \qquad \text{Consistency class F5 and F6} \qquad [12.14]$$

$$\bar{\sigma}_{h,\max} = 0.16 + \frac{0.8}{h_E} \leq 1 \qquad \text{SCC, where } h_E = t_E \cdot v \qquad [12.15]$$

Note that the predictions show that in the case of constant h_E, the pressure is lower for SCC than the highly workable, but vibrated, F5 and F6 concretes. Inserting Eq. 12.14 and Eq. 12.15 into Eq. 12.13 along with the value of γ_e = 25 kNm^3 results in the analytical models:

$$\bar{\sigma}_{h,\max} = 25kN/m^3 + 4.5 \cdot v \cdot t_E \leq \sigma_{h,E,hydro} \quad \text{Consistency class F5 and F6} \quad [12.16]$$

$$\bar{\sigma}_{h,\max} = 20kN/m^3 + 4.0 \cdot v \cdot t_E \leq \sigma_{h,E,hydro} \qquad \text{SCC} \qquad [12.17]$$

which are also shown in Fig. 12.24, together with measured formwork pressures.

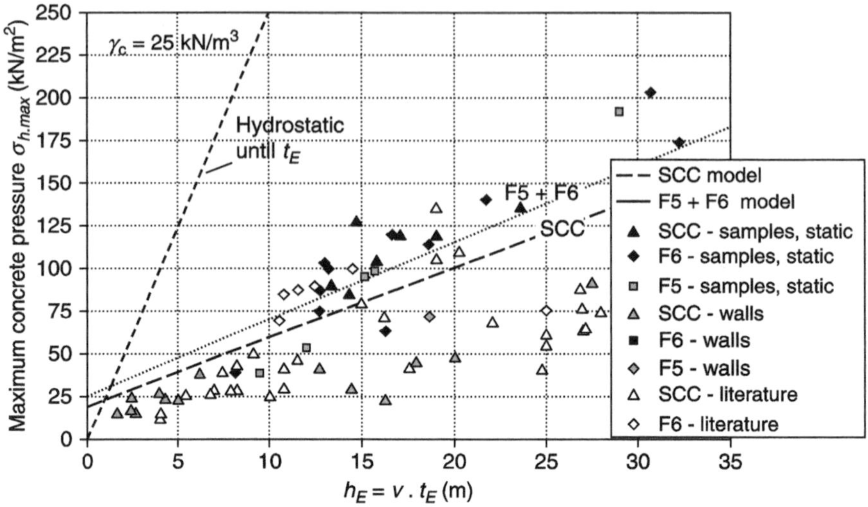

12.24 Lateral formwork pressures, $\sigma_{h,max}$, according to the analytical models (Eq. 12.15 and Eq. 12.16) and measured pressures (Proske and Graubner, 2010).

For the design of formwork and in order to reach a proposed model including the incorporation of a safety concept, a bilinear pressure distribution (Fig. 12.25) was chosen. The partial safety coefficient γ_F, according to a semi-probabilistic safety concept, has to be multiplied by the characteristic pressure (σ_{hk}) to obtain the design pressure (σ_{hd}). The partial safety factor for the ultimate limit state γ_F was predefined as equal to 1.5.

To incorporate all uncertainties such as varying casting rates and concrete setting times (found to be the most varying parameters) the following formulas for concrete pressure on formwork are proposed:

$$\sigma_{hk,\max} = h_s \cdot \gamma_c = (1[m] + 0.26 \cdot v \cdot t_E) \cdot \gamma_c \quad \text{SCC} \qquad [12.18]$$

$$\sigma_{hk,\max} = h_s \cdot \gamma_c = (1[m] + 0.30 \cdot v \cdot t_E) \cdot \gamma_c \quad \text{Consistency class F6} \qquad [12.19]$$

$$\sigma_{hk,\max} = h_s \cdot \gamma_c = (1[m] + 0.24 \cdot v \cdot t_E) \cdot \gamma_c \quad \text{Consistency class F5} \qquad [12.20]$$

Equation 12.20 for consistency class F5 predicts the lowest pressures, since the uncertainties of the model are the lowest for this concrete while for SCC the absence of vibration leads to lower predicted pressures than the more flowable and vibrated consistency class F6 concrete. These models are rather on the safer side and Proske and Graubner (2010) state that if the actual parameter variations are lower and if the workability loss or structural build-up is higher than assumed, the resulting formwork pressure is considerably lower than the models predict.

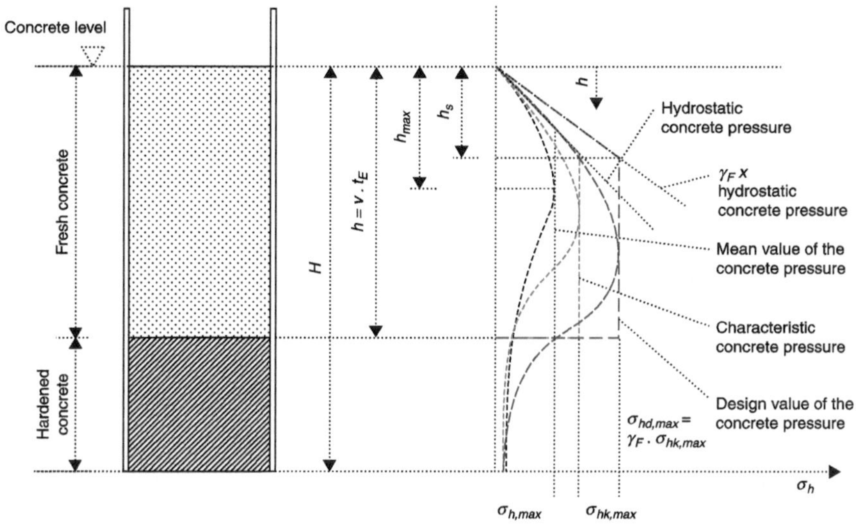

12.25 Distribution of lateral pressure from fresh concrete on formwork (Proske and Graubner, 2010).

12.6 Conclusions

A number of models to predict the lateral formwork pressure when casting with SCC have been presented. Of these, the approach by Proske and his German colleagues (Graubner *et al.*, 2008; Proske and Graubner, 2010) is the only one so far to have led to a national code also incorporating fluid concrete and SCC (DIN 18218: 2010–01). Khayat *et al.* (2010) compared their model with the one developed by Ovarlez and Roussel (2006) and found good mutual correlations. Thus, it seems fair enough to say that we now have models at hand that all work satisfactorily, but what is missing is an evaluation of the models in real field situations. A database with test results in the field will be necessary in order to include relevant safety factors to each model before they can be accepted as national codes.

Predicting formwork pressure when using SCC is still very much a topic of interest and, in many parts of the world, unpredictable formwork pressure is highlighted as the key obstacle for increased use of SCC. Greater use of SCC can considerably improve productivity, quality and work environment, and reduce costs and natural resource use when constructing concrete structures. Therefore, the reasons for untangling this knot are many. Technical committees within the American Concrete Institute (ACI), as well as within the International Union of Laboratories and Experts in Construction Materials, Systems and Structures (RILEM) are currently examining the question of formwork pressure and their goals are to find solutions to this problem, as well as harmonize the knowledge between both organizations.

12.7 References

ACI (1989), *Formwork for Concrete*, 5th edition, ACI SP–4, American Concrete Institute.

ACI Committee 238 (2008), *Report on Measurement of Workability and Rheology of Fresh Concrete*, ACI 238.1R–08, American Concrete Institute, 70 pp.

Amziane, S., Perrot, A. and Lecompte, T. (2008), 'A novel settling and structural build-up measurement method', *Measurement Science and Technology*, 19, Article: 105702, 8pp.

Assaad, J. (2004), *Formwork Pressure of Self-Consolidating Concrete – Influence of Thixotropy*, PhD thesis, Université de Sherbrooke, Quebec, Canada.

Assaad, J., Khayat, K.H. and Mesbah, H. (2003), 'Assessment of thixotropy of highly flowable and self-consolidating concrete', *ACI Materials Journal*, Vol. 100, No. 2, pp. 99–107.

Backsell, G. and Bydemar, B. (1968), *Dimensionering av vertikala formar* [Design of vertical concrete formwork], Report 44:1968, National Swedish Institute for Building Research.

Beitzel, M. (2009), *Formwork Pressure Behaviour of Fresh Concrete under Consideration of the Rheological Properties* [in German], PhD thesis, Institute for Concrete Structures and Building Materials, Karlsruhe Institute of Technology.

Billberg, P. (2003), 'Form pressure generated by self-compacting concrete', Proceedings 3rd International RILEM Symposium on Self-Compacting Concrete, RILEM PRO33 Reykjavik, Iceland, pp. 271–280.

Billberg, P. (2006), *Form Pressure Generated by Self-Compacting Concrete – Influence of Thixotropy and Structural Behaviour at Rest*, PhD thesis, Department of Structural Engineering, Royal Institute of Technology, Stockholm (downloadable from http://kth.diva-portal.org/smash/get/diva2:10913/FULLTEXT01).

Billberg, P. and Österberg, T. (2001), 'Thixotropy of self-compacting concrete', Proceedings 2nd International Symposium on Self-Compacting Concrete, Tokyo, Japan, pp. 99–108.

Clear, C.A. and Harrison, T.A. (1985), *Concrete Pressure on Formwork*, CIRIA Report 108, Construction Industry Research and Information Association.

Coussot, P. and Boyer, S. (1995), 'Determination of yield stress fluid behaviour from inclined plane test', *Rheology Acta*, Vol. 34, pp. 534–543.

DIN (1980) DIN 182 18 *Frischbetondruck auf lotrechte Schalungen* [Pressure of fresh concrete on vertical formwork], Deutsches Institut für Normung.

DIN (2010), DIN 18218: 2010–01 *Pressure of Fresh Concrete on Vertical Formwork* [in German], Beuth Verlag.

Ferraris, C., Zhang, M-H., Zhu, H., Picandet, V., Peltz, M. A. *et al.* (2007), *Measurement of Yield Stress for Concentrated Suspensions Using a Plate Device*, NIST Technical Note 1494., National Institute of Standards and Technology.

Gardner, N.J. (1985), 'Pressure of concrete on formwork – a review', *ACI Journal*, Vol. 82, pp. 744–753.

Graubner, C.-A., Beitzel, H., Beitzel, M., Bohnemann, C., Boska, E. *et al.* (2008), *Formwork Pressure Asserted by Highly Workable Concretes* [in German], F09–7–2008, TU Darmstadt, Institut für Massivbau.

Hurd, M.K. (2005), *Formwork for Concrete*, 7th edition, ACI-SP 4, American Concrete Institute.

Janssen, H.A. (1885), Versuche über Getreidedruck in Silozellen, *VDI Zeitschrift*, Vol. 39, pp. 1045–1049.

Khayat, K.H. and Omran, A.F. (2010), 'Evaluation of SCC formwork pressure', *Concrete International*, Vol. 32, No. 6, pp. 30–34.

Khayat, K.H. and Omran, A.F. (2011), 'Field monitoring of SCC formwork pressure and validation of prediction models', *Concrete International*, Vol. 33, Issue 6, pp. 33–39.

Khayat, K.H., Assaad, J., Mesbah, H. and Lessard, M. (2005), 'Effect of section width and casting rate on variations of formwork pressure of self-consolidating concrete', *Materials and Structures*, Vol. 38, pp. 73–78.

Khayat, K.H., Omran, A.F., Naji, S., Billberg, P. and Yahia, A. (2008), 'Test methods to evaluate form pressure of SCC', *Proceedings of the 3rd North American Conference on the Design and Use of Self-Consolidating Concrete* (SCC 2008), ed. Shah, S.P., Chicago, pp. 308–314.

Khayat, K.H, Omran, A.F and D'Ambrosia, M. (2010), 'Prediction of SCC formwork pressure in full-scale elements', Proceedings of SCC2010, Montreal, Canada, RILEM Bookseries, Vol. 1. pp. 231–242.

Koehler, E.P. (2004), *Development of a Portable Rheometer for Fresh Portland Cement Concrete*, MSc thesis, University of Texas at Austin, Austin, TX, USA.

Lange, D.A., Birch, B., Henchen, J., Liu, Y-S., Tejeda-Dominguez, F. and Struble, L. (2008), 'Modeling formwork pressure of SCC', *Proceedings of the 3rd North American Conference on the Design and Use of Self-Consolidating Concrete*, ed. Shah, S.P., Chicago, USA, pp. 295–300.

McCarthy, R., Billberg, P. and Silfwerbrand, J. (2007), 'An investigation and comparison of international design methods for vertical formwork', RILEM Proceedings PRO 54, SCC2007, 5th International RILEM Symposium on SCC, Ghent, Belgium pp. 479–484.

Mewis, J. (1979), 'Thixotropy – a general review', *Journal of Non-Newtonian Fluid Mechanics*, Vol. 6, pp. 1–20.

Omran, A. F. (2009), *Formwork Pressure Exerted by Self-Consolidating Concrete*, Doctoral thesis, Université de Sherbrooke, Quebec, Canada.

Ovarlez, G. and Roussel, N. (2006), 'A physical model for the prediction of lateral stress exerted by self-compacting concrete on formwork', *Materials and Structures*, Vol. 39, No. 2, pp. 269–279.

Perrot, A., Amziane, S., Ovarlez, G. and Roussel, N. (2009), 'SCC formwork pressure: influence of steel rebars', *Cement and Concrete Research*, 39, pp. 524–528.

Proske, T. (2007), *Formwork Pressure Asserted by Self-Compacting Concrete* [in German], PhD thesis, Technische Universität Darmstadt (downloadable from http://tuprints.ulb.tu-darmstadt.de/epda/000940/).

Proske, T. and Graubner, C.-A. (2010), 'Formwork pressure of highly workable concrete – experiments focused on setting, vibration and design approach', Proceedings of SCC2010, Montreal, Canada, RILEM Bookseries, Vol. 1, pp. 255–267.

Rodin, S. (1952), 'Pressure of concrete on formwork', *Proceedings of the Institution of Civil Engineers (London)*, Vol. 1, Part 1, No. 6, pp. 709–746.

Roussel, N. and Cussigh, F. (2008), 'Distinct-layer casting of SCC: the mechanical consequences of thixotropy', *Cement and Concrete Research*, Vol. 38, pp. 624–632.

Shunk, F.R. (1909), 'Pressure of concrete on forms', *Engineering News*, Vol. 62, pp. 288–289.

Tchamba, J. C., Amziane, S., Ovarlez, G. and Roussel, N. (2008), 'Lateral stress exerted by fresh cement paste on formwork: laboratory experiments', *Cement and Concrete Research*, 38, pp. 459–466.

Tejeda-Dominguez, F. (2005), *Laboratory and Field Study of Self Consolidating Concrete Formwork Pressure*, MSc thesis, University of Illinois at Urbana-Champaign.
Ueda, T. (ed.) (2002), *Standard Specifications for Concrete Structures – 2002*, JSCE Guidelines for Concrete No. 6, Japan Society of Civil Engineers.
Vanhove, Y., Djelal, C. and Magnin, A. (2004), 'Prediction of the lateral pressure exerted by self-compacting concrete on formwork', *Magazine of Concrete Research*, Vol. 56, No. 1, pp. 55–62.

13

Understanding the pumping of conventional vibrated and self-compacting concrete

D. FEYS, Université de Sherbrooke, Canada

Abstract: This chapter describes the pumping of concrete from a more theoretical point of view. It is shown that friction causes significantly higher pressure losses and must be avoided at all times. Based on several full-scale experimental studies, it is found that less flowable concretes, such as conventional vibrated concrete (CVC), move as a full plug in a pipe surrounded by a lubrication layer, while for more flowable concretes, like self-compacting concrete (SCC), a part of the bulk concrete is also sheared. For CVC, the properties of the lubrication layer relate well to the pumping parameters, while for SCC, the required pumping pressure is dominated by the concrete viscosity. The influence of other parameters is briefly discussed.

Key words: pressure, friction, pump, pipe, tribology.

13.1 Introduction

Concrete is the second-most used material in the world (after water). It offers a great flexibility to create structural elements of almost any shape because, if properly designed, concrete can fill the formwork completely. Concrete can be cast in a formwork in two different ways: by means of a bucket or by means of pumping. Pumping operations allow a faster and more continuous casting process and avoid the need to install a crane to lift and move the concrete bucket. Pumping is especially advantageous for small job sites, formworks with limited access or high-rise buildings. In May 2008, a new record of pumping concrete up to a height of 600 m in a single lift was set (Aldred, 2010). New developments in concrete technology can also lead to new casting techniques. For example, self-compacting concrete (SCC) can be pumped in the formwork from the bottom, which can be very advantageous when access is limited at the top of the formwork (Reymann and Khrapko, 2010).

Pumping of concrete has been carried out for decades and several manuals for concrete mix design, circuit lay-out and pumping operations have been developed based on practical experience (Schwing, 1983; Crepas, 1997; ACI Committee 304, 1998; ASTM-C33, 2003). For example, Fig. 13.1 shows a graph to determine the resulting pressure as a function of the discharge rate, pipe diameter, the equivalent length of the pipeline and the fresh properties of the concrete (spread). Despite the vast experience, problems still occur during pumping of which blocking is one of the main issues (Kaplan, 2001; Kaplan *et al.*, 2005). Blocking

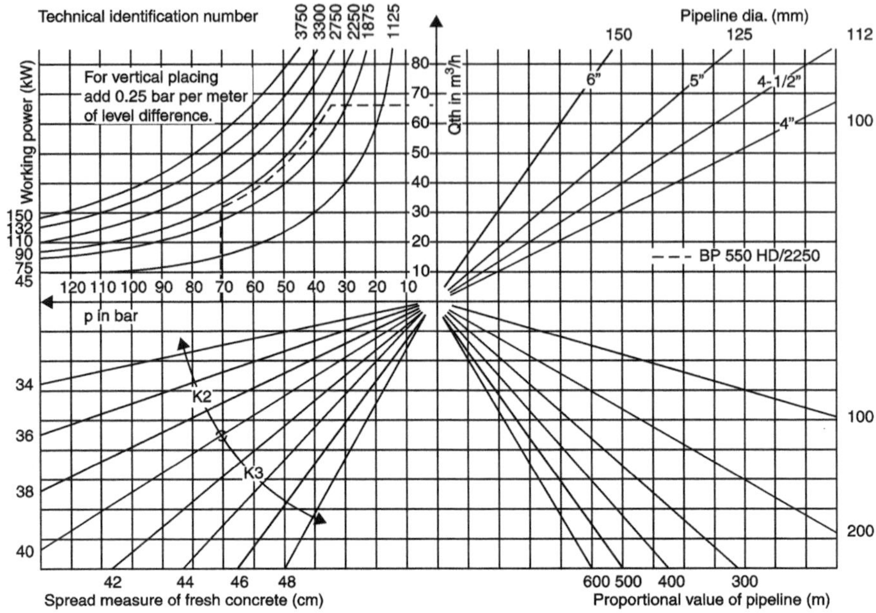

13.1 Determination of the required pumping pressure (horizontal left axis) as a function of the discharge rate (vertical axis at top), diameter of the pipeline (upper right), equivalent length of the circuit (lower right) and the spread of the concrete (lower left) (Schwing, 1983).

can lead to very high pressures, excessive movement of the pumping circuit and unfortunately also to injuries and casualties on the job site.

This chapter describes the pumping of concrete from a more theoretical point of view in order to provide a better understanding of the phenomena occurring during pumping operations. First, a distinction is made between flow and friction when concrete is moving through a pipeline. It is shown that friction must be avoided. With the focus on flow behaviour, the theoretical relationship between rheology and pumping is explained. A brief overview of recently conducted full-scale pumping experiments, measurement methods and different results is given and the experimental results are compared with theoretical formulae in case of straight sections. A detailed discussion is provided for both conventional vibrated concrete (CVC) and SCC, as major differences between these two concrete types have been observed. Finally, the influence of bends is briefly discussed.

13.2 Theoretical background

13.2.1 Flow or friction?

Concrete can be regarded as a suspension of sand and coarse aggregates in cement paste, or as a suspension of coarse aggregates in mortar. As explained in Chapter 1, the behaviour of a suspension is largely dependent on the concentration of solid particles. At a certain limit, the stress transfer switches from the hydrodynamic to the frictional type (Yammine *et al.*, 2008). In case of a hydrodynamic stress transfer, the movement of concrete is dependent on the shear rate in the interstitial liquid between the particles. In case of a frictional stress transfer, the movement is dominated by the contact between the particles and the normal force acting on them. In the first case, a sufficient amount of cement paste or mortar is available to lubricate the coarse aggregates, defining 'saturated' concrete. In the latter case, the coarse aggregates form a network through direct contacts and the concrete is called unsaturated (Fig. 13.2) (Ede, 1957).

The difference between saturated and unsaturated concrete has an enormous influence on the movement of concrete through pipes. In case of saturated concrete, the stress is transferred by means of hydrodynamic interactions. The movement of concrete can be regarded as a flow and the science of rheology can be applied. It is shown later in this chapter that the pressure loss per unit of length is directly related to the rheological properties. It is known that the rheological properties of liquids that do not exhibit a normal force during shearing (like water), are independent of the applied pressure (Macosko, 1994). As a result, the pressure loss per unit of length is constant over the entire length of a straight pipe section if no variations in pipe diameter or discharge rate occur. As a consequence, the total pressure increases linearly, in a straight horizontal pipe, from zero at the end (in fact, atmospheric pressure) to the maximum value at the outlet of the pump, as can be seen on the left side of Fig. 13.3 (Browne and Bamforth, 1977; Kaplan, 2001).

In case of unsaturated concrete, the movement is dominated by particle contacts and Coulomb's friction law must be applied. Coulomb's law states that the

13.2 Flow of concrete dominated by liquid interactions (left) and friction (right).

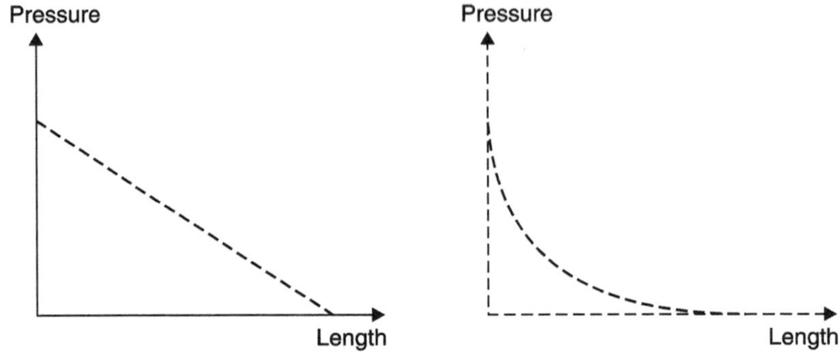

13.3 Evolution of total pressure as a function of the length of a straight, horizontal pipe. On the left, the evolution is linear for saturated concretes, while on the right, for unsaturated concretes, the relationship is exponential (after Browne and Bamforth, 1977).

tangential friction force is proportional to the friction coefficient between the two materials and the normal force. As a result, for the movement of concrete in pipes, the pressure loss per unit of length is dependent on the local pressure (Ede, 1957; Browne and Bamforth, 1977). The total pressure increases exponentially from the end to the beginning of the pipeline in case of a horizontal straight section (Fig. 13.3, right). Browne and Bamforth (1977) made theoretical calculations of the maximum distance concrete can be pumped, based on a certain set of properties of the concrete, the pump and the pipeline. If the concrete was saturated, the maximum pipeline distance was 250 m, while when unsaturated, the maximum pumping distance was reduced to 1.1 m! As a result, the unsaturated condition, corresponding to frictional stress transfer must be avoided in any case. More details can be found in Browne and Bamforth (1977).

From a practical point of view, concrete is defined as pumpable if it can be pumped through a pipeline without problems (ACI Committee 304, 1998) and if the properties of the concrete do not change significantly due to the pumping process (Jolin *et al.*, 2006). The pumpability of concrete is mainly related to the saturated or unsaturated condition. Practical guidelines define upper and lower limits for the grain size distribution of aggregates, maximum aggregate contents, minimum paste content, slump, etc, (Kasami *et al.*, 1979; Schwing, 1983; Sakuta *et al.*, 1989; Crepas, 1997; ACI Committee 304, 1998; ASTM-C33, 2003; Spiratos *et al.*, 2003; Jolin *et al.*, 2009). These guidelines are based on the on-site experience of practitioners, who in most cases try to avoid blocking of the pumpline.

Blocking mostly occurs as the consequence of a transition from liquid to solid interactions. Most blockings are observed during start-up (Kaplan, 2001). When inserting concrete in a pipeline, cement paste coats the walls of the pipes and fills the void in the seals between two pipes. Furthermore, especially for piston pumps, coarse aggregates move at each pumping stroke a little further ahead compared to the mortar due to inertia. As a result, a front of aggregates is created and the stress

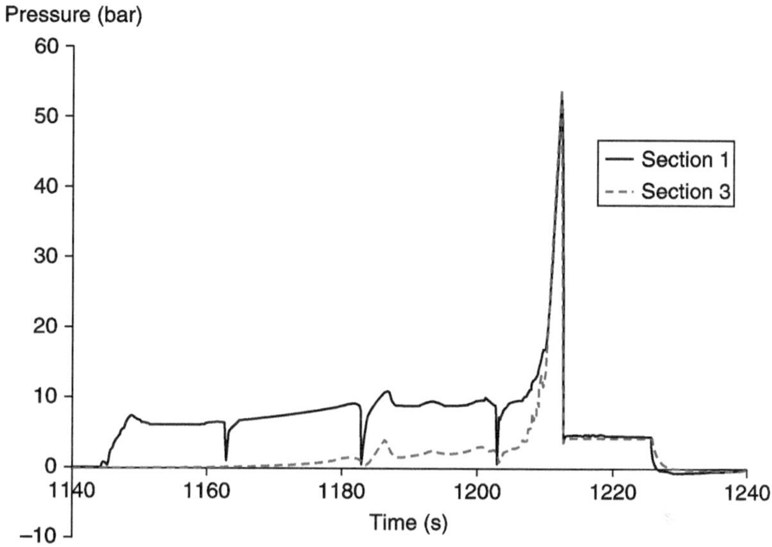

13.4 Transition from saturated to unsaturated concrete during the insertion of concrete in pipes, measured at two different sections. The transition occurs at t =1205 s and shows a rapidly increasing total pressure. At t = 1210 s, the pump can no longer deliver the pressure necessary to move the concrete and blocking occurred. From Feys (2009).

is transferred through friction. If the pump cannot deliver the pressure necessary to overcome friction, the flow stops and the pipe is blocked (Fig. 13.4). As a solution, a sufficient quantity of cement paste must be inserted in front of the concrete during start-up, and if the viscosity of this cement paste is sufficiently high, it can slow down the inertia effect acting on the coarse aggregates with each pumping stroke.

The transition from saturated to unsaturated concrete can also occur due to pressurised bleeding or water mobility under pressure (Fig. 13.5) (Browne and Bamforth, 1977; Kaplan, 2001; Kaplan *et al.*, 2005). In order to move concrete in a pipeline, a certain pressure gradient (pressure difference) to move the concrete in the pipes must be applied. This same pressure gradient causes also the water or cement paste to move between the aggregates, in the flow direction of the concrete. Water or cement paste separation with the flow direction causes zones in the pipes with a lack of water or cement paste. If insufficient cement paste is available in these zones, friction occurs and the pipeline can block. Pressurised bleeding can be avoided by optimising the grain size distribution (reducing the size of the voids), reducing the pumping pressure and increasing the viscosity of the cement paste. Viscosity-modifying agents (VMAs) (Khayat and Yahia, 1997; Khayat, 1998) are sometimes added to increase pumpability, especially if the aggregate grading is poor.

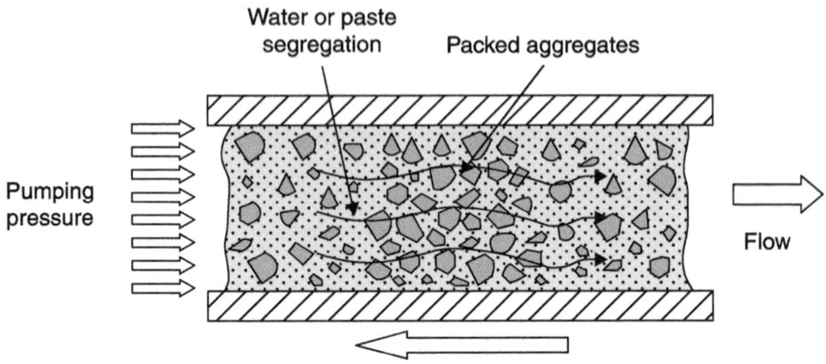

13.5 Principle of pressurised bleeding which creates zones with a higher content of aggregates due to water or paste movement in the direction of the flow. This can lead to a transition from saturated to unsaturated concrete (after Browne and Bamforth, 1977).

Several test methods to investigate pressurised bleeding or water mobility under pressure are described in the literature, but a detailed discussion of the equipment, modus operandi, advantages and disadvantages of each test is beyond the scope of this chapter. The interested reader is referred to specialised literature (Powers, 1968; Browne and Bamforth, 1977; Chalimo *et al.*, 1989; Sakuta *et al.* 1989; Kaplan, 2001).

For the remaining part of this chapter, the concrete is considered as saturated and the principles of rheology can be applied. For simplicity, use of the term 'friction' is minimized to avoid confusion with the unsaturated concrete flow condition.

13.2.2 Theoretical equations describing flow in pipes

In this section, the theoretical relationship between pressure loss and discharge rate as a function of the diameter of the pipe, length of the circuit and rheological properties of the material, is derived for Newtonian and Bingham liquids. The resulting equations are the Hagen–Poiseuille and Buckingham–Reiner equations respectively (Hagen, 1839; Poiseuille, 1840; Buckingham, 1921; Tattersall and Banfill, 1983; Macosko, 1994). The conditions for when to apply the Hagen–Poiseuille and Buckingham–Reiner equations are as follows:

- the flow is fully developed and isothermal
- the liquid is incompressible and homogeneous
- the flow is one-dimensional, with no radial or tangential flow component
- the flow must be steady
- no slippage at the wall is allowed
- a laminar flow condition is required

In a circular pipe, the equilibrium of forces acting on a certain section of concrete with length L results in the following equation:

$$\tau_w = \frac{\Delta p_{tot}}{L} \cdot \frac{R}{2} = \Delta p \cdot \frac{R}{2} \qquad [13.1]$$

where τ_w is the shear stress at the wall (Pa), Δp_{tot} is the total pressure loss over the section L (Pa), Δp is the pressure loss per unit of length (Pa/m), L is the length of the pipe (m) and R is the pipe radius (m) (Fig. 13.6). Furthermore, the shear stress evolves linearly from zero in the centre of the pipe to its maximum value τ_w at the wall, independent of the rheological properties of the material.

Based on the shear stress distribution and the rheological model, the shear rate at each distance r from the centre can be calculated. Integrating the shear rate to the radius r delivers the velocity profile, which is of quadratic form for a Newtonian liquid (Fig. 13.7). For a Bingham material, there exists a plug radius (r_{plug}) which defines the boundary between a zone with uniform velocity and a sheared zone, similar as for coaxial cylinders rheometers. For pipe flow, the zone with uniform velocity, also called the plug, is always located in the centre of the pipe and the plug radius is determined by:

$$r_{plug} = \frac{\tau_0}{\tau_w} \cdot R \qquad [13.2]$$

Integrating the velocity profile over the pipe area delivers the discharge rate, which is shown in the following equations (Eq. 13.3 and Eq. 13.4) for Newtonian and Bingham liquids respectively. These equations are generally known as the Hagen–Poiseuille and Buckingham–Reiner equations:

$$Q = \frac{1}{128} \cdot \frac{\pi \cdot \Delta p_{tot} \cdot D^4}{\eta \cdot L} \qquad [13.3]$$

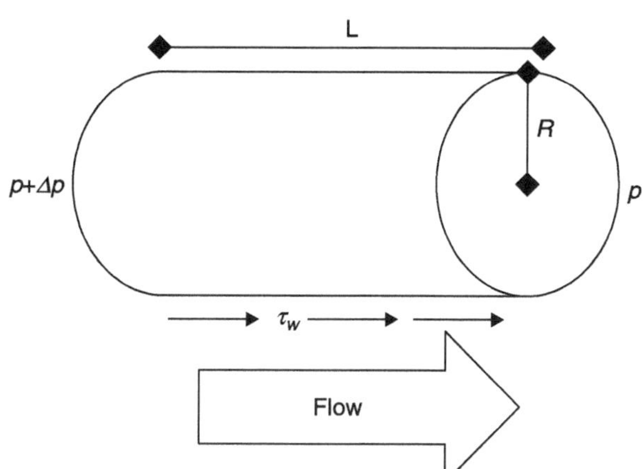

13.6 Equilibrium of forces for a material flowing through a pipe.

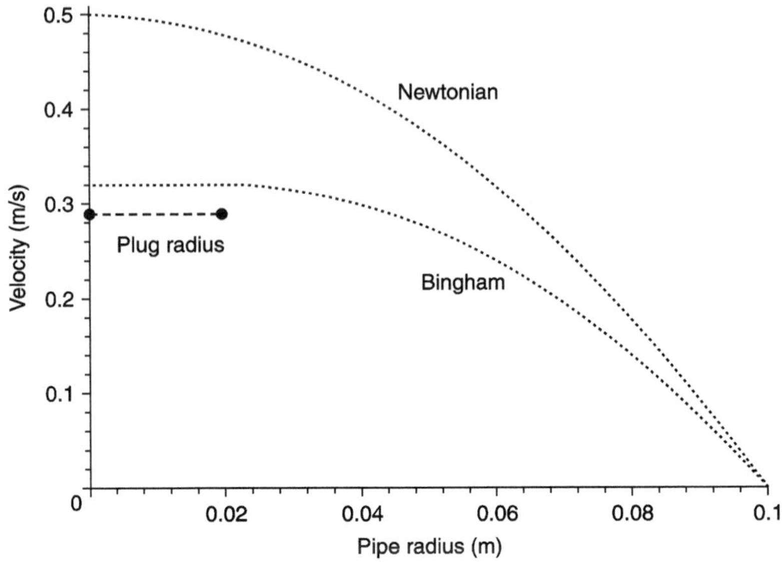

13.7 Theoretical velocity profile for a Newtonian liquid (i = 10 Pa·s; solid line) and a Bingham liquid (τ_0 = 20 Pa, μ_p = 10 Pa·s; dotted line) subjected to a wall shear stress τ_w = 100 Pa and a pipe radius of 0.1 m. A plug radius equal to 2 cm can be observed for the Bingham liquid.

$$Q = \frac{3 \cdot R^4 \cdot \Delta p_{tot}^{\ 4} + 16 \cdot \tau_0^4 \cdot L^4 - 8 \cdot \tau_0 \cdot L \cdot R^3 \cdot \Delta p_{tot}^{\ 3}}{24 \cdot \Delta p_{tot}^{\ 3} \cdot L \cdot \mu_p} \qquad [13.4]$$

where Q is the discharge rate (m^3/s), D is the pipe diameter (m), η is the Newtonian viscosity (Pa·s), τ_0 is the Bingham yield stress (Pa) and μ_p is the Bingham plastic viscosity (Pa·s).

From Eq. 13.3, it can be seen that the total pressure loss is linearly dependent on the discharge rate, Newtonian viscosity and length of the pipe, and inversely dependent on the radius or diameter to the power 4. Reducing the pipe diameter with a factor 2 will consequently increase the pressure with a factor 16, all other parameters remaining constant. Similar relationships for other rheological models can be calculated following this procedure, but the mathematical complexity increases significantly when adding an extra parameter in the rheological model (Feys *et al.*, 2008).

The equations above do not take into account the additional pressure needed to overcome a height difference. By means of the Bernoulli equation, it can be easily demonstrated that a height difference induces an additional hydrostatic pressure: $pg(h_2–h_1)$, in which p is the density (kg/m^3), g is the gravitational acceleration (m/s^2) and $h_2–h_1$ is the height difference between the outlet and inlet of the pipe (m). Note that the hydrostatic pressure is independent of the lay-out of the circuit;

only the height at inlet and outlet are of importance. On the other hand, in case of start-up of the pumping process or when the pipe empties due to gravity between two pumping strokes, the hydrostatic pressure becomes $\rho g(h_{max}-h_1)$, in which h_{max} is the maximum height of the pumping circuit (Lessard *et al.*, 1996; Chapdelaine, 2007).

13.3 Recent experimental research

This section describes some of the recent experimental research conducted by means of full-scale pumping tests. The main part of the discussion in the remaining sections is focussed on the results of Kaplan, Chapdelaine and Feys. The contributions of other authors are added where applicable.

13.3.1 Pumping circuits

Kaplan (2001) executed full-scale pumping tests on a horizontal 148 m long circuit at a concrete batching plant. The pump was a stationary piston pump. The loop circuit of Kaplan is divided in six straight sections of 23 m each with 180° bends in between (Fig. 13.8). The pipe diameter is 125 mm through the entire circuit. At the end of the circuit, the concrete flow is redirected in the reservoir of the pump by means of a grantry. The local pressure is measured at the beginning

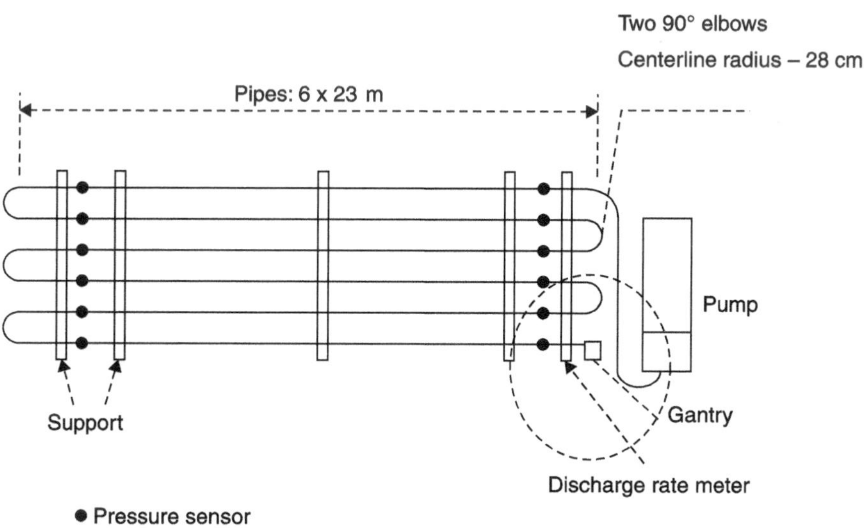

13.8 Schematic lay-out of the pumping circuit used in the investigation of Kaplan. The circuit measures 148 m in total and is divided into six sections of 23 m each, with 180° bends in between. At the beginning and end of each straight section, strain gauges are installed to measure the local pressure (after Kaplan, 2001).

and the end of each straight section. Attempts were made to measure the concrete pressure by means of sealed pressure sensors, but the plastic seals of the pressure sensors broke after several minutes of pumping due to the abrasive character of the concrete.

A second method for measuring the local pressure is to follow the deformation of the outer pipe wall by means of strain gauges (Browne and Bamforth, 1977; Kaplan, 2001). The sensitivity of this method depends on the thickness of the pipe wall, but fairly good results were obtained with this method. Furthermore, Kaplan (2001) showed that attaching strain gauges on the pipe wall in a longitudinal direction is not necessary, as the effect of the longitudinal deformation on the transversal deformation of the pipe is negligible. As a result, attaching strain gauges in the transversal direction is sufficient to monitor the pressure. The discharge rate was measured by means of an electromagnetic discharge meter installed at the end of the circuit. As a result, the filling coefficient of the pumping cylinders could be determined experimentally.

Chapdelaine (2007) used a low-discharge piston pump and measured the pressure loss over a 1.36 m long reducer from 100 to 50 mm inner diameter and over a 15 m long rubber hose with an inner diameter of 50 mm. The local pressure before and after the reducer were measured by means of pressure sensors with a metallic seal. In this way, the abrasive character of the concrete does not shorten significantly the life of the sensor. In a second part of Chapdelaine's work, the placing boom of a truck-mounted piston pump was equipped with five pressure sensors and measurements were performed on-site. The discharge rate was measured by counting the number of pumping strokes per unit of time. When the total volume of a pumping cylinder and the filling coefficient are known, the total discharge can be calculated. A method to relate the filling coefficient to the detailed pressure evolution with time has been proposed, of which more details can be found in Chapdelaine (2007).

Feys (2009) used a commercially available truck-mounted piston pump, without using the placing boom. The loop circuit consisted of steel pipes with an inner diameter of 100 mm, forming a 25, 81 or 105 m circuit (Fig. 13.9). Most experiments were conducted on the 25 m circuit, which contained a 12 m long horizontal straight section. This section was equipped with two pressure sensors (with a metallic seal) and three strain gauges were attached in transversal direction near the location of each pressure sensor. In the longer circuits, additional strain gauges were installed in other locations throughout the circuit. One of the measurement sections in the longer circuits enabled the determination of the pressure loss over a 90° or a 180° bend. The discharge rate was calculated by counting the number of pumping strokes per unit of time. The discharge rate was verified by means of a volumetric discharge measurement, during which the contents of a full pumping cylinder was pumped in a closed reservoir, equipped with a load cell, for all discharge rates applied (Feys, 2009).

(a)

(b)

13.9 (a) 25 m circuit employed by Feys, containing a straight horizontal section of 12 m equipped with two pressure sensors and 2 × 3 strain gauges. (b) Pumping tests in the long (105 m) circuit (Feys, 2009).

13.3.2 Concrete types studied

The three studies described above investigated the pumping of a wide range of concrete types which can be classified as plastic, fluid and self-compacting. The influence of different parameters on the pumping behaviour was studied. The main parameters investigated were:

- the sand-to-total aggregate ratio
- paste volume
- supplementary cementitious materials
- water to cementitious material (w/cm) or water to powder (w/p) ratio
- aggregate type
- initial slump or slump flow.

The detailed concrete compositions can be found in Kaplan (2001), Chapdelaine (2007) and Feys (2009).

13.3.3 Tests on fresh concrete: rheology and tribology

Apart from the standard tests on fresh concrete (such as slump, density and air content), the fresh concrete's rheological and tribological properties were determined. The BTRHEOM (Hu, 1995; Hu and de Larrard, 1996; Kaplan, 2001), IBB-rheometer with H-impeller (Chapdelaine, 2007) and Tattersall Mk-II rheometer (Tattersall and Banfill, 1983; Feys, 2009) were applied by Kaplan, Chapdelaine and Feys respectively. Kaplan modified the BTRHEOM and Chapdelaine modified the IBB-rheometer to obtain their respective tribometers.

A tribometer is used to measure the resistance to flow of concrete near a solid, smooth wall. In contrast to a rheometer, a tribometer does not contain a vane or ribs in the zone with the highest shear rates (which is the inner cylinder for a coaxial cylinders rheometer) as can be seen in Fig. 13.10. In the BTRHEOM-tribometer, the rotating inner cylinder (steel) is in direct contact with the outer cylinder. To prevent the flow of concrete inside the inner cylinder, a rubber seal is attached. The resistance induced by this rubber seal must be subtracted from the flow resistance of the concrete (Kaplan 2001).

Chapdelaine (2007) replaced the H-impeller of the IBB-rheometer by a smooth steel or rubber inner cylinder (Fig. 13.10). The movement of the concrete at the centre and at the outer wall of the bowl is prevented by specially designed ribs. The distance between the inner cylinder and the bottom of the bowl is 50 mm, and as a result, no rubber seal is needed. In the tribometer of Chapdelaine, the material at both the inside and the outside of the inner cylinder is sheared.

The testing procedure for a tribology test is very similar to that for a rheological test. Both Kaplan and Chapdelaine measured the torque during the stepwise decrease of the rotational velocity. Similarly as for the Reiner–Riwlin equation, the intercept with the T-axis (G) and the inclination of the T-N curve (H) are

(a) (b)

13.10 Tribometer developed by Chapdelaine (a,b). It is clearly visible that no ribs are provided at the inner cylinder. From Chapdelaine (2007).

transformed into yield stress ($\tau_{0,i}$, Pa) and viscous constant (η, Pa s/m): the properties of the lubrication layer (also called slippage layer) (Kaplan, 2001; Chapdelaine, 2007). In contrast to rheology, the shear stress is not expressed as a function of the shear rate, but as a function of the velocity, as can be seen in Eq. 13.5:

$$\tau = \tau_{0,i} + \eta \cdot v \qquad [13.5]$$

The basic condition for tribology measurements is that the bulk concrete is not sheared. All shearing is concentrated in the lubrication layer near the wall. In case of SCC, due to its low yield stress, this condition is not evidently fulfilled. The bulk SCC is likely to be sheared at higher velocities, contributing to the measured torque. As a result, tribology measurements on SCC are far more complicated and no measurement and data transformation procedure is yet available.

More details on the importance of the lubrication layer are given in Sections 13.4.1 and 13.4.2.

13.4 Pressure loss in straight sections

13.4.1 Theory versus experiments

One of the main goals of all research is to link the relationship between pressure loss and discharge rate with the properties of the concrete. In a first attempt, the Buckingham–Reiner equation is used to predict the pressure loss as a function of the discharge rate and the rheological properties of the material. Although the theoretical relationship between pressure loss and discharge rate has the same form as the experimental one, theory overestimates significantly the pressure for a certain discharge rate (Kaplan, 2001; Feys, 2009). The reason for this can be found in the conditions to apply the Buckingham–Reiner equation (see Section 13.2.2). It can be questioned as to whether the no-slip condition is fulfilled and whether the material remains homogeneous during the flow.

Certainly, the material cannot remain homogeneous during the flow in pipes, as the geometrical wall effect induces a lower concentration of aggregates near the wall. Furthermore, coarse particles have a tendency to move away from the zones with highest shear rates, which induces a particle migration from the wall to the centre. As a result, a layer with lower rheological properties is created near the wall of the pipe, which is called the lubrication or slippage layer. Different approaches exist in literature to treat the effect of the lubrication layer. In the recent studies of Kaplan and Chapdelaine, the properties of the lubrication layer are characterised by means of tribology. Other authors have simplified the lubrication layer into an apparent slippage near the wall, introducing a so-called slipping velocity v_{slip} (Ede, 1957; Morinaga, 1973; Browne and Bamforth, 1977; Sakuta *et al.*, 1989; Chalimo *et al.* 1989). In this case, the total velocity at a certain distance r from the centre is composed of the slipping velocity v_{slip}, which is assumed constant over the entire pipe section, and the shearing velocity v_{shear}, which varies with the distance from the centre. Note that the shear velocity can also be equal to zero.

13.4.2 Estimated velocity profile

As mentioned in Section 13.2.2, the plug radius r_{plug} defines the boundary between a zone of concrete flowing at uniform velocity, at the centre of the pipe, and a zone of concrete being sheared. The plug radius depends on the ratio of the material yield stress and the wall shear stress. The larger the yield stress, the smaller the sheared zone, and in theory, there should be no flow if the wall shear stress is smaller than the concrete yield stress.

For conventional vibrated concrete (CVC), however, the wall shear stress is often lower than the yield stress, and still the concrete is being pumped. This is due to the lubrication layer in which all shearing takes place. Consequently, the concrete flows as a large plug surrounded by the lubrication layer (Fig. 13.11). The pumpability of CVC is related to its ability to form the lubrication layer. If insufficient cement paste (or mortar) is available, the shearing takes place in the concrete itself. Furthermore, due to the very high concentration of coarse particles, CVC can show dilatancy (a volume increase) during shearing (Wallevik, 2009). This would provoke a transition from flow to friction, and blocking as a consequence, if no lubrication layer is present.

Highly workable concretes, and especially self-compacting concrete (SCC), have a low yield stress, possibly inducing shearing of the concrete itself. Although no experimental measurements of the velocity profile can be executed (unless two concretes with different colours set in the pipes (Vassiliev, 1993; Jacobsen *et al.*, 2009)), it is estimated that the theoretical determination of the plug radius by means of Eq. 13.2 is valid. On the other hand, a lubrication layer is also present, as the theoretical calculations overestimate the experimental results. The velocity profile for SCC flowing through pipes is assumed to be composed of a (small)

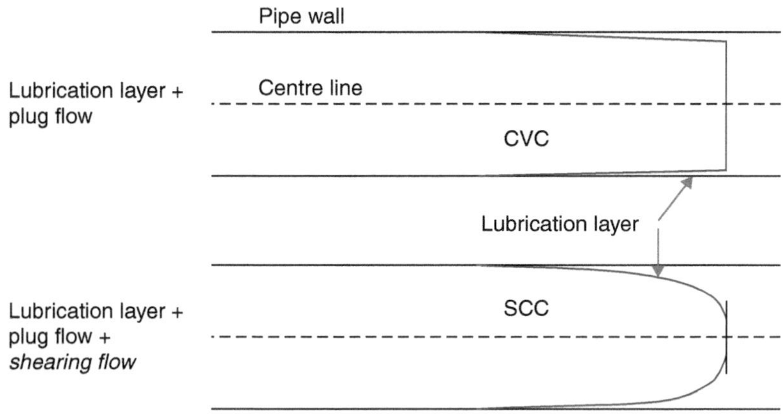

13.11 Estimated velocity profile during pumping of conventional vibrated concrete (CVC, top) and self-compacting concrete (SCC, bottom), showing that the bulk SCC is sheared while CVC is not.

plug in the centre, a lubrication layer at the wall and a zone in which the bulk concrete is sheared (Fig. 13.11).

13.4.3 Parameters influencing pressure during pumping

Rheological and tribological properties

Apart from the length and the diameter of the pipes, the relationship between pressure loss and discharge rate is influenced by the fresh properties of the concrete. The practical guidelines advise increasing the slump or spread of the concrete in order to decrease the pressure and facilitate the pumping process (Fig. 13.1). On the other hand, SCC is found to require higher pressures to be pumped, despite its very high slump. As discussed previously, a certain paste volume (or mortar volume) is required to create the lubrication layer to make the concrete pumpable (Jolin *et al.*, 2009). But once the concrete is pumpable, friction and blocking are avoided, and the rheological properties of the concrete and its lubrication layer determine the pressure loss–discharge curve.

Kaplan (2001) found that the pressure loss–discharge curve is bi-linear if Bingham is applicable as a rheological model for the concrete (Fig. 13.12). The distinction between the two linear parts of the curve is determined whether the bulk concrete is sheared or not. In the left part of the curve, the concrete flows as a full plug surrounded by the lubrication layer, while in the right part, a part of the bulk concrete is also sheared, causing an additional contribution to the total discharge rate. As a result, the inclination of the second part is always lower than the inclination of the first part of the curve. The transition from part 1 to part 2 occurs when the wall shear stress (related to the pressure loss) is approximately equal to the bulk concrete yield stress.

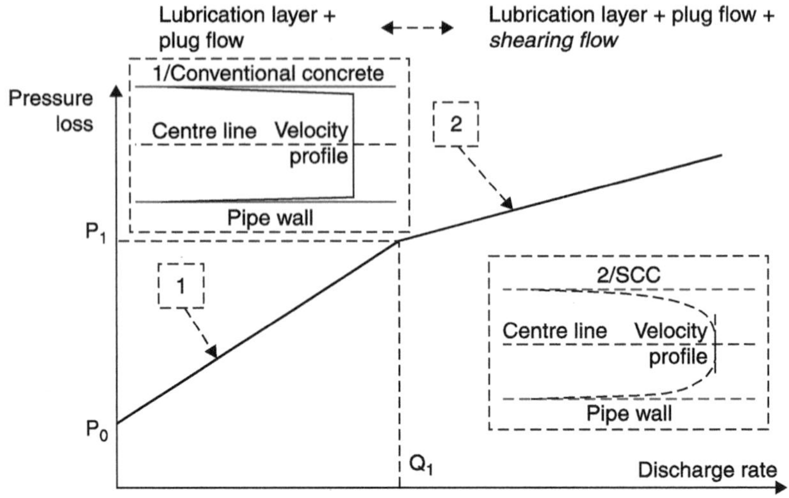

13.12 Graphical representation of Eq. 13.6 and Eq. 13.7. In part 1 of the curve, shearing only takes place in the lubrication layer, which corresponds mainly to the flow of CVC in pipes. In part 2 of the curve, the bulk concrete is also sheared, corresponding mainly to the flow of SCC in pipes (after Kaplan, 2001).

Based on this distinction, two theoretical equations were developed to predict the pressure loss–discharge rate relationship based on the properties of the lubrication layer for the first part (Eq. 13.6) and based on the properties of the lubrication layer and the bulk concrete for the second part of the curve (Eq. 13.7) (Kaplan, 2001). The properties of the lubrication layer in the equations are measured by means of a tribometer (as explained in Section 13.3.3).

$$p = \frac{2L}{R}\left(\frac{Q}{3600\pi R^2 k_r}\eta + \tau_{0,i}\right) \qquad [13.6]$$

$$p = \frac{2L}{R}\left(\frac{\dfrac{Q}{3600\pi R^2 k_r}\eta - \dfrac{R}{4\mu_p}\tau_{0,i} + \dfrac{R}{3\mu_p}\tau_0}{1 + \dfrac{R}{4\mu_p}\eta}\eta + \tau_{0,i}\right) \qquad [13.7]$$

where p is the total pressure (Pa), L is the total length of the pipeline (m), R is the radius of the pipe (m), Q is the real output discharge rate of the pump (m^3/h), $_{kr}$ is the filling coefficient of the pumping cylinders (–), $\tau_{0},_i$ and η are the yield stress (Pa) and viscous constant (Pa s/m) of the lubrication layer respectively, while τ_0 and μp are respectively the yield stress (Pa) and plastic viscosity (Pa s) of the bulk concrete. More information on the determination of the filling coefficient k_r can be found in Kaplan (2001) and Chapdelaine (2007).

The main part of the experimental work of Kaplan and Chapdelaine was carried out on CVC, and as a result, much of the data are situated in the left part of the curve. The yield stress of the CVC is sufficiently high to prevent the bulk concrete from being sheared. Both Kaplan (2001) and Chapdelaine (2007) obtained a very good relationship between the predicted and experimentally measured pressure losses (Fig. 13.13).

In the work of Feys (2009), mainly SCC was applied and it was found that the yield stress was in most cases lower than the wall shear stress, situating the results in part 2 of Fig. 13.12. As a consequence, the bulk concrete itself is also sheared and, if both the rheological and tribology properties of the concrete are measured, the prediction by means of Eq. 13.7 can be compared with the experimental results. Unfortunately, no tribology measurements were executed, but a good correlation between the pressure loss and the SCC viscosity was found at each

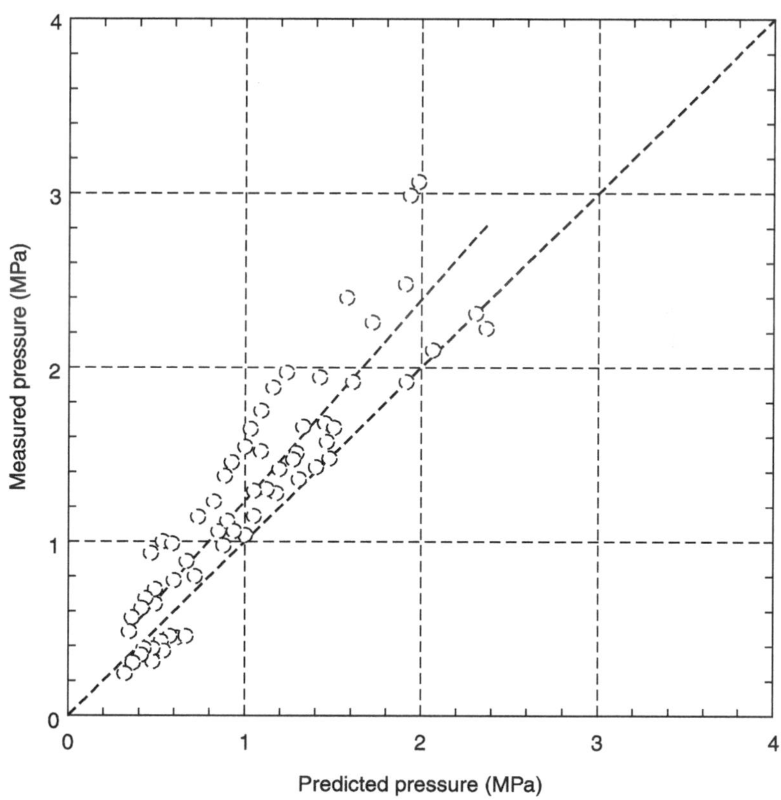

13.13 The predicted pressure (horizontal axis) by means of Eq. 13.6 and measured pressure (vertical axis) for all tests executed by Chapdelaine show a very good agreement for CVC. A similar result was obtained by Kaplan (Chapdelaine, 2007).

discharge rate (Fig. 13.14). The SCC employed by Feys also showed shear-thickening behaviour (Feys *et al.*, 2009), which is reflected in the non-linearity of the pressure loss–discharge rate curve for SCC, while the curve for CVC is linear as it is a Bingham material (Fig. 13.15).

For highly workable concretes, and especially SCC, the practical guidelines showing the increase in pressure loss with decrease in slump are no longer valid,

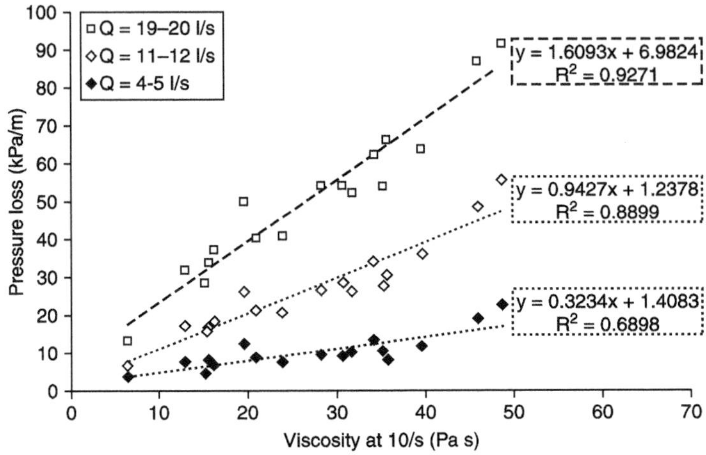

13.14 For SCC, the pressure loss is related to the concrete viscosity, dependent on the discharge rate applied (feys, 2009).

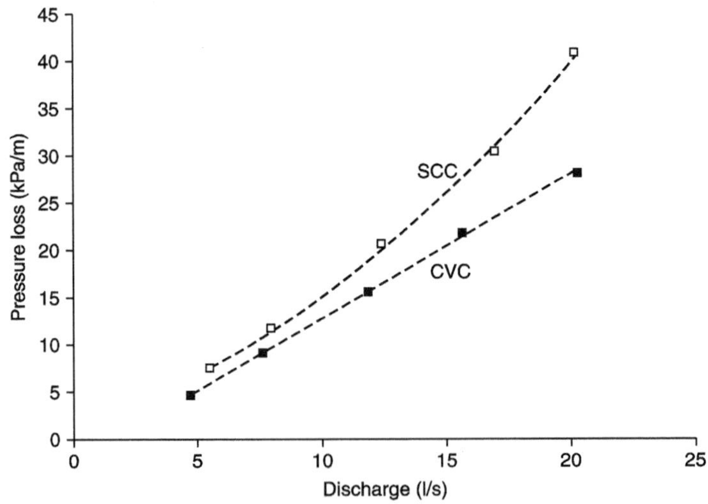

13.15 The type of rheological behaviour is reflected in the pressure loss–discharge rate curve, as the CVC and SCC were found to be Bingham and shear-thickening respectively (feys, 2009).

as the pressure losses are mainly dependent on the viscosity of the concrete (and its lubrication layer). Lowering the pressure during pumping of SCC can thus be achieved by lowering its viscosity. This has been proven by tests executed on a 250 m circuit by Jodeh and Nassar (2009). Two high-strength SCCs with equal slump flow (750 mm) were employed, but SCC 2 required a lower pressure to be pumped (92 bar compared to 250 bar for SCC 1) due to its lower V-funnel flow time (20 s and 10 s for SCC 1 and SCC 2 respectively). On the other hand, too low viscosity can cause pressurised bleeding and can result in blocking of the pipeline.

Total pressure

Kaplan (2001) executed a special test in which he wanted to verify the influence of high pressure on the pressure loss of concrete flowing through pipes. In theory, there should be no influence if the concrete is saturated (see Section 13.2). At the end of the pumping circuit, Kaplan installed a very narrow section with a valve, increasing the total pressure in the pipe. After the entire circuit was filled with water, a separator was inserted and concrete was pumping in behind. The test was stopped when the concrete reached the narrow section at the end (as it could not pass it). While the pressures applied during this 'high-pressure' test were up to 100 bar in total, which is up to 60 bar higher than during the normal tests, no additional pressure loss was observed in the high pressure tests compared to the normal testing procedure. This proves that rheological approach can be applied to study the pumping of saturated concretes.

Pipe material

Chapdelaine (2007) showed that the tribological properties of concrete depend on the material used as an inner cylinder of the tribometer. He mentioned that rubber provokes approximately 25% higher resistance than steel, indicating that higher pressures are needed to pump the same concrete through a rubber hose compared to a steel pipe. This type of results has been confirmed by Jacobsen *et al.* (2009) who visualised the velocity profile of concrete flowing through steel, rubber and acrylic pipes by using a grey and black concrete. The tests were conducted in an inclined pipe, while the flow was induced by gravity. They also measured the amount of material sticking to the wall when the pipe was empty. It was found that more material remained stuck to the rougher pipes (rubber and steel) and that the concrete had the highest flow rate in smooth pipes, although this conclusion was less clear when silica fume was incorporated in the mixture (Jacobsen *et al.*, 2009). The authors also state the need for further investigation to validate the importance of the pipe material in full-scale experiments.

Temperature

During pumping, the temperature of the concrete increases and the rate of increase with time ($\Delta T/\Delta t$) depends on the pressure loss per unit of length (Feys, 2009). A higher pressure loss will thus cause a faster increase in temperature. An increase in temperature accelerates the chemical reactions of cement with water, reduces the dormant period and causes a faster increase of yield stress and plastic viscosity with time, potentially reducing the workability of the concrete significantly so it can no longer be processed adequately on the job site (Petit *et al.*, 2006, 2009). The effect of temperature is most likely to occur in hot weather environment and during the construction of high-rise buildings, as the concrete stays in the pipes for a long time.

13.5 Pressure loss in bends and reducers

It is known from fluid mechanics that changing the flow direction requires additional energy. Therefore, concrete that passes through a bend or a reducer should require an additional pressure loss compared to a straight section. The practical guidelines use the term 'equivalent pipe length' to take into account the additional contributions from bends and reducers. The calculation is executed by simple rules of thumb, for example, a bend of 90° is equivalent to 3 m of straight pipes (Schwing, 1983, ACI Committee 304, 1998).

Kaplan (2001) and Chapdelaine (2007) came to the conclusion that, for their configurations of the pumping circuits and the (conventional vibrated) concretes pumped, the bends and reducers did not cause a significant additional pressure loss compared with a straight section. On the other hand, Feys (2009) found that 90° and 180° bends cause significantly larger pressure losses than a straight section for SCC, but the scatter on the results was too large to conclude some practical numbers.

The difference between the results of these studies is remarkable and reveals the complexity of the problem. The pressure loss in bends and reducers is influenced by a large number of parameters, amongst which the geometry of the bend or reducer, the flow velocity, the rheological properties of the concrete, inertia of coarse aggregates and the occurrence of a shearing or a plug flow are believed to be the most significant. Further research is needed to investigate the exact effect of bends and reducers on the total pressure needed to pump concrete.

13.6 Conclusions

Friction between coarse aggregates causes a significantly higher pressure loss than the hydrodynamic stress transfer in the cement paste, leading to many problems during the pumping of concrete. Friction can occur during start-up due to a lack of cement paste at the concrete front, or during the pumping process due to pressurised bleeding. Comparison of theoretical predictions by the Buckingham–Reiner equation and full-scale experimental results indicate the existence of a lubrication layer near the pipe wall, which is characterised by lower rheological properties. For less flowable concretes, like conventional vibrated concrete, concrete flows as a plug surrounded by the lubrication layer. Therefore, the pumping pressures are very well related to the properties of this lubrication layer. For more flowable concretes, such as self-compacting concrete, the velocity profile is assumed to be composed of a small plug in the centre of the pipe, a lubrication layer at the wall and a zone in which the bulk concrete is also sheared. Experiments have shown that the pressure losses are well related with the concrete viscosity, thus defining a limit of the established practical guidelines. Furthermore, it is shown that:

- the pressure loss is independent of the applied pressure if friction does not occur
- the velocity gradient in the lubrication layer depends on the pipe material
- temperature rises faster with increasing pressure loss.

Results on the influence of bends and reducers appear to be contradictory, but it can be concluded that their influence does not only depend on the geometry, but also on the rheological properties and composition of the concrete.

13.7 Acknowledgement

This chapter is based on large experimental investigations carried out in this field and reported in literature. Therefore, the following researchers, authors of key information used in this chapter, must be acknowledged: Roger D. Browne, Phillip B. Bamforth, Denis Kaplan and Frédéric Chapdelaine, along with those who assisted these researchers and the author of this chapter in their work.

13.8 References

ACI-Committee 304 (1998), *Placing concrete by pumping methods*, American Concrete Institute, Farmington Hills.

Aldred J (2010), 'Burj Khalifa – a new high for high-performance concrete', *Proc. Inst. Civ. Eng.: Civil Eng.*, 163, 66–73.

ASTM C 33 (2003), *Standard Specification for Concrete Aggregates*, American Society for Testing and Materials, Philadelphia.

Browne R D and Bamforth P B (1977), 'Tests to establish concrete pumpability', *ACI Journal*, 74, 193–203.

Buckingham E (1921), 'On plastic flow through capillary tubes', *Proc. Am. Soc. Testing Mater.*, 21, 1154–1161.

Chalimo T, Touloupov N and Markovskiy M (1989), *Peculiarities of Concrete Pumping* [in Russian], Edition Stroikniga, Minsk.

Chapdelaine F (2007), *Etude Fondamentale et Pratique sur le Pompage du Béton* [in French], PhD dissertation, University of Laval.

Crepas R A (1997), *Pumping Concrete, Techniques and Applications,* 3rd edition, Crepas and Associates, Elmhurst.

Ede A N (1957), 'The resistance of concrete pumped through pipelines', *Mag. Concr. Res.*, 9, 129–140.

Feys D, Verhoeven R and De Schutter G (2008), 'Extension of the Poiseuille formula for shear-thickening materials and application to self-compacting concrete', *Appl. Rheol.*, 18, 62705.

Feys D (2009), *Interactions between Rheological Properties and Pumping of Self-Compacting Concrete*, PhD dissertation, University of Ghent.

Feys D, Verhoeven R and De Schutter G (2009), 'Why is fresh self-compacting concrete shear thickening?', *Cem. Concr. Res.*, 39, 510–523.

Hagen G (1839), Ueber die Bewegung des Wassers in engen cylindrischen Rohren, (in German), *Poggendorfs Ann. Phys. Chem.*, 46, 423–442.

Hu C (1995), *Rhéologie des bétons fluides* [in French], PhD dissertation, Ecole Nationale des Ponts et Chaussées, Paris.

Hu C and de Larrard F (1996), 'The rheology of fresh high-performance concrete, *Cem. Concr. Res.*, 26, 283–294.

Jacobsen S, Haugan L, Hammer T A and Kalogiannidis E (2009), 'Flow conditions of fresh mortar and concrete in different pipes', *Cem. Concr. Res.*, 39, 997–1006.

Jodeh S A and Nassar G E (2009), *Pumpability assessment of C90 SCC*, Proceedings 2nd International Conference on Advances in Concrete Technology in the Middle-East: Self-Consolidating Concrete, Abu Dhabi, pp. 155–176.

Jolin M, Chapdelaine F, Gagnon F and Beaupré D (2006), *Pumping concrete: a fundamental and practical approach*, Proceedings 10th Conference on Shotcrete for Underground Support, Whistler.

Jolin M, Burns D, Bissonnette B, Gagnon F and Bolduc L-S (2009), *Understanding the pumpability of concrete*, Proceedings 11th Conference on Shotcrete for Underground Support, Davos.

Kaplan D (2001), *Pompage des bétons* [in French], PhD dissertation, Laboratoire Central des Ponts et Chaussées, Paris.

Kaplan D, de Larrard F and Sedran T (2005), 'Avoidance of blockages in concrete pumping process', *ACI Mater. J.*, 102, 183–191.

Kasami H, Ikeda T and Yamane S (1979), *On workability and pumpability of superplasticized concrete*, Proceedings 1st CANMET/ACI Conference on Superplasticizers in Concrete, Ottawa, pp. 67–86.

Khayat K H (1998), 'Viscosity-enhancing admixtures for cement-based materials – an overview', *Cem. Concr. Comp.*, 20, 171–188.

Khayat K H and Yahia A (1997), 'Effect of welan gum-high-range water reducer combinations on rheology of cement grout', *ACI Mat. J.*, 94, 365–372.

Lessard M, Baalbaki M and Aïtcin P-C (1996), 'Effect of pumping on air characteristics of conventional concrete, *Transp. Res. Rec.*, 1532, 9–14.

Macosko C W (1994), *Rheology Principle, Measurements and Applications*, Wiley-VCH, New York.

Morinaga M (1973), *Pumpability of concrete and pumping pressure in pipeline*, Proceedings RILEM Seminar on Fresh Concrete: Important Properties and their Measurement, Vol. 7, Leeds, pp. 1–39.

Petit J-Y, Khayat K H and Wirquin E (2006), 'Coupled effect of time and temperature on variations of yield value of highly flowable mortar', *Cem. Concr. Res.*, 36, 832–841.

Petit J-Y, Khayat K H and Wirquin E (2009), 'Coupled effect of time and temperature

on variations of plastic viscosity of highly flowable mortar', *Cem. Concr. Res.*, 39, 165–170.
Poiseuille J L M (1840), 'Recherches experimental sur le mouvement des liquids dans les tubes de très-petits diameters' [in French], *CR Acad. Sci. Paris*, 11, 961–967, 1041–4049.
Powers T (1968), *The Properties of Fresh Concrete*, John Wiley and Sons, London.
Reymann J and Khrapko M (2010), *A new vertical precast production system*, Proceedings 6th International RILEM Conference on SCC and 4th North-American Conference on SCC, Vol. II, Montréal, pp. 621–626.
Sakuta M, Kasanu I, Yamane S and Sakamoto A (1989), *Pumpability of Fresh Concrete*, Takenaka Technical Research Laboratory, Tokyo.
Schwing (1983), *Pumping, Concrete and Concrete Pumps – A Concrete Placing Manual*, Karl Ernst v. Eckardstein, 133pp.
Spiratos N, Pagé M, Mailvaganam N, Malhotra V M and Jolicoeur C (2003), *Superplasticizers for Concrete: Fundamentals, Technology, and Practice*, Supplementary Cementing Materials for Sustainable Development Inc., Ottawa.
Tattersall G H and Banfill P F G (1983), *The Rheology of Fresh Concrete*, Pitman, London.
Vassiliev V (1953), *Flow Regime in a Concrete Pipe* [in Russian], Edition No. 7, pp. 42–44 [cited in Kaplan, 2001].
Wallevik O H (2009), *Introduction to Rheology of Fresh Concrete*, course compendium, Reykjavik, ICI Rheocenter course – Reykjavik University.
Yammine J, Chaouche M, Guerinet M, Moranville M and Roussel N (2008), 'From ordinary rheology concrete to self-compacting concrete: a transition between frictional and hydrodynamic interactions', *Cem. Concr. Res.*, 38, 890–896.

Index

FSC
www.fsc.org
MIX
Papier | Fördert gute Waldnutzung
FSC® C083411

Kolibri 360 GmbH
Ferdinand-Jühlke-Straße 7
99095 Erfurt, Deutschland
produktsicherheit@kolibri360.de